Multidimensional Chromatography

Multidimensional Chromatography

LUIGI MONDELLO
Dipartimento Farmaco-chimico, Università degli Studi di Messina, Italy
ALASTAIR C. LEWIS
School of Chemistry and School of the Environment, University of Leeds, UK
KEITH D. BARTLE
School of Chemistry, University of Leeds, UK

JOHN WILEY & SONS, LTD

Chemistry Library

Other Wiley Editorial Offices

John Wiley & Sons, Inc., 605 Third Avenue,
New York, NY 10158-0012, USA

WILEY-VCH Verlag GmbH, Pappelallee 3,
D-69469 Weinheim, Germany

John Wiley & Sons Australia, Ltd, 33 Park Road, Milton,
Queensland 4064, Australia

John Wiley & Sons (Asia) Pte Ltd, 2 Clementi Loop #02-01,
Jin Xing Distripark, Singapore 129809

John Wiley & Sons (Canada) Ltd, 22 Worcester Road,
Rexdale, Ontario M9W 1L1, Canada

Library of Congress Cataloging-in-Publication Data

Mondello Luigi.
    Multidimensional chromatography / Luigi Mondello, Alastair C. Lewis, Keith D. Bartle.
        p.   cm
    Includes bibliographical references and index.
    ISBN 0-471-98869-3
    1. Chromatographic analysis.   I. Lewis, Alastair C.   II. Bartle, Keith D.   III. Title.

    QD79.C4 M65   2001
    543'.089—dc21

                                                            2001046684

British Library Cataloguing in Publication Data

A catalogue record for this book is available from the British Library

ISBN 0 471 98869 3

Typeset in 10/12pt Times by Thomson Press (India) Limited, New Delhi
Printed and bound in Great Britain by Biddles Ltd, Guildford and King's Lynn.
This book is printed on acid-free paper responsibly manufactured from sustainable forestry, in which at least two trees are planted for each one used for paper production.

CONTENTS

CONTRIBUTORS

Bartle, Keith D.
School of Chemistry, University of Leeds, Woodhouse Lane, Leeds LS2 9JT, UK

Beens, Jan
Department of Analytical Chemistry and Applied Spectroscopy, Faculty of Sciences, Free University de Boelelaan 1083, 1081 HV Amsterdam, The Netherlands

Chester, Thomas L.
The Procter & Gamble Company, Miami Valley Laboratories, PO Box 538707, Cincinnati, OH, 45253-8707, USA

Corradini, Claudio
Consiglio Nazionale delle Ricerche, Instituto di Cromatografia, Area della Ricerca di Roma, Via Salaria Km. 29 300, 00016-Monterotondo Scalo, Roma, Italy

Degen, Martha M.
Department of Chemistry, Oregon State University, Gilbert Hall 153, Corvallis, OR, 97331-4003, USA

de Jong, G. J.
Department of Analytical Chemistry and Toxicology, University of Groningen, Antonius Deusinglaan 1, 9713 AV Groningen, The Netherlands

Dugo, Giovanni
Dipartimento Farmaco-chimico, Facoltà di Framacia, Università degli Studi di Messina, Viale Annunziata, 98168-Messina, Italy

Dugo, Paola
Dipartimento di Chimica Organica e Biologica, Facoltà di Scienze, Università degli Studi di Messina, Salita Sperone, 98165-Messina, Italy

Kazakevich, Yuri V.
Department of Chemistry, Seton Hall University, 400 South Orange Avenue, South Orange, NJ, 07079, USA

Lanças, Fernando M.
Laboratory of Chromatography, Institute of Chemistry, University of São Carlos, Av. Dr Carlos Botelho 1465, 13560-970 São Carlos (SP), Brasil

Lewis, Alastair C.
School of Chemistry and School of the Environment, University of Leeds, Woodhouse Lane, Leeds LS2 9JT, UK

LoBrutto, Rosario
Department of Chemistry, Seton Hall University, 400 South Orange Avenue, South Orange, NJ, 07079, USA

Marcé, R. M.
Department of Analytical Chemistry and Organic Chemistry, Universitat Rovira i Virgili, Imperial Tarraco 1, 43005-Tarragona, Spain

Marriott, Philip J.
Chromatography and Molecular Separation Group, Department of Applied Chemistry, Royal Melbourne Institute of Technology, GPO Box 2467V, Melbourne 3001, Victoria, Australia

Mondello, Luigi
Dipartimento Farmaco-chimico, Facoltà di Farmacia, Università degli Studi di Messina, Viale Annunziata, 98168-Messina, Italy

Nyiredy, Sz.
Research Institute for Medical Plants, H-2011 Budakalàsz, PO Box 11, Hungary

Remcho, Vincent T.
Department of Chemistry, Oregon State University, Gilbert Hall 153, Corvallis, OR, 97331-4003, USA

Snow, Nicholas H.
Department of Chemistry, Seton Hall University, 400 South Orange Avenue, South Orange, NJ, 07079, USA

Somsen, G. W.
Department of Analytical Chemistry and Toxicology, University of Groningen, Antonius Deusinglaan 1, 9713 AV Groningen, The Netherlands

PREFACE

Separation Science is a mature and unified subject in which now conventional chromatographic and electrically driven processes are applied in the analysis of mixtures of compounds ranging from permanent gases to proteins. The boundaries between previously distinct techniques are increasingly blurred and it is becoming very evident that is a single theory may be applicable to chromatography whatever the physical state of the mobile phase. Gas, liquid and supercritical fluid chromatography can be regarded as special cases of the same procedure, while capillary electrochromatography combines liquid chromatography with electrophoresis.

Separation science is now very focused on reducing not only timescales for analyzis, but also the size and physical nature of the analytical device, Miniaturisation of entire analytical procedures provides a strong driving force for these trends in unifying theory and practice, and is a process likely to continue, as separations using microfluidic devices are developed. In spite of these many advances however, the complexity of many naturally occurring mixtures exceeds the capacity of any single method, even when optimized to resolve them. For many years therefore, intense effort has been concentrated on coupling separations methods together to increase resolution, and these have proceeded parallel with advances in coupling separation methods with spectroscopy. As our ability to isolate components in mixtures has increased, so has our appreciation for the shear complexity of compounds found in nature, Even separation systems with the capacity to isolate many thousands of species, are found to be inadequate when applied to commonplace mixtures such as diesel fuel. We clearly have some way to go in realising separation systems that can provide truly universal and complete separations.

Recent advances in multidimensional separation methods have been rapid and we considered that the time was appropriate to bring together accounts by leading researchers who are developing and applying multidimensional techniques. These authors have emphasized underlying theory along with instrumentation and practicalities, and have illustrated techniques with real-world examples. We hope that the eader will be as excited as we are by this combined account of progress. We thank all our contributors for their significant efforts in producing chapters of high scientific quality. We are especially indebted to Katya Vines of John Wiley who guided the project through its early stages and more recently to Emma Dowdle who brought it to completion.

KEITH BARTLE, Leeds
ALLY LEWIS, Leeds
LUIGI MONDELLO, Messina

Part 1

General

1 Introduction

K.D. BARTLE

University of Leeds, Leeds, UK

1.1 PREAMBLE

The natural world is one of complex mixtures: petroleum may contain 10^5–10^6 components, while it has been estimated that there are at least 150 000 different proteins in the human body. The separation methods necessary to cope with complexity of this kind are based on chromatography and electrophoresis, and it could be said that separation has been the science of the 20th century (1, 2). Indeed, separation science spans the century almost exactly. In the early 1900s, organic and natural product chemistry was dominated by synthesis and by structure determination by degradation, chemical reactions and elemental analysis; distillation, liquid extraction, and especially crystallization were the separation methods available to organic chemists.

Indeed, great emphasis was placed on the presentation of compounds in crystalline form; for many years, early chromatographic procedures for the separation of natural substances were criticized because the products were not crystalline. None the less, the invention by Tswett (3) of chromatographic separation by continuous adsorption/desorption on open columns as applied to plant extracts was taken up by a number of natural product researchers in the 1930s, notably by Karrer (4) and by Swab and Jockers (5). An early example (6) of hyphenation was the use of fluorescence spectroscopy to identify benzo[*a*]pyrene separated from shale oil by adsorption chromatography on alumina.

The great leap forward for chromatography was the seminal work of Martin and Synge (7) who in 1941 replaced countercurrent liquid–liquid extraction by partition chromatography for the analysis of amino acids from wool. Martin also realized that the mobile phase could be a gas rather than a liquid, and with James first developed (8) gas chromatography (GC) in 1951, following the gas-phase adsorption–chromatographic separations of Phillips (9).

Early partition chromatography was carried out on packed columns, but in 1958 Golay, in a piece of brilliant inductive reasoning (10), showed how a tortuous path through a packed bed could be replaced by a much straighter path through a narrow open tube. Long, and hence highly efficient columns for GC, could thus be fabricated from metal or glass capillaries, and remarkable separations were soon

Multidimensional Chromatography, edited by L. Mondello, A. C. Lewis and K. D. Bartle
©2002 John Wiley & Sons Ltd.

demonstrated. None the less, practical difficulties associated with capillary column technology generally restricted open tubular GC to a minority of applications until the fused silica column revolution in 1979. Dandeneau and Zerenner realized (11) that manufacturing methods for fibre-optic cables could be applied to make robust and durable capillary tubes with inactive inner surfaces. Lee *et al.* then delineated (12) the chemistry underlying the coating of such capillaries with a variety of stationary phases, and the age of modern high-resolution GC was born. Small diameter fused-silica capillaries were also found by Jorgenson and Lukacs (13) to be suitable for electrodriven separations since the heat generated could be readily dissipated because of the high surface-area-to-volume ratio. The invention of capillary supercritical fluid chromatography (SFC) in 1981 by Lee, Novotny and, co-workers (14) also depended on the availability of fused-silica capillary columns.

Liquid chromatography, however, took a different course, largely because slow diffusion in liquids meant that separations in open tubes necessitated inner diameters which were too small to make this approach practical. On the other hand, greatly increased efficiencies could be achieved on columns packed with small silica particles with bonded organic groups, and the technology for such columns was made available following the pioneering work of Horvath *et al.* (15) and Kirkland (16), thus giving rise to high performance liquid chromatography (HPLC). Even so, the available theoretical plate numbers (N) are limited in HPLC at normally accessible pressures and a different separation principle is therefore made use of. Since the resolution, R, for the separation of two compounds with retention factors k_1 and k_2 is given by:

$$R = \frac{\sqrt{N}}{4} \left(\frac{\alpha - 1}{\alpha} \right) \left(\frac{k_2}{1 + k_2} \right) \tag{1.1}$$

where α, the selectivity, is k_2/k_1, it follows that increased resolution based on column efficiency can only be achieved by very large increases in column length, because of the square-root dependence of R on N. However, a small increase in α has a major influence on R, and selectivity is therefore the principal means of achieving separation in HPLC through the tremendous variety of differently bonded stationary phase groups.

1.2 PACKED CAPILLARY COLUMN AND UNIFIED CHROMATOGRAPHY

Small-diameter packed columns offer (17) the substantial advantages of small volumetric flow rates (1–20 (μL min^{-1})), which have environmental advantages, as well as permitting the use of 'exotic' or expensive mobile phases. Peak volumes are reduced (see Table 1.1), driven by the necessity of analysing the very small (picomole) amounts of substance available, for example, in small volumes of body fluids, or in the products of single-bead combinatorial chemistry.

The increasing use of microcolumns has moved chromatography towards unification. Giddings was the first to point out (18) that there was no distinction between

Table 1.1 Comparison of packed columns for analytical chromatography

Column internal diameter	Volumetric flow rate	Injection volume	UV-detector cell volume	Sensitivity improvement[a]
4.6 mm ("conventional")	1 mL min^{-1}	20 μL	8 μL	(1)
1 mm ('microbore')	50 μL min^{-1}	1 μL	400 nL	21
250 μm ("micro")	3 μL min^{-1}	60 nL	25 nL	340
75 μm ('nano')	300 nL min^{-1}	5 nL	3 nL	3760

[a] Values are expressed relative to (conventional) 4.6 mm column.

chromatographic separation modes, which are classified according to the physical state of the mobile phase (GC, SFC or HPLC) but which move towards convergence as microcolumns are employed. Towards the end of the 1980s, the concept arose of using a single chromatographic system to carry out a range of separation modes, namely the unified chromatograph. Such an instrument (19) has been used (20) (Table 1.2) in the analysis of the complete range of products derived from petroleum, from gases to vacuum residues and polymers, with either open-tubular or packed-capillary columns, and gas, supercritical or liquid mobile phases.

Recently, Chester has described (21) how a consideration of the phase diagram of the mobile phase shows that a one-phase region (Figure 1.1) is available for the selection of the mobile phase parameters, and that the boundaries separating

Table 1.2 Application of unified chromatography in petroleum analysis (20)

Mode	Sample	Column[a]	Detection[b]
GC	Petroleum gases	Packed capillary (ODS)	FID
	Gasoline		
	Kerosene	Open tubular	FID
	Diesel fuel		
SFC	Petroleum wax		FID
	Atmospheric and vacuum residues	Open tubular	
	Lube oil additives		
	Aromatic fractions	Packed capillary (ODS)	UV, FID
	Lube oil additives	Packed capillary (diol)	FID
GC–SFC (sequential)	Crude oil, etc.		
	Gasoline, diesel fuel in lube oil	Open tubular	FID
HPLC and SFC/HPLC (sequential)	Aromatic fractions		
	Polymers	Packed capillary (SiO$_2$)	UV

[a] ODS, octadecylsilyl silica.
[b] FID, flame-ionization detector.

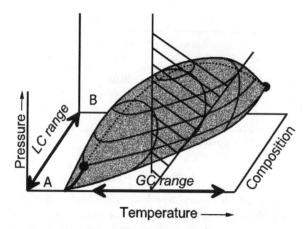

Figure 1.1 Phase diagram of the chromatographic mobile phase (after reference (21)), where the plane describes changing the composition and pressure at constant temperature.

individual techniques are totally arbitrary. By varying the pressure, temperature and composition, solute–mobile phase interactions can be varied so as to permit the elution of analytes ranging from permanent gases to ionic compounds; the dependence of the solute diffusion coefficient in the mobile phase on pressure, temperature and composition also influences mass transfer and therefore has an important bearing on the choice of an appropriate mobile phase.

The provinces of the common chromatographic separation modes are shown in Figure 1.1, with GC and HPLC practice corresponding to two of the areas; SFC with a single (here carbon dioxide) mobile phase is carried out on the front face of the diagram. In principle, however, any part of the phase diagram outside the two-phase region (shaded in Figure 1.1) may be employed. Figure 1.2 shows a microcolumn chromatogram obtained with simultaneous change of the pressure and composition of a carbon-dioxide mixture mobile phase, as indicated by the plane in Figure 1.1. Table 1.3 summarizes some of the uses of different regions of the phase diagram of the mobile phase.

1.3 RESOLVING POWER OF CHROMATOGRAPHIC SYSTEMS

The peak capacity, n, of a single-column chromatographic system generating N theoretical plates is given by:

$$n = \frac{\sqrt{N}}{4R} \ln\left(\frac{t_2}{t_1}\right) + 1 \qquad (1.2)$$

for a retention window from time t_1 to t_2. Some values of n for commonly used chromatographic separation methods are listed in Table 1.4, where it is immediately clear that there is a considerable mismatch between the capabilities of even very long GC columns or small-particle HPLC columns and the requirements for the analysis of mixtures commonly met in petroleum, natural product or biological chemistry. For example, a GC chromatogram of gasoline on a 400 m long capillary column developing 1.3×10^6 plates in an 11 h analysis with a peak capacity of 1000 still showed (22) considerable overlap. In the case of HPLC, even if the current predictions of the high plate numbers that might be possible with electrodriven capillary electrochromatography (CEC) (23) or with very high pressures and very small monodisperse

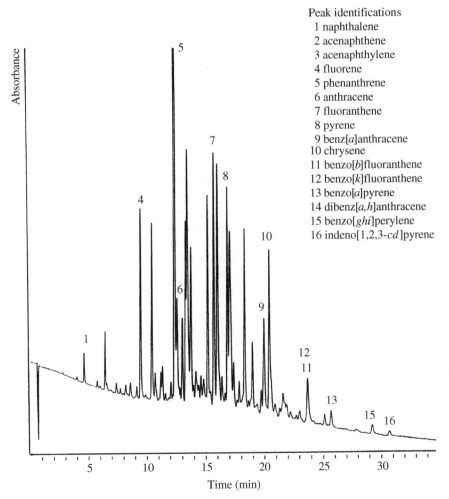

Peak identifications
1 naphthalene
2 acenaphthene
3 acenaphthylene
4 fluorene
5 phenanthrene
6 anthracene
7 fluoranthene
8 pyrene
9 benz[a]anthracene
10 chrysene
11 benzo[b]fluoranthene
12 benzo[k]fluoranthene
13 benzo[a]pyrene
14 dibenz[a,h]anthracene
15 benzo[ghi]perylene
16 indeno[1,2,3-cd]pyrene

Figure 1.2 Chromatogram of coal-tar oil obtained by using the following conditions: column, Waters Spherisorb PAH 5 mm in 250 μm id × 30 cm fused silica; column oven temperature, 100°C; UV detector wavelength to 254 nm; mobile phase, 100 to 300 bar CO_2 and 0.10 to 1.00 μL min^{-1} methanol over 30 minutes.

Table 1.3 Uses of different regions of the mobile phase diagram (cf. Figure 1.1)

Use	Reference
Change mobile phase during run for wide-ranging mixtures	D. Ishii, *J. Chromatogr. Sci.* **27**, 71 (1989); K. D. Bartle and D. Tong, *J. Chromatogr. A.* **703**, 17 (1995)
Faster diffusion available in enhanced fluidity (CO_2-based) mobile phases	S. V. Olesik, *Anal. Chem.* **63**, 1812 (1991)
Better solubility and faster diffusion available in high-temperature μLC	R. Trones, A. Iveland and T. Greibrokk, *J. Microcolumn Sep.* **7**, 505 (1995)
Solvating-gas chromatography	C. Shen and M. L. Lee, *Anal. Chem.* **69**, 2541 (1997)
High-pressure GC	S. M. Shariff, M. M. Robson and K. D. Bartle, *J. High Resolut. Chromatogr.* **19**, 527 (1996)

particles (24, 25) come to fruition for routine applications, full resolution of real mixtures will still not be possible.

The limitations of one-dimensional (1D) chromatography in the analysis of complex mixtures are even more evident if a statistical method of overlap (SMO) is applied. The work of Davis and Giddings (26), and of Guiochon and co-workers (27), recently summarized by Jorgenson and co-workers (28) and Bertsch (29), showed how peak capacity is only the maximum number of mixture constituents which a chromatographic system may resolve. Because the peaks will be randomly rather than evenly distributed, it is inevitable that some will overlap. In fact, an SMO approach can be used to show how the number of resolved simple peaks (s) is related to n and the *actual* number of components in the mixture (m) by the following:

$$s = m \exp\left(-\frac{2m}{n}\right) \qquad (1.3)$$

Table 1.4 Peak capacities in modern high-resolution chromatography[a]

Method	Column Length	Theoretical Plates	Peak Capacity[b]
GC	50 m	2×10^5	260
HPLC	25 cm (5 μm particles)	2.5×10^4	90
CEC	25 cm (3 μm particles)	6×10^4	140
	50 cm (3 μm particles)	1.2×10^5	200
	50 cm (1.5 μm particles)	2×10^5	260

[a] Calculated from equation (1.2) using $R = 1$.
[b] $K = 10$.

The fraction of the peaks resolved (s/m) also represents the probability, p, that a component will be separated as a single peak, so that:

$$P = \exp\left(-\frac{2m}{n}\right) \tag{1.4}$$

The values of n and the corresponding N which are necessary to resolve 50–90% of the constituents of a mixture of 100 compounds are listed in Table 1.5, thus making clear the limitations of one-dimensional chromatography. For example, to resolve over 80 % of the 100 compounds by GC would require a column generating 2.4 million plates, which would be approximately 500 m long for a conventional internal diameter of 250 μm. For real mixtures, the situation is even less favourable; to resolve, for example, 80 % the components of a mixture containing all possible 209 polychlorinated biphenyls (PCBS) would require over 10^7 plates.

1.4 TWO-DIMENSIONAL SEPARATIONS

A considerable increase in peak capacity is achieved if the mixture to be analysed is subjected to two independent displacement processes with axes z and y orientated at right angles, and along which the peak capacities are, respectively, n_z and n_y. For the orthogonality criterion to be satisfied, the coupled separations must be based on different separation mechanisms; the maximum peak capacity is then $n_z \times n_y$ (Figure 1.3), and the improvement in resolving power is spectacular. Thus, a peak capacity of 200 in the first dimension and one of 50 in the second, as is quite possible in comprehensive two-dimensional (2D) GC, yields a total peak capacity of 10^4, which would require in one dimension a plate number (30) of approximately 4×10^8 plates in a 250 μm id column of 80 km in length! The peak capacity of 10^4 of the two-dimensional system would permit resolution of 98 of the 100 components in the mixture discussed above, and in principle 200 of the 209 PCBs. If, however, the two separations are correlated, as for example, might hold for the separation of the polycyclic aromatic hydrocarbons (PAHs,) naphthalene, phenanthrene, chrysene, etc., by normal phase HPLC coupled to non-polar GC, there is little improvement over either method applied singly, and the retention pattern in two dimensions is

Table 1.5 Peak capacity and corresponding plate numbers required to resolve a given fraction of a 100-component mixture

Fraction of peaks resolved	Required peak capacity	Number of theoretical plates
0.5	290	250000
0.6	390	460000
0.7	560	950000
0.8	900	2430000
0.9	1910	10950000

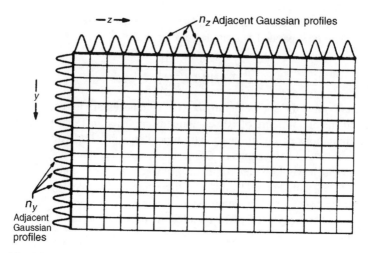

Figure 1.3 Peak capacity of a 2D system (reproduced with permission from reference (30)).

diagonal (Figure 1.4) (31). Such a system would, however, be effective for alkylated PAHs, where the two separations are less correlated (more similar retentions to the parent PAHs in HPLC, but better separated from the parents in GC) (Figure 1.4).

Giddings pointed out (32) that separated compounds must remain resolved throughout the whole process. This situation is illustrated in Figure 1.5, where two secondary columns are coupled to a primary column, and each secondary column is fed a fraction of duration Δt from the eluent from the first column. The peak capacity of the coupled system then depends on the plate number of each individual separation and on Δt. The primary column eliminates sample components that would otherwise interfere with the resolution of the components of interest in the secondary columns. An efficient primary separation may be wasted, however, if Δt is greater than the average peak width produced by the primary column, because of the recombination of resolved peaks after transfer into a secondary column. As Δt increases, the system approaches that of a tandem arrangement, and the resolution gained in one column may be nullified by the elution order in a subsequent column.

Two-dimensional separations can be represented on a flat bed, by analogy with planar chromatography, with components represented by a series of 'dots'. In fact, zone broadening processes in the two dimensions result in elliptically shaped 'spots' centred on each 'dot'. Overlap of the spots is then possible, but Bertsch (30) also showed how the contributors to the overall resolution, R, along the two axes, R_z and R_y contribute to the final resolution according to the following:

$$R \simeq \left(R_z^2 + R_y^2 \right)^{1/2} \qquad (1.5)$$

If the resolution is greater than 1 along either axis, then the final resolution will be also greater than 1. It follows that the isolation of a component in a two-dimensional system is much more probable than in a linear system because two displacements similar to that of another component are much less likely than for a single displacement.

The coupling of chromatographic techniques is clearly attractive for the analysis of complex mixtures, and numerous combinations have been proposed and developed (Figure 1.6). Truly comprehensive two-dimensional hyphenation is generally achieved by frequent sampling from the first column into the second, with a very rapid analysis. The interface is crucial here, and is designed so that components separated in the first dimension are not allowed to recombine; a variety of multiport valving arrangements have been used, but transfer between columns is most efficient if some kind of modulation is employed. The best example so far is the thermally modulated injection of very small samples from a primary GC column into a second GC column (33, 34).

More commonly, a fraction, based on chemical type, molecular weight or volatility, is 'heart-cut' from the eluent of the primary column and introduced into a secondary column for more detailed analysis. If the same mobile phase is used in both dimensions, fractions may be diverted by means of pressure changes – an approach first used in 1968 in GC-GC by Deans (35), and applied by Davies *et al.* in SFC–SFC (36). If the mobile phases are different, valves are employed, and special

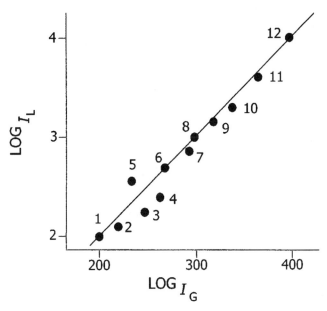

Figure 1.4 Two-dimensional plot of HPLC (log I_L) and GC (log I_G) retention indexes: (1) naphthalene; (2) 2-methylnaphthalene; (3) 2,3-dimethylnaphthalene; (4) 2,3,6-trimethyl-naphthalene; (5) biphenyl; (6) fluorene; (7) dibenzothiophen; (8) phenanthrene; (9) 2-methylphenanthrene; (10) 3,6-dimethylphenanthrene; (11) benzo[*a*]fluorene; (12) chrysene (data replotted from reference (31)).

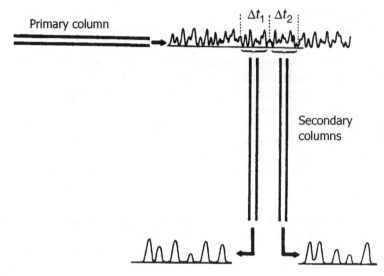

Figure 1.5 Representation of a coupled column system consisting of a primary column and two secondary columns (reproduced with permission from reference (30)).

arrangements may be necessary in order to eliminate large volumes of a liquid or a supercritical fluid. In HPLC–GC, this is achieved (31, 37, 38) by the use of a retention gap, pre-column and early solvent-vapour exit so that HPLC fraction with volumes of the order of hundreds of microlitres may be transferred to a GC column.

1.5 THE ORIGINS OF MULTIDIMENSIONAL CHROMATOGRAPHY

The main origin of multidimensional chromatography lies in planar chromatography. The development of paper chromatography, i.e. the partition between a liquid moving by capillary action across a strip of paper impregnated with a second liquid

		2nd Dimension			
		GC	SFC	HPLC	Electrically driven (CE, CEC)
1st Dimension	GC	√√	*	*	*
	SFC	√	√	*	*
	HPLC	√	√	√√	√√

Figure 1.6 Scope of chromatographic hyphenation: √ 'heart-cut', systems; √√ comprehensive versions.

(e.g. water), proceeded in parallel with the development of liquid–liquid partition chromatography on columns, and in 1944 Martin and co-workers (39) discussed the possibility of different eluents in different directions. Kirchner *et al.* pioneered (40) two-dimensional thin-layer chromatography (TLC) in the early 1950s before it was put on a firm footing by Stahl (41). A variety of hyphenated chromatography–electrophoresis techniques were demonstrated, but the most important planar separation was high resolution 2D gel electrophoresis, reported by O'Farrell in 1975 (42). Here, up to 1000 proteins from a bacterial culture were separated by using isoelectric focusing in one direction and sodium dodecylsulphonate-polyacrylamide gel electrophoresis in the second. Two-dimensional gel electrophoresis is still commonly used today in protein and DNA separation.

Most developments in the past two decades, however, have involved coupled column systems which are much more amenable to automation and more readily permit quantitative measurements, and such systems form the subject of this present book. A review on two-dimensional GC was published (43) in 1978 (and recently updated (29)), and the development by Liu and Phillips in 1991 of comprehensive 2D GC marked a particular advance (33). The fundamentals of HPLC–GC coupling have been set out (37) with great thoroughness by Grob. Other work on a number of other aspects of multidimensional chromatography have also been extensively reviewed (44, 45).

ACKNOWLEDGEMENTS

This chapter is based, in part, on a paper read before the 'Seventh International Symposium on Hyphenated Techniques in Chromatography,' held in Brugge in Belgium, in February 2000. I am indebted to the many colleagues who have worked in my Laboratory at Leeds on multidimensional chromatography, especially Tony Clifford, Nick Cotton, Ilona Davies, Paola Dugo, Grant Kelly, Andy Lee, Ally Lewis, Luigi Mondello, Peter Myers, Mark Raynor, Bob Robinson, Mark Robson and Daixin Tong.

REFERENCES

1. L. S. Ettre, 'Chromatography: The separation technique of the 20th century', *Chromatographia* **51**: 7 (2000).
2. L. S. Ettre and A. Zlatkis (Eds), *75 Years of Chromatography – a Historical Dialogue,* Elsevier, Amsterdam (1979).
3. M. Tswett, 'Physikalisch-Chemische Studier über das chlorophyll. Die absorptionen', *Ber. Dtsch. Botan. Ges* **24**: 316 (1906).
4. P. Karrer, 'Purity and activity of Vitamin A', *Helv. Chim. Acta* **22**: 1149 (1939).
5. G. M. Swab and K. Jockers, 'Inorganic chromatography I', *Angew. Chem.* **50**: 546 (1937).
6. I. Berenblum and R. Schoental, 'Carcinogenic constituents of shade oil,' *Brit. J. Exp. Path.* **24**: 232 (1943).
7. A. J. P. Martin and R. L. M. Synge, 'A new form of chromatogram employing two liquid phases. I: A theory of chromatogaphy', *Biochem. J.* **35**: 1158 (1941).

8. A. T. James and A. J. P. Martin, 'The separation and micro estimation of volatile fatty acids', *Biochem. J.* **50**: 679 (1952).

9. C. G. S. Phillips, 'The chromatography of gases and vapours', *Disc. Faraday. Soc.* **67**: 241 (1949).

10. M. J. E. Golay, 'Theory of chromatography in open and coated tubular columns with round and rectangular cross-section,' in *Gas Chromatography Amsterdam 1958* (Amsterdam Symposium), Desty D. H. (Ed.), Butterworths Scientific Publications, London, pp. 36–55 (1958).

11. R. D. Dandeneau and E. H. Zerenner, 'An investigation of glasses for capillary chromatography', *J. High. Resolut. Chromatogr. Chromatogr. Commun.* **2**: 351 (1979).

12. M. L. Lee, F. J. Yang and K. D. Bartle, 'Column technology,' in *Open Tubular Column Gas Chromatography. Theory and Practice,* John Wiley & Sons, New York, Ch. 3, pp. 50–99 (1984).

13. J. W. Jorgenson and K. D. Lukacs, 'Zone electrophoresis in open-tubular glass capillaries,' *Anal. Chem.* **53**: 1298 (1981).

14. M. Novotny, S. R. Springston, P. A. Peaden, J. C. Fjeldsted and M. L. Lee, 'Capillary supercritical fluid chromatography,' *Anal. Chem.* **53**: 407A (1981).

15. C. Horváth, W. Melander and I. Molnár, 'Solvophobic interactions in liquid chromatography with non-polar stationary phases', *J. Chromatogr.* **125**: 129 (1976).

16. J. J. Kirkland, 'High speed liquid-partition chromatography with chemically bonded organic stationary phases,' *J. Chromatogr. Sci.* **9**: 206 (1971).

17. M. Novotny, 'Recent advances in microcolumn liquid chromatography', *Anal. Chem.* **60**: 500A (1988).

18. J. C. Giddings, *Dynamics of Chromatography*, Marcel Dekker, New York (1965).

19. D. Tong, K. D. Bartle, A. A. Clifford and R. E. Robinson, 'Unified chromatograph for gas chromatography, supercritical fluid chromatography and micro-liquid chromatography,' *Analyst* **120**: 2461 (1995).

20. D. Tong, K. D. Bartle and R. E. Robinson, 'Unified chromatography in petrochemical analysis', *J. Chromatogr. Sci.* **31**: 77 (1993).

21. T. L. Chester, 'Chromatography from the mobile-phase perspective', *Anal. Chem.* **69**: 165A (1997).

22. T. A. Berger, 'Separation of a gasoline on an open tubular column with 1.3 million effective plates', *Chromatographia* **42**: 63 (1996).

23. M. G. Cikalo, K. D. Bartle, M. M. Robson, P. Myers and M. R. Euerby, 'Capillary electrochromatography', *Analyst* **123**: 87R (1998).

24. J. E. MacNair, K. C. Lewis and J. W. Jorgenson, 'Ultrahigh-pressure reversed-phase liquid chromatography in packed capillary column', *Anal. Chem.* **69**: 983 (1997).

25. J. A. Lippert, B. Xin, N. Wu and M. L. Lee, 'Fast ultrahigh-pressure liquid chromatography: on-column UV and time-of-flight mass spectrometric detection', *J. Microcolumn. Sep.* **11**: 631 (1999).

26. J. M. Davis and J. C. Giddings, 'Statistical theory of component overlap in multicomponent chromatograms', *Anal. Chem.* **55**: 418 (1983).

27. M. Martin, D. P. Herman and G. Guiochon, 'Probability distributions of the number of chromatographically resolved peaks and resolvable components in mixtures', *Anal. Chem.* **58**: 2200 (1986).

28. T. F. Hooker, D. J. Jeffery and J. W. Jorgenson, 'Two-dimensional Separations', in *High Performance Capillary Electrophoresis,* M. G. Khaledi (Ed.), John Wiley & Sons, New York, pp. 581–612 (1998).

29. W. Bertsch, 'Two-dimensional gas chromatography: concept, instrumentation and applications – Part 1: fundamentals., conventional two-dimensional gas chromatography, selected applications', *J. High. Resolut. Chromatogr.* **22**: 647 (1999).

30. W. Bertsch, 'Multidimensional gas chromatography', in *Multidimensional Chromatography. Techniques and Applications,* H. J. Cortes (Ed.), Marcel Dekker, New York, pp. 75–110 (1990).

31. I. L. Davies, M. W. Raynor, J. P. Kithinji, K. D. Bartle, P. T. Williams and G. E. Andrews, LC – SFE – GC – SFC interfacing', *Anal. Chem.* **60**: 683A (1988).

32. J. C. Giddings, 'Two-dimensional separation: concept and promise', *Anal. Chem.* **56**: 1258A (1984).

33. Z. Liu and J. B. Phillips, 'Comprehensive two-dimensional gas chromatography using an on-column thermal modulator interface', *J. Chromatogr. Sci.* **29**: 227 (1991).

34. J. B. Phillips, J. Beens and U. A. Th Brinkman, in *Hyphenation: Hype and Fascination,* Brinkman UA Th (Ed.), Elsevier, Amsterdam, pp. 331 – 347 (1999).

35. R. R. Deans, 'A new technique for heart cutting in gas chromatography', *Chromatographia* **1**: 18 (1968).

36. I. L. Davies, B. Xu, K. E. Markides, K. D. Bartle and M. L. Lee, 'Multidimensional coupled supercritical fluid chromotography', *J. Microcolumn. Sep.* **1**: 71 (1989).

37. K. Grob, *On-Line Coupled LC–GC,* W. Bertsch, W. G. Jennings and P. Sandra (Series Eds), Hüthig, Heidelberg, Germany (1991).

38. L. Mondello, P. Dugo, G. Dugo, A. C. Lewis and K. D. Bartle, 'High performance liquid chromatography coupled on-line with high resolution gas chromatography. State of the art', *J. Chromatogr.* **842**: 373 (1999).

39. R. Consden, A. H. Gordon and. A. J. P. Martin, 'A partition chromatographic method using paper', *Biochem. J.* **38**: 224 (1944).

40. J. G. Kirchner, J. M. Miller and G. Keller, 'Separation and identification of some terpenes by a new chromatographic technique', *Anal. Chem.* **23**: 420 (1951).

41. E. Stahl, 'Thin-layer chromatography II. Standardisation, detection, documentation and application', *Chem. Ztg.* **82**: 323 (1958).

42. P. H. O'Farrell, 'High resolution two-dimensional electrophoresis of proteins', *J. Biol. Chem.* **250**: 4007 (1975).

43. W. Bertsch, 'Methods in gas chromatography: two dimensional techniques', *J. High Resolut. Chromatogr.* **1**: 1, 85, 289 (1978).

44. H. J. Cortes (Ed.), *Multidimensional Chromatography. Techniques and Applications,* Marcel Dekker, New York (1990).

45. U. A. Th Brinkman (Ed.), *Hyphenation: Hype and Fascination,* Elsevier, Amsterdam (1999).

2 Coupled High Performance Liquid Chromatography with High Resolution Gas Chromatography

L. MONDELLO
Università di Messina, Messina, Italy

2.1 INTRODUCTION

The analysis of complex matrices, such as natural products, food products, environmental pollutants and fossil fuels, is today a very important area of separation science. The latest developments in chromatographic techniques have yielded highly efficient systems, used with specific detectors to obtain high selectivity and or sensitivity.

Sometimes, the resolving power attainable with a single chromatographic system is still insufficient for the analysis of complex matrices. An approach commonly used to obtain greater resolution is multidimensional chromatography.

In the specific case of high performance liquid chromatography coupled with high resolution gas chromatography (HPLC–HRGC), the selectivity of the LC separation is combined with the high efficiency and sensitivity of GC separation, thus giving a relatively high peak capacity. Off-line coupling of LC and GC is frequently used because of the ease of collecting and handling liquids, but this technique is long and laborious, and involves numerous steps with the risk of contamination, formation of artifacts and possible loss of sample. On-line coupling of LC and GC presents a number of advantages: the amount of sample required is less, there is no sample work-up, no evaporation or dilution is necessary, and fully automated sample pre- or post-treatments are possible. The disadvantages of the on-line system are that the system is more difficult to operate, the initial set-up is expensive, and interfaces are relatively complicated. The main problem to be solved in on-line LC–GC coupling is the transfer of amounts of liquid from LC to GC, where the latter operates in a different physical state. Different approaches have been studied for allowing the introduction of large amount of solvent into the GC column. These techniques must selectively remove the solvent, thus leaving the solute in a sharp band at the entrance to the separation column.

Multidimensional Chromatography, edited by L. Mondello, A. C. Lewis and K. D. Bartle
©2002 John Wiley & Sons Ltd.

2.2 TRANSFER TECHNIQUES

This chapter will describe some of the direct transfer techniques used for both normal phase and reversed phase eluents. An overview of the various techniques (direct and indirect) for the analysis of water or water-containing eluents will also be given. In fact the nature of the eluent greatly influences the choice of the transferring technique, as will be explained in the discussion for each section. Figure 2.1 summarizes the concepts of eluent evaporation (with some subclasses), thus allowing the transfer of large LC fractions to the gas chromatograph.

2.2.1 RETENTION GAP TECHNIQUES

The retention gap method (1, 2) represents the best approach in the case of qualitative and quantitative analysis of samples containing highly volatile compounds. The key feature of this technique is the introduction of the sample into the GC unit at a temperature below the boiling point of the LC eluent (corrected for the current inlet pressure), (see Figure 2.2). This causes the sample vapour pressure to be below the carrier gas inlet pressure, and has two consequences, as follows:

- volatile components are reconcentrated by the solvent effects, primarily solvent trapping (3);

- the high-boiling compounds are spread by band broadening in space.

A layer of condensed eluent is built up ahead of the evaporation site which acts as a thick layer of retaining stationary phase, thus blocking the further movement of all but the most volatile compounds into the column. Solvent evaporation, therefore proceeds from the rear towards the front of the sample layer (see Figure 2.2).

These effects refer to the reconcentration obtained with an uncoated inlet. In fact, the term retention gap means a column inlet of a retention power lower than that of the analytical column. This retention gap is placed in front of the analytical column, thus allowing different reconcentration mechanisms to occur.

Together with this solvent effect, another effect, called phase soaking, occurs in the retention gap technique: if a large volume of solvent vapour has saturated the carrier gas, the properties of the stationary phase can be altered by swelling (thicker apparent film), a change in the viscosity or changed polarity. The consequence is that the column shows an increased retention power, which can be used to better retain the most volatile components.

During the evaporation process, band broadening in space spreads the high-boiling compounds. Two retention gap effects can reconcentrate the bands, namely phase-ratio focusing and cold trapping, generally known as stationary phase focusing effects.

Phase ratio focusing is based on the higher migration speed of components through the retention gap compared to that through the analytical column. Reconcentration depends on the ratio between the retention power in the pre- and in

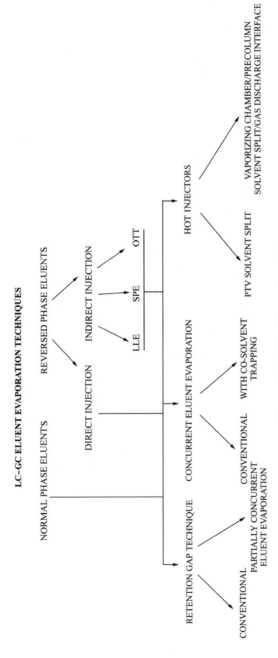

Figure 2.1 LC–GC transfer techniques.

the main column. Cold trapping occurs when the solutes reach an analytical column which is maintained at a temperature too low for them to migrate, so they accumulate and concentrate until the oven temperature is increased sufficiently to allow them to move through the column.

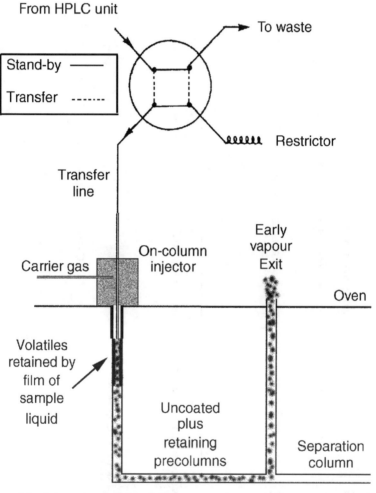

Figure 2.2 Schematic representation of an on-column interface. The eluent leaving the HPLC detector enters the valve and in the stand-by position, leaves it to go to waste. When the valve is switched on, the eluent is pumped through the transfer line into the inlet of the on-column injector. The liquid floods the capillary wall, thus creating a layer that will retain the solutes. Evaporation occurs from the rear part of the solvent so refocusing the chromatographic band. At the end of the transfer, the valve is switched off, and the eluent again flows to waste.

The retention gap technique, due to the solvent effects explained above, allows the analysis of compounds eluting immediately after the solvent peak. The limitation of this technique is the need for long retention gaps and long analysis times, since the solvent has to be completely evaporated prior to starting the elution of components of interest. It is possible to find in the literature many applications which use the classical retention gap for the transfer from HPLC to HRGC. In particular, Figure 2.3 (a) (4) shows the LC chromatogram of a bovine urine sample after formation of the dipentaflorobenzyl ether of diethylstilbestrol (DES), which is a synthetic estrogen.

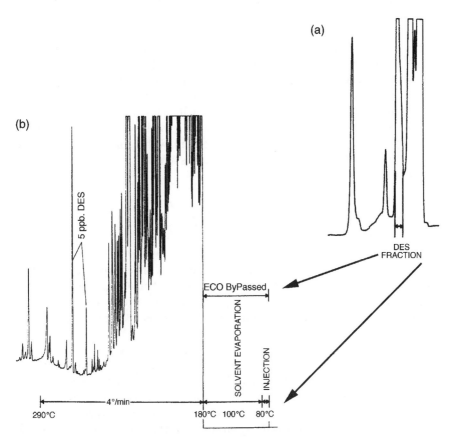

Figure 2.3 (a) LC chromatogram of a bovine urine sample of DES carried out on a 100 × 3 mm id glass column, packed with 5 μm silica gel Spherisorb S-5-W, with cyclohexane/1% THF as eluent at a flow rate of 260 μl/min. (b) GC–ECD (electron-capture detector) chromatogram of the transferred fraction coming from a urine sample spiked with 5 ppb of DES (two isomers). Reprinted from *Journal of Chromatography, 357*, K. Grob *et al.*, 'Coupled HPLC–GC as a replacement for GC–MS in the determination of diethylstil bestrol in bovine urine', pp 416–422, copyright 1986, with permission from Elsevier Science.

Transfer of an LC fraction of 300 μl volume occurred by the conventional retention gap technique. In fact, Figure 2.3(b) shows the GC chromatogram obtained after the transfer of the LC fraction.

An additional technique, called 'partially concurrent eluent evaporation,' has been introduced in order to overcome some drawbacks of the retention gap technique. In this case, a large amount of solvent is evaporated during introduction (concurrently), yet still producing a zone flooded by the eluent, thus providing solvent trapping (5). This technique allows the use of shorter retention gaps or larger transfer volumes. In theory, an early vapour exit (6) should be placed between the uncoated precolumn and the main column, but in practice a short section of the main column (7) is placed between the precolumn and the vapor exit, so rendering the closure of the vent less critical.

In particular, Figure 2.4(a) (7) shows the LC chromatogram of fat extracted from an irradiated chicken. Irradiation of foods containing fat produces radiolysis fragments of triglycerides, such as acids, propanediol esters, alkenes, aldehydes and methyl esters. The alkane/alkene fraction transferred to the gas chromatograph correspond to the first 200 μl eluted after the dead volume. The second fraction of interest was that of the aldehydes, eluted after a few minutes. The two fractions (Figure 2.4(b)), both of about 200 μl, were transferred by partially concurrent eluent evaporation by using a precolumn of 12 m × 0.50 mm id, with a 3 m × 0.32 mm id fused silica section of the separation column serving as a retaining precolumn before the solvent vapour exit.

2.2.2 LOOP-TYPE INTERFACES (CONCURRENT ELUENT EVAPORATION)

The retention gap techniques, essential for the analysis of very volatile components, are often replaced by concurrent eluent evaporation techniques, due to their simplicity and the possibility of transfering very large amount of solvent. In this case, the solvents are introduced into an uncoated inlet at temperatures at or above the solvent boiling point.

In this way, the liquid can be transferred at a speed corresponding to the evaporation speed. The fraction to be analysed is contained in a loop (see Figure 2.5), connected to a switching valve. By opening the valve, the sample in the loop is driven by the carrier gas into the GC unit (8), instead of the LC pump. An early vapour exit is usually placed after a few metres of the deactivated precolumn (9) and a short piece (3–4 m) of the main column (retaining precolumn). This valve is opened during solvent evaporation in order to reduce the amount of solvent that would reach the detector, and at the same time, to increase the solvent evaporation rate (6).

When the sample solvent evaporates at the front end of the liquid, volatile compounds co-evaporate with the solvent and start moving through the main column. In this way, volatile components can be lost through the early vapour exit or, if venting is delayed, the most volatile compounds reach the detector even before the end of

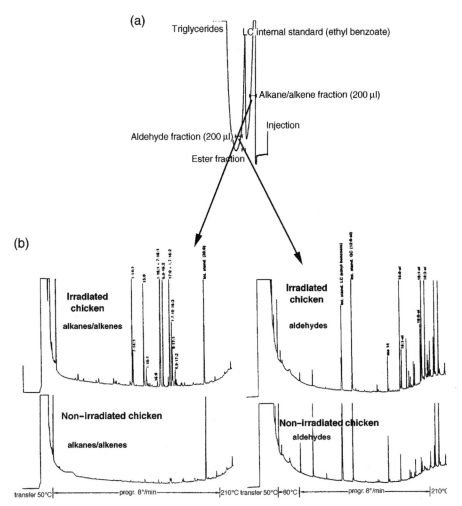

Figure 2.4 (a) LC chromatogram of fat extracted from an irradiated chicken. The alkane/alkene and the aldehyde fractions (200 μm each) were transferred to the GC unit in order to determine the effects of irradiation. (b) GC chromatograms of the fractions transferred of a non-irradiated and an irradiated chicken dose of 5 KGY. Reprinted from *Journal of High Resolution Chromatography*, 12, M. Biedermann *et al.*, 'Partially concurrent eluent evaporation with an early vapor exit; detection of food irradiation through coupled LC–GC analysis of the fat', pp. 591–598, 1989, with permission from Wiley-VCH.

solvent evaporation. In practice, the first properly shaped peaks are eluted at temperatures only 40–120 °C above the transfer temperature.

The main advantages of this technique are that short retention gaps are sufficient (3–20 m), the evaporation rate is faster than that obtained with transfers at a temperature below the boiling point of the solvent, and the requirement that solvent

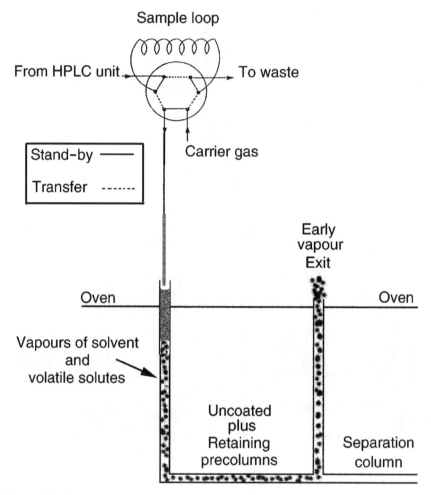

Figure 2.5   Schematic representation of a loop-interface scheme for concurrent eluent evaporation. The sample is first loaded in a loop and then, after switching the valve, directed by the carrier into the GC column. The solvent evaporates from the front end of the liquid, thus causing band broadening. Since the column is not flooded, very large amount of liquid can be introduced.

wets the retention gap is less important. Since its first reported use in 1985 (10), this technique has been largely used for the analysis of sterols, waxes, higher aliphatic and triterpene alcohols, contaminants of water and foods products, and fuel products. Figure 2.6, for example, shows the results of transferring a 10ml volume of *n*-hexane containing C_{14}–C_{26} alkanes (about 1 ppb each) (8), introduced at a rate of 120 μl/ min through an on-column interface under conditions of concurrent eluent evaporation.

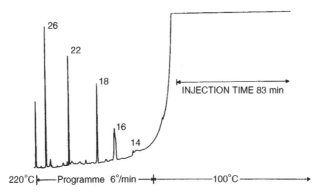

26

22

18

INJECTION TIME 83 min

16

14

220°C ├——Programme 6°/min ——┼—————100°C —————▶

Figure 2.6 Gas chromatogram of a 10 ml test sample containing C_{14}–C_{26} alkanes in *n*-hexane (about 1 ppb each): the carrier gas (H_2) inlet pressure was 2.5 bar for a 22 m × 0.32 mm id separation column coupled with a 2 m × 0.32 mm id uncoated precolumn (no vapour exit). Reprinted from *Journal of High Resolution Chromatography*, **9**, K. Grob *et al.*, 'Concurrent solvent evaporation for on-line coupled HPLC–HRGC', pp. 95–101, 1986, with permission from Wiley-VCH.

The main drawback of this technique is represented by the loss of volatile compounds. A solution to this problem is represented by the so-called co-solvent trapping technique used during concurrent eluent evaporation. In practice, a small amount of a higher boiling co-solvent is added to the main solvent at such a concentration that some co-solvent is left behind as a liquid, while some is co-evaporated with the main solvent. The co-solvent remaining forms a barrier of liquid film, which evaporates from the rear to the front, as in the retention gap technique. Volatile components are retained by the solvent trapping effect, and released when the evaporation of the co-solvent is completed. Although this technique shows some similarities to the retention gap technique, the main difference is represented by the interface used: the latter technique uses an on-column interface, while concurrent eluent evaporation is designed for a loop-type interface.

While partially concurrent eluent evaporation is easier to use, and is preferred for the transfer of normal phase solvents, concurrent eluent evaporation with co-solvent trapping is the technique of choice for transfer of water-containing solvents, because wettability is not required.

Figure 2.7 (11) shows a gas chromatogram obtained by co-solvent trapping and concurrent eluent evaporation after injecting 500 μl of diluted gasoline. The main solvent was *n*-pentane with 5% of *n*-heptane as co-solvent. It is noteworthy that without the co-solvent, higher-boiling compounds could be lost.

2.3 VAPORIZATION WITH HOT INJECTORS

Together with the techniques described above, other techniques using hot injectors for the transfer of large-volumes in capillary gas chromatography have been developed. Transfer of large-volume solvents in a programmed temperature vaporizing

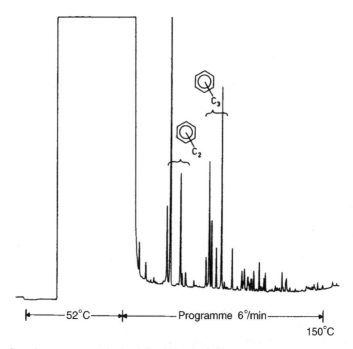

Figure 2.7   Gas chromatogram obtained for 500 µl of diluted gasoline in *n*-pentane introduced by concurrent eluent evaporation, using *n*-heptane as the co-solvent. Reprinted from
Journal of High Resolution Chromatography, **11**, K. Grob and E. Müller, 'Co-solvent effects
for preventing broadening or loss of early eluted peaks when using concurrent eluent evaporation in capillary GC. Part 2: *n*-heptane in *n*-pentane as an example', pp. 560–565, 1988, with
permission from Wiley-VCH.

(PTV) injector is carried out by opening the split valve during the transfer of the
fraction. An additional purge time is used to remove the remaining solvent from the
bed. The solvent is selectively eliminated, while the solutes are retained on the packing material. When solvent evaporation is concluded, the split line is closed before
the chamber is heated, thus allowing splitless transfer of the solutes into the column.
PTV, with solute trapping in packed beds, can be successfully applied to the transfer
of large volumes, although this method does present some disadvantages. The main
problem of injecting into a hot vaporising chamber occurs when thermally labile
compounds need to be analysed. In fact, the temperature of the injector would
decrease due to the large amount of solvent evaporating inside the chamber, and
therefore higher temperatures are necessary. Moreover, since the packed bed has a
high retention power, the chamber has to be heated above the column temperature in
order to release the solutes. However, this technique shows some advantages when
compared to the techniques that use uncoated precolumns for large sample introduction. First, wettability is not important for the retention of the liquid, and secondly,
the packing materials used for the liners are more stable than deactivated silica
tubing. In addition, packed beds retain more liquid per unit internal volume and the

PTV chamber is more easily heated than a capillary column. The introduction of large volumes with the use of such an injector in solvent-split mode was first investigated in the 1970s. Staniewski, Cramers and co-workers (12–17) demonstrated that by reducing the injector temperature and by increasing the purge gas during the solvent elimination it was easier to remove solvent, and showed that by using chambers packed with porous glass beads instead of Tenax TA or Thermotrap recovery of solutes may be improved.

A different approach was recently introduced by Sandra and co-workers. (18, 19) for the transfer of large volumes from a liquid chromatograph, to a gas chromatograph. This interface consist of a flow cell that was made by modifying an autosampler vial. The solvent coming from the HPLC unit which contains the solutes is continuously sampled by the flow cell. When the fraction of interest is inside the vial, a large-volume injection is made using the PTV device in the solvent-vent mode. This technique was successfully applied to the analysis of pesticides, as shown in Figure 2.8 (19).

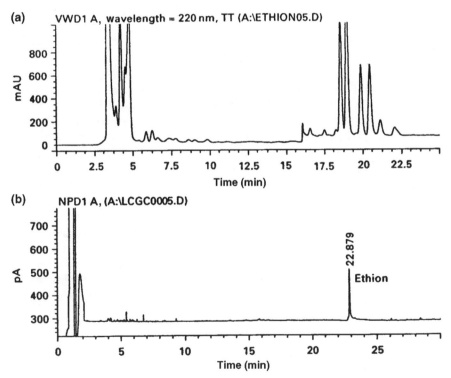

Figure 2.8 (a) HPLC fractionation of orange oil on Lichrosorb 100 diol. (b) LC–GC-NPD analysis of peel orange oil (from Florida), contaminated with ethion. Reprinted from Proceedings of the 20th International Symposium on Capillary Chromatography, F. David *et al.*, 'On-line LC–PTV–CGC: determination of pesticides in essential oils', 1998, with permission from Sandra P.

2.4 TRANSFER OF WATER-CONTAINING SOLVENT MIXTURES

A large number of samples to be analysed by gas chromatography have an aqueous matrix. Figure 2.1 shows that for the analysis of reversed-phase eluents both direct and indirect methods have been used. The direct injection of water-containing solvent mixtures in capillary GC would be attractive for accelerating an analysis but has many difficulties. The main drawbacks are as follows: the high boiling point and molar enthalpy of vaporising, thus requiring extensive solvent evaporation; due to its small molecular weight, water forms a very large volume of vapour per volume of liquid (about six time more than hexane), and elimination of this vapour volume via the column is tedious and time-consuming; condensed water destroys the deactivation of the precolumn due to the hydrolysis of siloxane bonds; the extremely high surface tension of water does not allow it to wet deactivated capillary surfaces, which thus makes water a poor solvent for the formation of the solvent film that is essential for solute trapping.

2.4.1 DIRECT INJECTION BY USING A RETENTION GAP

On-column injection of large volumes of aqueous samples has achieved considerable attention in the field of on-line reversed phase LC–GC. The main problem in direct introduction of water, as mentioned above, is the poor wettability of the uncoated precolumns. In 1989, Grob and Li (20) tested several fused silica and glass precolumns deactivated by using different methods and concluded that the transfer of aqueous solvents by retention gap techniques was not achievable because it was impossible to find a precolumn that at the same time was both water-wettable and inert. This problem was tentatively solved by using an organic solvent with a higher boiling point than that of water. In fact, these same authors (21) investigated the wettability of phenyl- and cyanosilylated precolumns with mixtures of organic solvents and water, by using the retention gap technique for transferring water and mixtures of water with organic solvents. Their results demonstrated that, depending on the organic solvent being used, mixtures of such solvents with high concentrations of water still wet the precolumns (e.g. 70% water and 30% 1-propanol), although water did not evaporate together with the organic component. Azeotropically boiling mixtures, e.g. 28% water and 72% of 1-propanol, demonstrated that wettability of such precolumns is possible and thus allows the introduction of water by the retention gap technique.

2.4.2 DIRECT INJECTION BY USING CONCURRENT SOLVENT EVAPORATION

As mentioned above, concurrent solvent evaporation does not need good wettability of the solvent used for the LC–GC transfer. However, due to the large amount of vapour released and the high temperatures needed for concurrent solvent evaporation of water, this technique is limited to high-boiling analytes. As an illustration of this, the determination of atrazine in tap water is shown in Figure 2.9 (22). This method is based on enrichment of the atrazine from 10 ml of water on a small LC column packed with silica-C18, and desorption with methanol–water (60:40) + 5% 1-propanol to the GC column by concurrent eluent evaporation using a loop-type interface. Both the retaining precolumn and the separation column were coated with Carbowax 20 M which had a very high retention for atrazine. This was necessary to obtain the high elution temperature required for atrazine. In fact, when transferring the LC fraction at 112 °C the atrazine peak was perfectly shaped only when eluted at about 250°C.

2.4.3 DIRECT INJECTION BY USING CONCURRENT SOLVENT EVAPORATION WITH A CO-SOLVENT

A partial solution to the problem of producing sharp peaks at low elution temperatures is to add a small amount of a higher-boiling co-solvent to the main solvent. As suggested by Grob and Muller (23, 24), butoxyethanol can be used as a suitable cosolvent for aqueous mixtures in such cases.

2.4.4 DIRECT INTRODUCTION OF WATER VIA A VAPORIZER CHAMBER/PRECOLUMN SOLVENT SPLIT/GAS DISCHARGE INTERFACE

Recently, the direct introduction of water-containing eluents via a vaporizer chamber/precolumn solvent split/gas discharge interface has been reported (25, 26). Water and water-containing eluents were driven into a vaporizer chamber at 300°C by the LC pump (Figure 2.10). This high temperature permitted evaporation of water at a rate up to around 200 μl/min. The vapours were then removed through a retaining precolumn and a early vapour exit, driven by the flow of carrier gas (discharge). The vaporizing chamber consisted of a 1 mm id glass tube, packed with a 2 cm plug of Carbofrit and internally coated with polyimide. Solvent/solute separation occurred in the retaining precolumn, and special attention was given to the oven temperature during the transfer, being held close to the temperature at which recondensation occurs (the dew point). This method was successfully applied to the determination of phthalates in drinking water. Figure 2.11(a) shows a liquid chromatogram, obtained on a column packed with C-18 (5 μm) bonded silica (1 cm × 3 mm i.d.) of a water sample spiked with dibutyl phthalate (DBP) and diethylhexyl phthalate (DEHP). After sample enrichment, 10 ml of the fraction was transferred to the gas chromatograph, driven by the LC eluent (water/methanol 15:85) and by reducing the flow rate to 100 μl/min. The LC–GC–MS(EI) chromatogram of the treated water containing 55 and 40 ng/l of DBP and DHEP, respectively is shown in Figure 2.11(b).

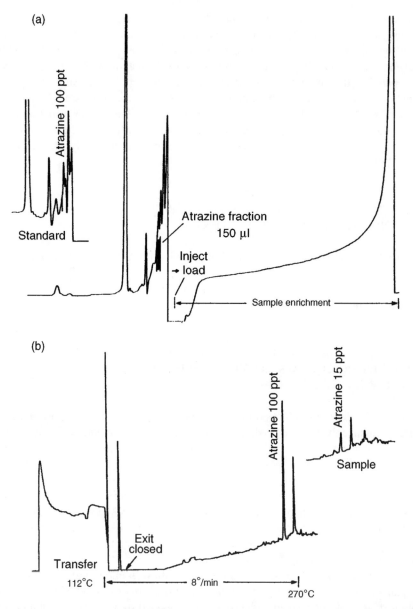

Figure 2.9   (a) Liquid chromatogram (UV detection at 220 nm) obtained after initial sample enrichment (10 ml of water). At the point indicated, shortly before the content of the sample loop was completely transferred, the injection valve was switched, thus starting elution. The atrazine fraction is marked. On the left, part of the chromatogram of another tap water sample spiked with 100 ppt of atrazine is shown. (b) Gas chromatogram of tap water spiked with 100 ppt and 15 ppt of atrazine. Reprinted from *Journal of Chromatography A*, **473**, K. Grob and Z. Li, 'Coupled reversed-phase liquid chromatography – capillary gas chromatography for the determination of atrazine in water', pp. 423–430, copyright 1989, with permission from Elsevier Science.

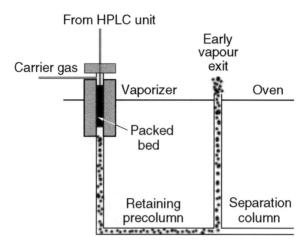

Figure 2.10   Schematic representation of a vaporizing chamber/precolumn solvent split/gas discharge interface, where the vaporizer is packed and heated at a suitable temperature for solvent evaporation. The vapour exit can be positioned at the end of the retention gap.

2.5   INDIRECT INTRODUCTION OF WATER

2.5.1   SOLID PHASE EXTRACTION

A method for coupling reversed phase LC to GC without introducing water into the GC unit uses precolumns or solid-phase extraction (SPE) cartridges for the solutes enrichment, followed by drying and extraction with organic solvent. Normally, a large volume of water (1 ml) containing the organic compounds is passed through the precolumn. The organic compounds are then eluted by an organic solvent into the gas chromatograph, after flushing of the precolumn with nitrogen to remove the residual water. This method, mostly used by Brinkman and co-workers was first developed (27) by using an SPE cartridge (4 mm × 1 mm id) inserted in a six-port valve and packed with C-18 bonded silica.

Subsequently, Vreuls and co-workers (28–31) used a small LC column installed on a 6- or 10-port valve. These columns usually had a length of 2–10 mm and an id of 1–4.6 mm (32). A large volume (1–10 ml) of aqueous sample containing the organic compounds is then passed and concentrated on a polymer-packed column. After several washings with water, the column must be dried with nitrogen. The organic compounds are then eluted by using an organic solvent such as hexane, or ethyl acetate (e.g. 50 μl) which is introduced on-line into the GC unit. Due to the large amount of organic solvent introduced into the gas chromatograph, large-volume introduction techniques are required, such as those discussed above. A typical on-line SPE–GC set-up is shown in Figure 2.12 (32).

Drying carried out with purging of a precolumn is time consuming, and an interesting alternative is the use of a drying cartridge positioned between the SPE and GC units. This small cartridge should be reconditioned during the GC run by heating. In

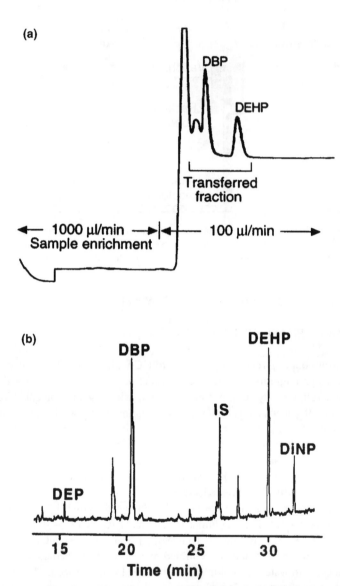

Figure 2.11   (a) HPLC chromatogram obtained for a sample of drinking water spiked with dibutyl phthalate (DBP) and diethylhexyl phthalate (DEHP). (b) The LC–GC/MS chromatogram of the same sample after sample enrichment. Reprinted from *Journal of High Resolution Chromatography*, **20**, T. Hyötyläinen *et al.*, 'Reversed phase HPLC coupled on-line to GC by the vaporizer/precolumn solvent split/gas discharge; analysis of phthalates in water', pp. 410–416, with permission from Wiley-VCH.

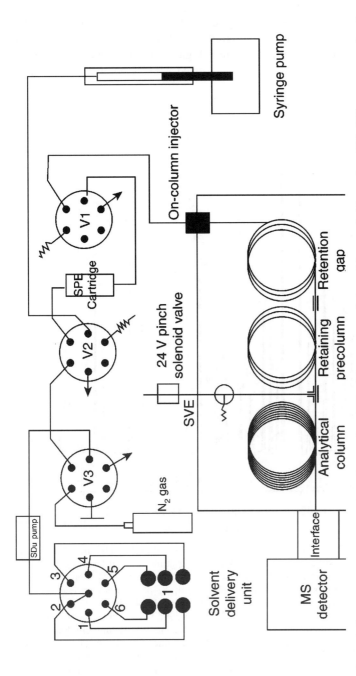

Figure 2.12 Schematic representation of an on-line SPE–GC system consisting of three switching valves (V1–V3), two pumps (a solvent-delivery unit (SDU) pump and a syringe pump) and a GC system equipped with a solvent-vapour exit (SVE), an MS instrument detector, a retention gap, a retaining precolumn and an analytical column. Reprinted from *Journal of Chromatography*, *A* **725**, A. J. H. Louter *et al.*, 'Analysis of microcontaminants in aqueous samples by fully automated on-line solid-phase extraction–gas chromatography–mass selective detection', pp. 67–83, copyright 1996, with permission from Elsevier Science.

a recent paper (33), three drying agents and four solvents were tested by using a mixture of 24 microcontaminants ranging widely in polarity and volatility. Silica was preferred among the drying agents investigated (sodium sulfate, silica and molecular sieves), while methyl acetate was the preferred solvent after examining pentane, hexane, ethyl acetate and methyl acetate. In fact, Figure 2.13 shows the on-line SPE-drying–GC–MS chromatogram of tap water spiked with 24 pollutants at low-ng/l detection levels.

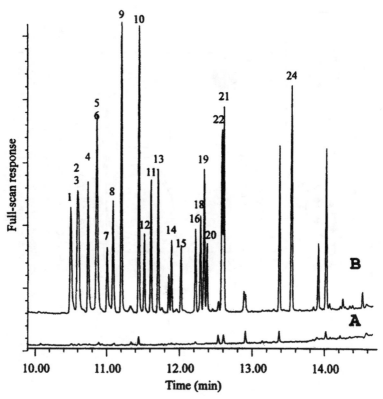

Figure 2.13 On-line SPE–GC–MS (full scan; *m/z* 35–435) chromatograms of 10 ml of (Amsterdam) tap water without (A) and with (B) spiking at the 0.5 μg/l level. Peak assignment: 1, 3,4-dichlorobenzene; 2, dimethylphthalate; 3, 1,3-dinitrobenzene; 4, 4-butoxyphenol; 5, acenaphthene; 6, 3-nitroanalinine; 7, 1-naphthenelol; 8, pentachlorobenzene; 9, 2,5-diethoxyaniline; 10, diethylphthalate; 11, 1-nitrophthalene; 12, 1,2,4,6-bis-*O*-(1-methylethylidine)-a-L-sorbofuranose; 13, ributylphosphate; 14, trifluralin; 15, 1,4-dibutoxybenzene; 16, hexachlorobenzene; 17, dimethoate; 18, simazine; 19, atrazine; 20, trichloroethylphosphate; 21, phenantrene, 22, diazinon (internal standard) 23, caffeine; 24, metolachlor. Reprinted from *Journal of High Resolution Chromatography,* **21**, T. Hankemeier *et al.*, 'Use of a drying cartridge in on-line solid-phase extraction–gas chromatography–mass spectrometrys, pp. 450–456, 1998, with permission from Wiley-VCH.

2.5.2 LIQUID–LIQUID EXTRACTION

This method relies on the on-line extraction of an organic compound from an aqueous sample by using an organic solvent. In automated liquid–liquid extraction (LLE), the solvent (organic) is injected via a small segmentor into the water sample, which flows through a small-diameter fused silica tube. Phase separation is achieved by using a semipermeable PTFE membrane (34–36) or a sandwich-type phase separator. For the first case (34), the configuration adopted is illustrated in Figure 2.14. The water sample was delivered by either an HPLC pump, syringe or a pressurized flask, and then extracted with *n*-pentane. Separation of the two phases was achieved by using a membrane separator (Figure 2.15), based on a semipermeable PTFE membrane, which has different wetting characteristic for the phases. This system was used to analyse volatile organic trace compounds, as shown in Figure 2.16 where a water sample was spiked with 200 ppb of a naphtha fraction (34).

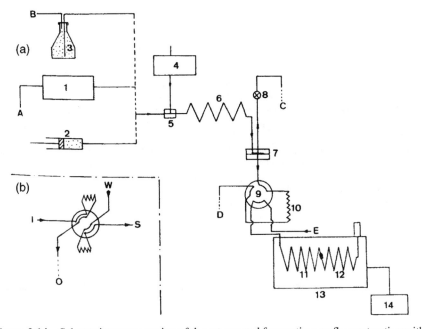

Figure 2.14 Schematic representation of the set-up used for continuous flow extraction with on-line capillary GC: (A) 1 and 4 HPLC pumps; 2, syringe pump; 3, pressurized flask with water sample; 5, segmentor; 6, fused-silica capillary tube; 7, phase separator; 8, needle valve; 9, six-port valve; 10, sample loop; 11, empty precolumn; 12, GC column; 13, GC oven; 14, recorder/integrator; A, water sample; B, pressurized nitrogen, C, extracted water; D, extracted drain; E, carrier gas: (B) I, sample; O, sample drain; W, clean water stream; S, segmentor. Reprinted from *Journal of Chromatography, A* **330**, J. Roeraade, 'Automated monitoring of organic trace components in water. I. Continuous flow extraction together with on-line capillary gas chromatography', pp. 263–274, copyright 1985, with permission from Elsevier Science.

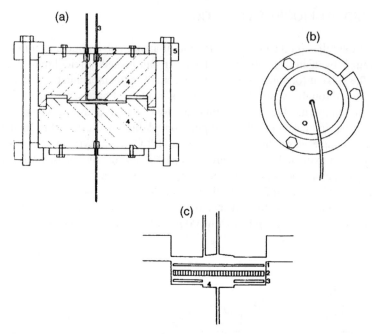

Figure 2.15   Schematic representation of the phase separator: (A) 1, cavity with silicone rubber O-ring; 2, stainless-steel compression plate; 3, fused-silica tubing; 4, stainless-steel body; 5, aluminium ring: (B) bottom view of the separator (top view when solvents with a lower density than that of water are employed): (C) enlarged view of the separator cavities; 1, Fluoropore membrane; 2, stainless-steel screen; 3, Fluoropore washer; 4, cavity for the organic phase (diameter 4 mm; depth 0.1 mm). Reprinted from *Journal of Chromatography A*, **330**, J. Roeraade, 'Automated monitoring of organic trace components in water. I. Continuous flow extraction together with on-line capillary gas chromatography', pp. 263–274, copyright 1985, with permission from Elsevier Science.

Another approach is where membranes are not used in the construction material of the sandwich-phase separator (36). In this configuration, water was pumped through the system by using a membrane pump (1 ml/min) with the organic phase (isooctane) being delivered by a second membrane pump (again at a rate of 1 ml/min). The extraction coil was a 6.4 m long PTFE tube with phase separation being accomplished by the use of a sandwich-phase separator (Figure 2.17) (36,37). with this arrangement, 100 µl of isooctane could be injected by fully concurrent solvent evaporation. By using such a system the determination of 0.1 pg/l of hexachlorocyclohexanes in water was possible (Figure 2.18).

A similar on-line LLE system, equipped with an on-line derivatization system, has also been reported. Brinkman and co-workers, for example, have presented a complete automated on-line system that allows both the LLE and alkylation of organic acids (38). More recently, this same group have reported the coupling of

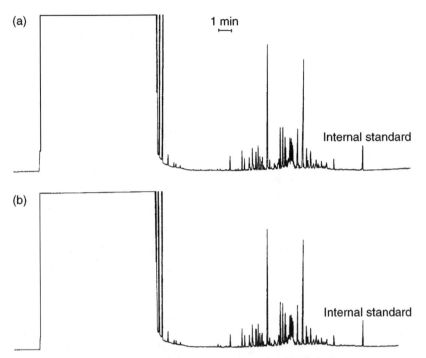

Figure 2.16 Chromatograms of a pentane extract of a water sample containing 200 ppb of a naphtha fraction: (a) sample extracted by using a continuous flow system, where a pressurized bottle was employed as the sample-delivery system; (b) batch-extracted sample. Reprinted from *Journal of Chromatography, A* **330**, J. Roeraade, 'Automated monitoring of organic trace components in water. I. Continuous flow extraction together with on-line capillary gas chromatography', pp. 263–274, copyrigth 1985, with permission from Elsevier Science.

such a system with an atomic emission detector (AED) for the analysis of nitrogen-chlorine- and sulfur-containing pesticides in aqueous samples (39), as shown in Figure 2.19.

2.5.3 OPEN TUBULAR TRAPPING

The analysis of organic compounds from aqueous samples is also possible by using open tubular trapping (OTT) columns (13, 40–42). The extraction step involves sorption of the analytes from water (40) into the stationary phase of an open tube (5 m × 0.53 mm id) with a film of CP-Sil-5CB (5 μm film thick). Removal of water is achieved by purging the trap with nitrogen, and desorption of the analytes with an organic solvent (75 μl). Solvent elimination prior to transfer to the GC column is carried out by using a PTV injector and a multidimensional system (Figure 2.20).

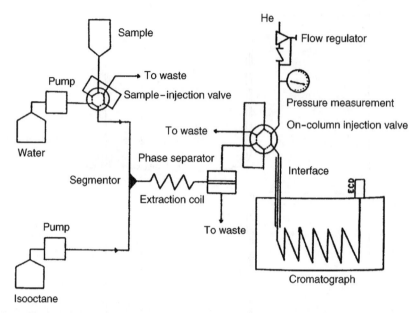

Figure 2.17   Schematic representation of the set-up used for on-line liquid–liquid extraction coupled with capillary GC when using a membrane phase separator. Reprinted from *Journal of High Resdution Chromatography*, **13**, E. C. Goosens *et al.*, 'Determination of hexachloro-cyclohexanes in ground water by coupled liquid–liquid extraction and capillary gas chromatography', pp. 438–441, 1990, with permission from Wiley-VCH.

Figure 2.21 shows the on-line extraction gas chromatogram of 2.25 ml of water spiked at 5 ppb levels with 14 different organic pollutants (40). In this case, the authors concluded that wall-coated open tubular traps (thick-film polysiloxane phases) can be used for the on-line extraction of organic compounds from water. However, when using swelling agents such as pentane, non-polar analytes can be trapped quantitatively, while for more polar compounds chloroform is the most suitable solvent.

2.6   CONCLUSIONS

Coupled liquid chromatography–gas chromatography is an excellent on-line method for sample enrichment and sample clean-up. Recently, many authors have reviewed in some detail the various LC–GC transfer methods that are now available (1, 43–52). For the analysis of normal phase eluents, the main transfer technique used is, without doubt, concurrent eluent evaporation employing a loop-type interface. The main disadvantage of this technique is co-evaporation of the solute with the solvent,

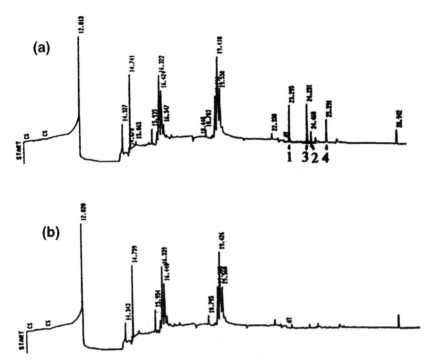

Figure 2.18 (a) Gas chromatogram of a standard solution of various hexachlorocyclohex-anes (HCHs) in water, obtained after on-line isooctane extraction: 1, α-HCH; 2, β-HCH; 3, γ-HCH; 4, δ-HCH. (b) Gas chromatogram obtained for a reference blank (distilled water) after the same on-line extraction treatment. Reprinted from *Journal of High Resolution Chromatography*, **13**, E. C. Goosens *et al.*, 'Determination of hexachlorocyclohexanes in ground water by coupled liquid–liquid extraction and capillary gas chromatography', pp. 438–441, 1990, with permission from Wiley-VCH.

thus leading to the loss of the more volatile components. If more-volatile compounds have to be analysed, then transfer from the LC unit to the GC unit is best achieved by using retention gap techniques. Due to the solvent effects explained above, this tech-nique allows the analysis of compounds which elute immediately after the solvent peak. The main drawback of this approach is that it is restricted to the transfer of small fractions, due to the limited capacity of the uncoated precolumns. Larger frac-tions can be transferred by partial concurrent evaporation, which still retains the advantages of the retention gap technique. Indirect injection of water-containing elu-ents seems to be the appropriate choice for the analysis of such samples (LLE, SPE and OTT). However, direct injection of water via a vaporizer chamber/precolumn solvent split/gas discharge interface seems to be a promising technique for transfer-ing reversed-phase eluents.

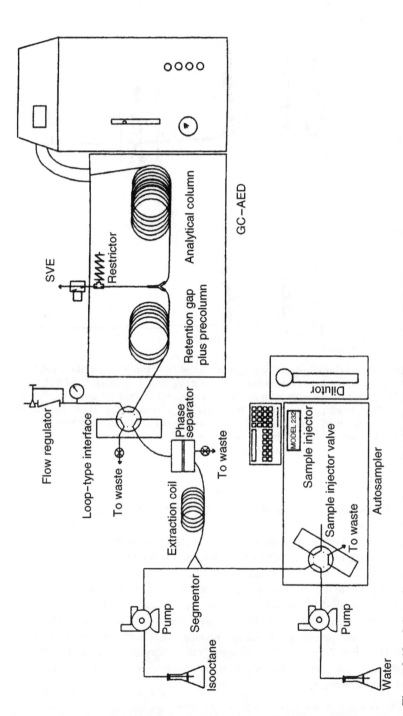

Figure 2.19 Schematic representation of an on-line liquid–liquid extraction–GC/AED system. Reprinted from *Journal of High Resolution Chromatography*, **18**, E. C. Goosens *et al.*, 'Continuous liquid–liquid extraction combined on-line with capillary gas chromatography–atomic emission detection for environmental analysis', pp. 38–44, 1995, with permission from Wiley-VCH.

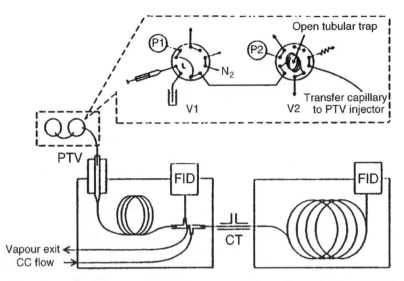

Figure 2.20 Schematic representation of the set-up used for on-line extraction–GC: V1 and V2, valves; P1 and P2, syringe pumps; L, sample loop; CC flow, countercurrent flow; CT, cold trap. Reprinted from *Journal of High Resolution Chromatography*, **16**, H. G. J. Mol *et al.*, 'Use of open-tubular trapping columns for on-line extraction–capillary gas chromatography of aqueous samples', pp. 413–418, 1993, with permission from Wiley-VCH.

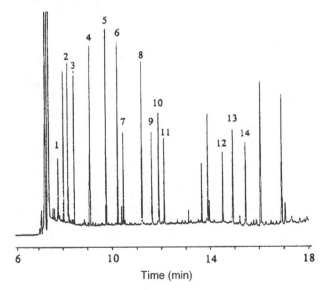

Figure 2.21 A gas chromatogram of a sample of river water (2.25 ml) spiked at 5 ppb levels with: 1, toluene; 2, ethylbenzene; 3, methoxybenzene; 4, *p*-dichlorobenzene; 5, dimethylphenol; 6, dimethylaniline; 7, chloroaniline; 8, indole; 9, dichlorobenzonitrile; 10, trichlorophenol; 11, dinitrobenzene; 12, trifluranil; 13, atrazine; 14, phenanthrene. Reprinted from *Journal of High Resolution Chromatography*, **16**, H. G. J. Mol *et al.*, 'Use of open-tubular trapping columns for on-line extraction–capillary gas chromatography of aqueous samples', pp. 413–418, 1993, with permission from Wiley-VCH.

REFERENCES

1. K. Grob, *On-Line Coupled LC–GC*, W. Bertsch, W. G. Jennings and P. Sandra (Series Eds), Hüthig, Heidelberg, Germany (1991).
2. K. Grob, D. Fröhlich, B. Schilling, H. P. Neukom and P. Nägeli, 'Coupling of high-performance liquid chromatography with capillary gas chromatography',*J. Chromatogr.* **295**: 55–61 (1991).
3. K. Grob, *On-Column Injection in Capillary Gas Chromatography*, W. Bertsch, W. G. Jennings and P. Sandra (Series Eds), Hüthig, Heidelberg, Germany (1991).
4. K. Grob, H. P. Neukom and R. Etter, 'Coupled HPLC–GC as a replacement for GC–MS in the determination of diethylstilbestrol in bovine urine', *J. Chromatogr.* **357**: 416–422 (1986).
5. F. Munari, A. Trisciani, G. Mapelli, S. Trestianu, K. Grob and J. M. Colin, 'Analysis of petroleum fractions by on-line micro HPLC–HRGC coupling, involving increased efficiency in using retention gaps by partially concurrent solvent evaporation', *J. High Resolut. Chromatogr. & Chromatogr. Commun.* **8**: 601–606 (1985).
6. T. Noy, E. Weiss, T. Herps, H. van Cruchten and J. Rijks, 'On-line combination of liquid chromatography and capillary gas chromatography. Preconcentration and analysis of organic compounds in aqueous sample', *J. High Resolut. Chromatogr. & Chromatogr. Commun.* **11**: 181–186 (1988).
7. M. Biedermann, K. Grob and W. Meier, 'Partially concurrent eluent evaporation with an early vapor exit; detection of food irradiation through coupled LC–GC analysis of the fat', *J. High Resolut. Chromatogr.* **12**: 591–598 (1989).
8. K. Grob, C. Walder and B. Schilling, 'Concurrent solvent evaporation for on-line coupled HPLC–HRGC', *J. High Resolut. Chromatogr. & Chromatogr. Commun.* **9**: 95–101 (1986).
9. K. Grob and J. M. Stoll, 'Loop-type interface for concurrent solvent evaporation in coupled HPLC–GC. Analysis of raspberry ketone in a raspberry sauce as an example', *J. High Resolut. Chromatogr. & Chromatogr. Commun.* **9**: 518–523 (1986).
10. K. Grob and B. Schilling, 'Coupled HPLC–capillary GC–state of the art and outlook', *J. High Resolut. Chromatogr. & Chromatogr. Commun.* **8**: 726–733 (1985).
11. K. Grob and E. Müller, 'Co-solvent effects for preventing broadening or loss of early eluted peaks when using concurrent eluent evaporation in capillary GC. Part **2**: *n*-heptane in *n*-pentane as an example', *J. High Resolut. Chromatogr. & Chromatogr. Commun.* **11**: 560–565 (1988).
12. J. Staniewski and J. A. Rijks, 'Potentials and limitations of the linear design for cold temperature programmed large volume injection in capillary GC and for LC–GC interfacing' in *Proceedings of the 13th International Symposium on Capillary Chromatography,* Riva del Garda, Italy, Sandra P. (Ed.), Hüthing, Heidelburg, Germany, pp. 1334–1337 (1991).
13. J. Staniewski, H.G. Janssen, C. A. Cramers and J. A. Rijks, 'Programmed-temperature injector for large-volume sample introduction in capillary gas chromatography and for liquid chromatography–gas chromatography interfacing', *J. Microcolumn Sep.* **4**: 331–338 (1993).
14. H. G. J. Mole, J. Staniewski, J. A. Rijks, H. G. Janssen, C. A. Cramers and R. T. Ghijsen, 'Use of an open tubular trapping column as interface in on-line coupled reversed phase LC–capillary GC' in *Proceedings of the 14th International Symposium on Capillary Chromatography,* Baltimore, MD, USA, Sandra P. (Ed.), Hüthing, Heidelburg, Germany, pp. 645–653(1992).
15. J. Staniewski, H. G. Janssen, J. A. Rijks and C. A. Cramers, 'Introduction of large volumes of methylene chloride in capillary GC with electron capture detection', in

Proceedings of the 15th International Symposium on Capillary Chromatography, Riva del Garda, Italy, Sandra P. (Ed.), Hüthing, Heidelburg, Germmany, pp. 401–405 (1993).

16. J. Staniewski, H. G. Janssen and C. A. Cramers, 'A new approach for the introduction of large sample volumes in capillary GC for LC–GC interfacing', in *Proceedings of the 15th International Symposium on Capillary Chromatography,* Sandra P. (Ed.), Hüthing, Heidelburg, Germmany, pp. 808–813 (1993).

17. J. Staniewski and J. A. Rijks, 'Potential and limitations of differently designed programmed-temperature injector liners for large volume sample introduction in capillary GC,' *J. High Resolut. Chromatogr.* **16**: 182–187 (1993).

18. F. David, P. Sandra, D. Bremer, R. Bremer, F. Rogles and A. Hoffmann, 'Interface for HPLC/GC coupling', *Labor. Praxis.* **21**: 82–86 (1997).

19. F. David, R. C. Correa and P. Sandra, 'On-line LC–PTV–CGC: determination of pesticides in essential oils', in *Proceedings of the 20th International Symposium on Capillary Chromatography,* Riva del Garda, Italy, Sandra P. and Rocktraw (Eds), Hüthing, Heidelburg, Germany (CD-ROM) (1998).

20. K. Grob and Z. Li, 'Introduction of water and water-containing solvent mixtures in capillary gas chromatography. I. Failure to produce water-wettable precolumns (retention gaps)', *J. Chromatogr.* **473**: 381–390 (1998).

21. K. Grob and Z. Li, 'Introduction of water and water-containing solvent mixtures in capillary gas chromatography. II. Wettability of precolumns by mixtures of organic solvents and water retention gas techniques', *J. Chromatogr.* **473**: 391–400 (1989).

22. K. Grob and Z. Li, 'Coupled reversed-phase liquid chromatography–capillary gas chromatography for the determination of atrazine in water', *J. Chromatogr.* **473**: 423–430 (1989).

23. K. Grob and E. Müller, 'Introduction of water and water-containing solvent mixtures in capillary gas chromatography. IV. Principles of concurrent solvent evaporation with co-solvent trapping', *J. Chromatogr.* **473**: 411–422 (1989).

24. K. Grob, 'Concurrent eluent evaporation with co-solvent trapping for on-line reversed-phase liquid chromatography–gas chromatography. Optimization of conditions', *J. Chromatogr.* **477**: 73–86 (1989).

25. T. Hyötyläinen, K. Grob, M. Biedermann and M-L. Riekkola, 'Reversed phase HPLC coupled on-line to GC by the vaporizer/precolumn solvent split/gas discharge; analysis of phthalates in water', *J. High Resolut. Chromatogr.* **20**: 410–416 (1997).

26. T. Hyötyläinen, K. Jauho and M-L. Riekkola, 'Analysis of pesticides in red wines by on-line coupled reversed phase liquid chromatography with a vaporizer/precolumn solvent split/gas discharge interface', *J. Chromatogr.* **813**: 113–119 (1997).

27. E. Noorozian, F. A. Maris, M. W. F. Nielen, R. W. Frei, G. J. de Jong and U. A. Th Brinkman, 'Liquid chromatographic trace enrichment with on-line capillary gas chromatography for the determination of organic pollutants in aqueous samples', *J. High Resolut. Chromatogr. & Chromatogr. Commun.* **10**: 17–24 (1987).

28. J. J. Vreuls, W. J. G. M. Cuppen, G. J. de Jong and U. A. Th Brinkman, 'Ethyl acetate for the desorption of a liquid chromatographic precolumn on-line into a gas chromatograph', *J. High Resolut. Chromatogr.* **13**: 157–161 (1990).

29. J. J. Vreuls, R. T. Ghijsen, G. J. de Jong and U. A. Th Brinkman, 'Drying step for introduction of water-free desorption solvent into a gas chromatograph after on-line liquid chromatographic trace enrichment of aqueous samples', *J. Chromatogr.* **625**: 237–245 (1992).

30. J. J. Vreuls, A. J. Bulterman, R. T. Ghijsen and U. A. Th Brinkman, 'On-line preconcentration of aqueous samples for gas chromatographic–mass spectrometric analysis', *Analyst* **117**: 1701–1705 (1992).

31 A. J. Bulterman, J. J. Vreuls, R. T. Ghijsen and U. A. Th Brinkman, 'Selective and sensitive detection of organic contaminants in water samples by on-line trace enrichment–gas chromatography–mass spectrometry', *J. High Resolut. Chromatogr.* **16**: 397–403 (1993).

32 A. J. H. Louter, C. A. van Beekvelt, P Cid Montanes, J, Slobodnik, J. J. Vreuls and U. A. Th Brinkman, 'Analysis of microcontaminants in aqueous samples by fully automated on-line solid-phase extraction–gas chromatography–mass selective detection', *J. Chromatogr* **725**: 67–83 (1996).

33 T. Hankemeier, A. J. H. Louter, J. Dallüge, J. J. Vreuls and U. A. Th Brinkman, 'Use of a drying cartridge in on-line solid-phase extraction–gas chromatography–mass spectrometry', *J. High Resolut. Chromatogr.* **21**: 450–456 (1998).

34 J. Roeraade, 'Automated monitoring of organic trace components in water. I. Continuous flow extraction together with on-line capillary gas chromatography', *J. Chromatogr.* **330**: 263–274 (1985).

35 E. Fogelqvist, M. Krysell and L-G. Danielsson, 'On-line liquid–liquid extraction in a segmented flow directly coupled to on-column injection into a gas chromatograph', *Anal. Chem.* **58**: 1516–1520 (1986).

36 E. C. Goosens, R. G. Bunschoten, V. Engelen, D. de Jong and J. H. M. van den Berg, 'Determination of hexachlorocyclohexanes in ground water by coupled liquid–liquid extraction and capillary gas chromatography', *J. High Resolut. Chromatogr.* **13**: 438–441 (1990).

37 C. de Ruiter, J. H. Wolf, U. A. Th Brinkman and R. W. Frei, 'Design and evaluation of a sandwich phase separator for on-line liquid/liquid extraction', *Anal. Chim. Acta* **192**: 267–275 (1987).

38 E. C. Goosens, M. H. Broekman, M. H. Wolters, R. E. Strrijker, D. de Jong and G. J. de Jong, 'A continuous two-phase reaction system coupled on-line with capillary chromatography for the determination of polar solutes in water', *J. High Resolut. Chromatogr.* **15**: 242–248 (1992).

39 E. C. Goosens, D. de Jong, G. J. de Jong, F. D. Rinkema and U A Th Brinkman, 'Continuous liquid–liquid extraction combined on-line with capillary gas chromatography–atomic emission detection for environmental analysis', *J. High Resolut. Chromatogr.* **18**: 38–44 (1995).

40 H. G. J. Mol, H.-G. Janssen and C. A. Cramers, 'Use of open-tubular trapping columns for on-line extraction–capillary gas chromatography of aqueous samples', *J. High Resolut. Chromatogr.* **16**: 413–418 (1993).

41 H. G. J. Mol, J. Staniewski, H.-G. Janssen and C. A. Cramers, 'Use of an open-tubular trapping column as phase-switching interface in on-line coupled reversed-phase liquid chromatography–capillary gas chromatography', *J. Chromatogr.* **630**: 201–212 (1993).

42 H. G. J. Mol, H.-G. Janssen, C. A. Cramers and U. A. Th Brinkman, 'On-line sample enrichment–capillary gas chromatography of aqueous samples using geometrically deformed open-tubular extraction columns', *J. Microcolumn Sep.* **7**: 247–257 (1995).

43 A. J. H. Louter, J. J. Vreuls and U. A. Th Brinkman, 'On-line combination of aqueous-sample preparation and capillary gas chromatography', *J. Chromatogr.* **842**: 391–426 (1999).

44 L. Mondello, P. Dugo, G. Dugo, A. C. Lewis and K. D. Bartle, 'High performance liquid chromatography coupled on-line with high resolution gas chromatography. State of the art', *J. Chromatogr.* **842**: 373–390 (1999).

45 L. Mondello, P. Dugo, G Dugo, K. D. Bartle and A. C. Lewis, 'Liquid chromatography–gas chromatography', in *Encyclopedia Separation Science*, Academic Press, London, pp. 3261–3268 (2000).

46 E. C. Goosens, D. de Jong, G. J. de Jong and U. A. Th Brinkman, 'On-line sample treatment–capillary gas chromatography', *Chromatographia* **47**: 313–345 (1998).

47 T. Hyötyläinen and M-L. Riekkola, 'Direct coupling of reversed-phase liquid chromatography to gas chromatography', *J. Chromatogr.* **819**: 13–24 (1998).

48 L. Mondello, G. Dugo and K. D. Bartle, 'On-line microbore high performance liquid chromatography–capillary gas chromatography for food and water analyses: a review', *J. Microcolumn Sep.* **8**: 275–310 (1996).

49 H. G. J. Mol, H.-G. M. Janssen, C. A. Cramers, J. J. Vreuls and U. A. Th Brinkman, 'Trace level analysis of micropollutants in aqueous samples using gas chromatography with on-line sample enrichment and large volume injection', *J. Chromatogr.* **703**: 277–307 (1995).

50 K. Grob, 'Developments of the transfer techniques for on-line high performance liquid chromatography–capillary gas chromatography', *J. Chromatogr.* **703**: 265–276 (1995).

51 M.-L. Riekkola, 'Applications of on-line coupled liquid chromatography–gas chromatography', *J. Chromatogr.* **473**: 315–323 (1989).

52 I. L. Davies, K. E. Markides, M. L. Lee, M. W. Raynor and K. D. Bartle, 'Applications of coupled LC–GC: a review', *J. High Resolut. Chromatogr.* **12**: 193–207 (1989).

3 Multidimensional High Resolution Gas Chromatography

A. C. LEWIS

University of Leeds, Leeds, UK

3.1 INTRODUCTION

The coupling of gas chromatography columns to enable multidimensional separations has been widely reported in many areas of industrial and environmental analysis. The application of multidimensional GC has been focused in essentially two areas: (i) increasing peak capacity of the separation system, and (ii) increasing the speed of analysis of the separation system. It was perhaps the former of these two that drove the early interest in two-dimensional GC couplings, and this still remains important today. Despite GC still being very much a developing technique, two-dimensional systems, were being applied to the analysis of crude oil and refinery products as early as the late 1960s (1). These early applications focused on achieving a higher degree of deconvolution with a two-column system for the characterization of feedstock and refinery fuels, and this over the past three decades has become a recurring application of two-dimensional gas chromatography.

In common with all multidimensional separations, two-dimensional GC has a requirement that target analytes are subjected to two or more mutually independent separation steps and that the components remain separated until completion of the overall procedure. Essentially, the effluent from a primary column is reanalysed by a second column of differing *stationary phase selectivity*. Since often enhancing the peak capacity of the analytical system is the main goal of the coupling, it is the relationship between the peak capacities of the individual dimensions that is crucial. Giddings (2) outlined the concepts of peak capacity product and it is this function that results in such powerful two-dimensional GC separations.

This present chapter will not focus on the statistical theory of overlapping peaks and the deconvolution of complex mixtures, as this is treated in more detail in Chapter 1. It is worth remembering, however, that of all the separation techniques, it is gas chromatography which is generally applied to the analysis of the most complex mixtures that are encountered. Individual columns in gas chromatography can, of course, have extremely high individual peak capacities, for example, over 1000 with a 10^6 theoretical plates column (3), but even when columns such as these are

Multidimensional Chromatography, edited by L. Mondello, A. C. Lewis and K. D. Bartle
©2002 John Wiley & Sons Ltd.

applied to moderately complex samples such as gasoline, they still suffer from incomplete resolution of the mixture. The analysts[1] most common approach to improve resolution is generally a modification of single-column physical parameters, increase the length, decrease the column internal diameter, or a combination of the two. For many complex sample analyses, however, changes such as these offer only slight improvements in resolving typical target compounds, which may well be isomeric or enantiomeric in nature. It is a well-known theory that a doubling in column length results in only a $\sqrt{2N}$ increase in the number of theoretical plates. What is more often required is not simply greater numbers of theoretical plates on the same column, but complementary selectivity, achieved by using a serially coupled secondary separation. The degree to which a multidimensional GC separation produces enhancement in peak capacity can be related to the degree of orthogonality between the stationary phase selectivities in each dimension. For any given application, therefore, single-column methods may be described as being reliant on column efficiency, whereas-two dimensional system depend on stationary phase selectivities.

Application of two-dimensional GC to increase the speed of analysis was pioneered initially by industrial and process applications, which required on-line high-speed analysis of only a single or very limited number of target analytes. In this mode, the primary GC column has taken on a role similar to the primary column in LC–GC, in that it is used more for sample pre-fractionation than high resolution separation. Through the use of pre-columns and backflushing, many applications have taken advantage of the power of two-dimensional GC, to allow the rapid analysis of relatively volatile components in a matrix of higher-molecular-weight species.

3.2 PRACTICAL TWO-DIMENSIONAL GAS CHROMATOGRAPHY

In many respects, the coupling of GC columns is well suited since experimentally there are few limitations and all analytes may be considered miscible. There are, however, a very wide variety of modes in which columns may be utilized in what may be described as a two-dimensional manner. What is common to all processes is that segments or bands of eluent from a first separation are directed into a secondary column of differing stationary phase selectivity. The key differences of the method lie in the mechanisms by which the outflow from the primary column is interfaced to the secondary column or columns.

In most two-dimensional GC applications reported from the late 1960s until the early 1990s, coupling was via the transfer of limited numbers of discrete fractions of eluent from one standard capillary column to a secondary one. In early work, this focused on using packed columns for at least one of the dimensions, although currently most new techniques report the coupling of two or more capillary columns. This mode is often described in literature as 'heart-cut', or linear in nature. Since both columns have roughly equivalent peak capacities, the time taken for the analysis by each dimension is significant with respect to the other. That is to say that only a single portion of primary column flow is reanalysed using the second column selectivity. This naturally has a very significant impact on the application of the technique, in that only a limited fraction of the total sample is analysed with the full resolution of a two-dimensional separation. This is not to say, however, that this limits the usefulness of the technique, as later example applications of two-dimensional GC will show.

Efforts have been made, however, to extend the range or extent of samples that can be analysed by using a two-dimensional separation when used in heart-cut mode. This has been reported to include the use of numerous parallel micro-traps to essentially store the primary column eluent fractions ready for second-column separation, and the use of parallel second-dimension columns.

The ultimate extension of two-dimensional GC was introduced in the early 1990s (4) and involves the reanalysis of all components from the primary dimension on a secondary column. To enable this, the peak capacity of the secondary column is often very much smaller than the primary thus allowing completion of the separation in a time that may be considered insignificant as a fraction of the time required to complete the primary stage. This is discussed in much further detail in Chapter 4.

3.2.1   EXPERIMENTAL CONFIGURATIONS

The most basic classification of GC couplings is into off-line and on-line interfacing. Off-line is described as the manual collection of effluent from a column prior to manual re-injection to a second column. While this can be relatively simple to perform, there are very significant problems in handling volatile species, and the potential for artefact generation exists. Perhaps more importantly, the reproducibility of manual handling of samples is poor and automation is clearly not practical. Far more commonly used is a direct on-line coupling between systems. This may be described as a system where the collection and transfer of fractions between columns is performed within a sealed analytical system. In practice, this is enabled by the automatic diverting of column flows via mechanical or pressure-driven switching devices. Automation and reproducibility are greatly increased, and the introduction of chemicals external to the procedure are eliminated.

Figure 3.1 shows several potential on-line modes of two-dimensional GC operation. These couplings demonstrate HRGC–HRGC performed by using a single heart-cut from the primary to the secondary column, multiple heart-cuts, transferred to multiple intermediate traps, and heart-cuts transferred to a multiple parallel secondary column configuration.

Although the ability to generate separation systems with significantly enhanced peak capacities is the most obvious practical usage of two-dimensional GC, there are several ancillary benefits which are often also achieved when analysis is performed using this approach.

- *Time for analysis.* The analysis of complex samples when performed on single columns generally requires very long separation periods, commensurate with the

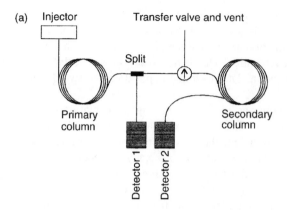

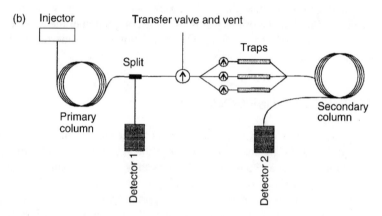

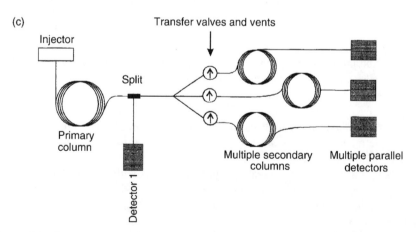

Figure 3.1 Two-dimensional gas chromatography instrumental configurations: (a) direct transfer heart-cut configuration; (b) multiple parallel trap configuration; (c) multiple parallel column configuration.

long column lengths used. Practically, the determination of a species of interest may take an extremely long period of time if single-column efficiency is the only tool used for its chromatographic isolation. The use of a pair of coupled columns may often result in the equivalent peak capacity being generated in a substantially shorter period of time. This, of course, is only really a useful option if the analysis required is focused on only a limited number of analytes, but does offer very significant time reduction possibilities if the methodology is optimized.

- *Sample capacity.* In common with the speed of analysis, the isolation of an analyte in a complex mixture using only a single dimension often results in the application of narrow internal diameter capillary columns. While it is possible when using large-volume sample techniques to introduce relatively large quantities of material on to capillary columns (5, 6), a limit may be reached when the target analyte is in the present of an excess complex matrix. In these instances, overloading of the column with interference material may well occur. Since the sample capacity q of a capillary column is related to the internal diameter by the following relationship:

$$q = C_m d_c^{\frac{3}{2}}(1 + k)(Lh)^{\frac{1}{2}} \tag{3.1}$$

where C_m- is the solubility in the mobile phase, d_c is the internal column diameter, k is the capacity factor, L is the column length and h is the height equivalent of a theoretical plate. There is clearly a very rapid decrease in available sample capacity when column internal diameters are reduced. This may be effectively overcome by either fractionation of unwanted matrix material or the generation of equivalent peak capacities using two coupled GC columns with wider internal diameters.

3.2.2 TWO-DIMENSIONAL GC APPARATUS

For all three of the examples of two-dimensional gas chromatography configurations shown in Figure 3.1, an interfacing unit is required between the primary and secondary columns. The types of instrumentation used for this purpose range from relatively simple manually operated valves, to more complex but flexible computer pressure and flow control systems. What is not indicated on the figure, but is, of course, a further operational parameter is that the two columns used may be operated at the same or independent temperatures, through the use of either single or multiple GC ovens. In what may be considered the simplest case, both GC columns are contained within a single GC oven, with a mechanical valve used to direct flow from the primary to the secondary column at the appropriate moment during the primary separation. This mode of operation highlights a major limiting factor in two-dimensional gas chromatography – that peak widths introduced to the second column from the first will critically limit the peak capacity of the second column. This arises since the peak width eluting from the primary column must be less than the peak width resulting from second column unless a refocusing or zone compression is performed. Peak widths of 10 s may be typical of eluting peaks in capillary GC, but, of course, an injection period of 10 s (which is essentially what is occurring between the primary and secondary columns) would be generally considered excessive. In practice, a number of mechanisms are used to enable refocusing and zone compression at the point of transfer, ranging from cryogenic and chemical micro-traps, to phase ratio refocusing at cooler oven temperatures, when the secondary column is held at temperatures independent of the first. An additional consideration of this simple two-dimensional system, beyond just that of peak dispersion, highlights that the sample will be in contact with a metallic or polymer surface of the valve, and that a significant pressure drop may exist with two columns in line. In addition, consideration must be made to match the flow rates, and sweep dead flows in the system along with precise and instantaneous transfer from the primary to the secondary column. The switching process itself must also not upset flows or general system or detector stability.

Solutions do exist to these problems. The engineering of mechanical valves is now significantly advanced and very low dead volume units which minimize band broadening are commercially available. Similarly, the inertness of the rotor material used to make the gas-tight seal between the metal body channels has been vastly improved. The extent to which the problems of band dispersion and inertness have been overcome can perhaps be assessed from some anecdotal reports of two-dimensional GC performed by using modern valves. A study in 1985 compared the Deans switch and mechanical valve performance for a range of reactive species, although the results obtained were largely inconclusive (7). More recent work by Mills and Guise (8) reported the successful analysis of free acids and anhydrides in a valve-based two-dimensional system with few problems. However contradictory evidence is available for some compounds such as aldehydes and alcohols, e.g. acetal and acetaldehyde were analysed successfully by Adam (9) whilst furaneol gave significant problems (10). It is clear, therefore that individual methodologies must be carefully examined when considering the use of mechanical valve interfaces. Given, however, the various deactivation techniques, such as silanization, which are now used for the production of stainless steel columns, future developments in deactivation, and high-temperature stability make this simple route worthy of consideration. Figure 3.2 shows the typical valve switching arrangements used for two-dimensional gas chromatography in both the heart-cut and analyse/monitor positions. In the analyse position, flow from the primary column passes into the two-position switching valve and is diverted to the first detector. This allows the progress of the primary separation to be actively monitored. At the moment of transfer, the valve positions are switched and flow exiting the first column is passed directly to the secondary column. On completion of the heart-cut, the valve is returned to its original position. The secondary separation is monitored on the second detector.

3.2.3 DEANS SWITCHING

The non-intrusive manipulation of carrier gas effluent between two columns clearly has significant advantages in two-dimensional GC. In addition, a pressure-driven switch between the columns introduces no extra band broadening to an eluting peak.

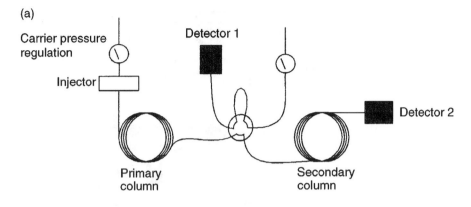

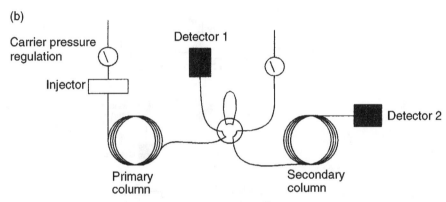

Figure 3.2 Valve switching interfaces in (a) heart-cut position and (b) primary column monitor/secondary column analyse position.

In 1968, Deans (11) introduced a basic principle of pressure switching which has subsequently been used extensively for heart-cutting, venting and backflushing within two-dimensional systems. The principle of operation is shown in Figure 3.3 in a diagram adapted from an earlier review by Bertsch (12). The process indicated highlights three distinct phases, i.e. that of survey or prefractionation, sample transfer and backflush of the primary column. The basis of the method is in the diversion of flows by using pressure balancing at junctions. The flow of carrier gas to each junction is controlled by solenoid valves, with the magnitude of pressure introduced being determined by the inlet and outlet pressures of the interacting devices and their individual flow resistances.

There are three phases in the operation of a Deans switch :

(a) Prefractionation-operation in the SURVEY position indicted in Figure 3.3 results in a balance of pressure such that flow from the primary column is diverted at the junction between the columns (marked 'A') towards Detector 1. This set of pressures prevents sample entering the second column, but does

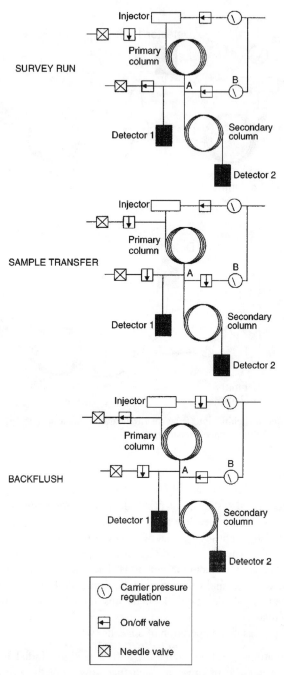

Figure 3.3 Survey, sample transfer and backflush positions used during the non-intrusive Deans heart-cut switching process.

provide the carrier gas for the second column. This is supplied as excess pressure at the junction A.

(b)   Sample transfer-The second pressure configuration results in both columns being coupled in a sequential manner. A minor portion of the primary eluent is split at junction A to go to Detector 1, with the majority passing directly on to the secondary column.

(c)   Analysis of fraction-once the sample transfer is complete, the third pressure configuration is adopted. The carrier gas flow through the secondary column is maintained by excess pressure at junction A, supplied from regulator B. Concurrently, the primary column is backflushed by also using the pressure supplied from regulator B.

Following the backflush of the primary column and separation of the analytes on the second column, the system can then be returned to its original prefractionation position, ready for the next sample injection.

The system can be made more sophisticated through the addition of fused silica restrictors acting as bleeds, thus preventing back diffusion of analytes towards the solenoid valves. Additional pre-concentration stages can also be provided between the primary and secondary columns to reduce the peak widths introduced to the secondary column.

Overall, the technical complexity of the Deans switch system is considerably greater than that of a mechanical switching valve and it is accepted that reliability and ease of use is reduced as the system complexity increases. For many compound types, however, the completely non-intrusive nature of the Deans method offers sufficient advantages to justify its application. However, the use of modern electronic pressure and flow controls integrated into the overall computer control of the chromatographic system does now make the operation of Deans switches significantly easier or more reliable than has been reported in its earlier applications.

3.2.4   INTERMEDIATE TRAPPING

Whilst the most simple form of two-column coupling results in a switch at the midpoint between two columns in a single oven, large numbers of more complex adaptations do exist. Perhaps the most significant additional modification that can be made is refocusing of analytes prior to the secondary separation. As highlighted earlier, the peak widths from the primary column separation fundamentally limit the resolving power of the second column, with refocusing being the key method in reducing this dispersion effect. Refocusing can be enabled by a number of means, including thermal focusing by using a cryogenic trap, cooled typically with CO_2 or liquid nitrogen, a thermally modulated retaining column, or reconcentration at the secondary column head prior to temperature programming. W. Bertsch highlighted a large number of refocusing advantages, which included (i) knowing the exact starting point of the second dimension, (ii) addition of several heart-cuts to form a composite prior to injection, (iii) the ability to connect high-flow packed columns with low-flow narrow capillary columns, (iv) multiple enrichment to enhance sensitivity, (v) reduction in band broadening, (vi) the addition of auxiliary inlets for the addition of standards, and (vii) the operation of analytical columns with a different carrier gas to that of the primary. While some of these potential advantages are perhaps little exploited in reality, they highlight the flexibility that refocusing at the midpoint can achieve. It is, of course, this zone compression that lies at the heart of the next generation of two-dimensional gas chromatographs which have utilised comprehensive chromatography.

The practical use of cryogenic traps varies greatly in sophistication. Fully automated cryogenic traps are available as add-on equipment for capillary GC systems. Couplings for GC–GC may be readily adapted from such automated preconcentration devices that were designed initially for headspace analysis. Far simpler intermediate refocusing can be achieved by using only a short length of capillary column dipped into a Dewar flask of liquid nitrogen. The low thermal mass of the capillary means that on withdrawal from the cryogenic liquid a very rapid release of trapped analytes occurs, thus introducing a very narrow injection band of analytes to the secondary column. Care must be taken, however, in ensuring complete retention of solute at the intermediate stage with the use of thin liquid film coatings aiding this process. The use of polysiloxane-coated capillaries dipped in liquid nitrogen is particularly effective since they remain as fluids even at low temperatures. Other commonly used phases, such as polyethylene glycol, however, are less effective as they often undergo phase solidification.

A variation on the method of thermal modulation is the use of a length of capillary column coated with a thick film of stationary phase. At ambient oven temperatures, this results in a retention of semivolatile analytes, which may be subsequently released to the secondary column once the trap is heated. The rapid cycling time possible with this methodology has resulted in its common application as the intermediate trap in comprehensive GC.

3.2.5 MULTIPLE OVENS

A very early application of a double-oven system (that is, primary and secondary columns held at independent temperatures) was reported in 1973 by Fenimore *et al.* (13). In this particular application, the need to use a two-oven system was driven by detector stability considerations, requiring accurate and stable control of the secondary column temperature. Later systems began to use independent temperature controls of both ovens to enable more rapid and higher resolution separations. Refocusing achieved through cooled on-column transfer to a secondary column for temperature programming introduces a further area for two-dimensional GC method refinement. The use of multiple ovens held at differing temperatures for primary and subsequent columns requires a significant additional investment in equipment. It does, however, offer a number of significant advantages over simple single-column methods, as follows:

(a) Columns with high-temperature phases can be used in combination with those of limited stability. This is most relevant when a coupling is made between highly stable cross-linked methyl polysiloxane and high-polarity (but low thermal stability) wax or porous layer open tubular (PLOT)-type columns.

(b) The ability to maintain the ovens at two independent isothermal temperatures allows accurate retention index behaviour to be established.

(c) Optimization of stationary phase selectivity by temperature adjustment of coupled columns.

(d) Analyte refocusing on transfer from the primary to the secondary column.

3.3 PRACTICAL EXAMPLES OF TWO-DIMENSIONAL CHROMATOGRAPHY

It is through observing examples of actual applications that the best understanding of GC–GC separation principles can be achieved. Over the past 30 years, there have been essentially three main areas where two-dimensional gas chromatography has been applied:

- petroleum, fuels, feedstocks and combustion analysis;
- flavours, fragrance and food;
- environmental contaminants.

Multidimensional techniques are used extensively in all of these areas, with GC–GC being only one of the many commonly used hyphenated methods. Reviews of each of these application areas is discussed in greater detail in Chapters 10, 13 and 14. The remaining sections of this present chapter, however, will use some selected GC–GC applications to demonstrate how such techniques in particular have been applied in practice.

What is common to all of these areas is that the relevant number of published GC–GC papers is very small when compared to those concerning single-column and GC–MS methods. While approximately 1000 papers per year are currently published on single-column GC methods and, in recent years, nearly 750 per year on GC–MS techniques, only around 50 per annum have been produced on two-dimensional GC. Of course, this may not be a true reflection of the extent to which two-dimensional GC is utilized, but it is certainly the case that research interest in its application is very much secondary to that of mass spectrometric couplings. A number of the subject areas where two-dimensional methods have been applied do highlight the limitations that exist in single-column and MS-separation analysis.

A large amount of fuel and environmentally based analysis is focused on the determination of aliphatic and aromatic content. These types of species are often notoriously difficult to deconvolute by mass spectrometric means, and resolution at the isomeric level is almost only possible by using chromatographic methods. Similarly, the areas of organohalogen and flavours/fragrance analysis are dominated by a need to often quantify chiral compounds, which in the same way as aliphatic isomers yield highly similar mass fragmentation patterns. In these cases, it is only through high resolution separations that these goals may be achieved.

3.3.1 PETROLEUM, FUELS, FEEDSTOCKS AND COMBUSTION ANALYSIS METHODOLOGIES

As mentioned at the start of this chapter, it was the analysis of crude oil fractions that was at the forefront of multidimensional GC development. The need to accurately quantify components within such mixtures was of tremendous importance not only commercially, but also in diagnosing environmental consequences and impact. As mixtures, fuel type samples are obvious candidates for multidimensional chromatography. They naturally contain very large numbers of aliphatic compounds, with often a staggering degree of sample complexity and isomeric abundance. There are, for example, around 4×10^7 potential isomers (14) in the diesel fraction range $C_{10}-C_{25}$. Within matrices such as these, however, there exist large numbers of groups of compounds of particular interest to the analyst, such as heterocompounds containing nitrogen, sulfur and oxygen, as well as amounts of polycyclic aromatic materials. For example, their presence may be significant in reducing catalytic convertor performance, or in increasing the emission of NO_x, SO_2 or harmful carcinogenic compounds to the atmosphere.

In almost all fuel and feedstock analysis, the peak capacity of a single column is far from sufficient to resolve all of the species present and analysis of the PIONA (paraffins; isoparaffins; olefins; naphalenes; aromatics) fraction is a particular challenge. It was PIONA analysis that highlighted the first significant industrial application in 1971 (15) of two-dimensional GC, with a method that has now found general acceptance (16). The ASTM standard methods of PIONA analysis can be highly complex in nature, using between two to five parallel secondary GC columns, with valve transfers, independent temperature controls and multiple re-injections. These standard GC–GC methods still do not produce a complete baseline separation of the mixtures, but do offer a highly targeted analysis of a large number of key species. In general, this type of GC–GC analysis system is optimized for condensed-phase, relatively involatile material. More volatile mixtures, such as exhaust emissions or refinery gases, are still sufficiently complex to display incomplete resolution on a single column. In cases such as these it is essential to isolate the gas-phase species from the less volatile material in order to avoid column and detector contamination. Early GC–GC applications used packed columns in combination with liquid-film capillary columns (17, 18). More recently, backflushed liquid-film pre-columns have been used with PLOT-type secondary columns (19). In these configurations, the high-retention PLOT columns require, that involatile material is completely excluded, or otherwise irreversible adsorption may occur.

The production of petrochemicals from feedstocks has also been an area of wide application of two-dimensional GC. A detailed knowledge of the composition of feedstocks and intermediates in the manufacturing process is central to obtaining products with the correct compositional and combustion characteristics. A typical example of the use of gas chromatography when applied to petrochemical analysis can be found in reference (20), while reviews of such applications are also available in selected issues of the *Journal of High Resolution Chromatography* (21). What is evident from the many reported applications of two-dimensional GC in the areas of hydrocarbon and related analysis is the accuracy of quantitation when compared to that of GC – MS. Since resolution is achieved by chromatographic means alone, well characterized detectors such as the flame ionization detector (FID) may be used in the place of the less easily quantified mass spectrometric detection (22).

Of considerable commercial interest has been the isolation of various alkyl-substituted polycyclic aromatic compounds where these serve as indicators of the geological history of sedimentary rock samples, highly useful in the exploration for oil reserves. The isolation of alkyl-substituted aromatics, however, is a typical problem to which multidimensional GC may be applied. Large numbers of substituted aromatic isomers exist, and their deconvolution by using mass spectrometry alone is still unclear. The most accurate method to directly quantify individual isomers is therefore to resolve each one chromatographically. Schäfer and Höltkemeir (20) in 1992 presented a two-dimensional method that had sufficient selectivity and resolution to individually isolate 1,4-dimethylnaphthalene from 2,3-dimethylnaphthalene without the use of mass spectrometry. This method effectively used a Deans switch to transfer fractions of crude oil between a primary low-polarity column and a 50 % phenyl/25% cyanopropyl/25% methyl polysiloxane secondary column. Figure 3.4 shows a series of chromatograms obtained from the recombination of heart-cuts made from this primary separation. In this case, while a single column of moderate polarity would have enabled some resolution between the target species, to achieve complete isolation would have required an impractical length of column, and hence analysis time.

The analysis of combustion products presents problems of complexity similar to that of feedstock and raw fuel analysis. A highly complex matrix of aliphatic material often exists (as unburnt fuel in the combustion exhaust), whilst the species of interest, for example, carcinogens or mutagens are often at very low concentrations. A classic example of multidimensional GC is its use in the analysis of flue-cured tobacco essential oil condensate.

The chromatograms reported by Gordon *et al.* (23) shown in Figure 3.5 illustrate the huge complexity of even small heart-cuts made from the primary separation. Once again, a Deans-type switch was used for sample transfer. For the primary chromatogram, each cut is seen to contain only a handful of peaks, yet when a further secondary separation is performed (based on polarity rather than boiling point) a large numbers of extra species can be isolated. The huge complexity in even the second-dimension chromatogram required that the second column was temperature programmed, and a two-oven approach was therefore applied. In the case of the tobacco condensate it becomes questionable, even with a second separation with full temperature programming, whether the analytical system has sufficient capacity, and that possibly a higher dimension was required to truly characterize the sample. In this

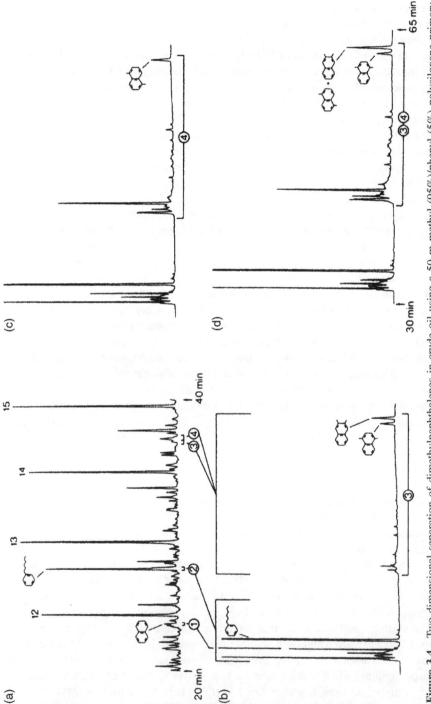

Figure 3.4 Two-dimensional separation of dimethylnaphthalenes in crude oil using a 50 m methyl (95%)/phenyl (5%) polysiloxane primary column and a 50 m methyl (50%)/phenyl (25%)/cyanopropyl (25%) polysiloxane secondary column. The top trace indicates the primary separation monitor, while the following chromatograms indicate individual heart-cut secondary analysis. Reproduced from R.G. Schäfer and J. Höltkemeir, *Anal. Chim. Acta*, 1992, **260**, 107 (20).

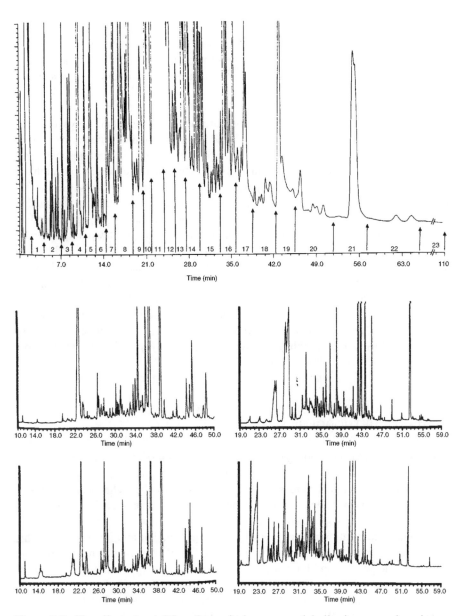

Figure 3.5 Two-dimensional GC analysis of tobacco essential oil using non-polar primary and polar secondary separations. The top trace indicates the primary separation, with the four resulting heart-cut chromatograms shown below being obtained on the transfer of approximately 1–2 min fractions of primary eluent. Reproduced from B.M. Gordon *et al. J. Chromatogr. Sci.* 1988, **26**, 174 (23).

example, it was essential that further chemical information was available to produce even group-type classifications in such a mixture, with the addition of a mass spectrometer generally offering the required extra analytical dimension.

A number of environmental applications of two-dimensional GC have arisen as a direct result of the impact of combustion emissions on the environment. The analysis of aromatic species in air may be achieved in a similar manner to petrochemical analysis, although sample pre-concentration becomes essential as the abundances are proportionally lower. Similarly, the measurement of carcinogenic polycyclic aromatic compounds, in particular nitrogen-and oxygen-containing derivatives, is of particular epidemiological interest. The isolation of species in the atmosphere presents similar problems to analysis in fuels, in that a high-concentration aliphatic matrix often masks the target analytes. Multidimensional methods using both GC–GC and LC–GC offer excellent resolution with simple quantitation when using flame ionization detectors, with a number of such applications being covered in Chapter 13. An application worth mentioning from a technical viewpoint, however, is where two-dimensional GC has been used to reduce the analysis time and utilizes isothermal conditions. In particular, a valve-transfer system has been reported for the determination of isoprene and dimethyl sulfide in air, which involves the combination of carrier gas pressure programming with non-polar and PLOT columns (24). Although limited applications have been reported in environmental analysis, GC–GC offers great potential for deconvoluting much of the complexity in these samples, and as well as enabling the use of simple analytical systems which allow high-speed repetitive measurements.

3.3.2 FLAVOURS, FRAGRANCE AND FOOD ANALYSIS METHODOLGIES

In common with the previous application category, this grouping is of great commercial importance to a number of global industries. The determination of species central to the flavour and fragrance industries is highly complex in both chemical complexity and human response. A recurring theme, however, in this type of analysis is that very low concentrations of specific compounds must be isolated from a potentially complex, high-concentration matrix to enable both quantitation and assessment. In this field, assessment may be structural or functional, through the use of a mass spectrometer, or by impact on the human senses through organoleptic assessment, respectively. The use of organoleptic methods (that is, the use of the human nose to 'detect' components in a mixture) often highlights problems of resolution in GC separations of complex mixtures. While ideally each strongly detected peak on a chromatogram would have a correlated peak on the detector, this is rarely the case. Often multiple elutions at a given retention time mean that organoleptic assessment merely narrows the field of possible compounds, and very commonly that the nose recognizes components that are not apparent on the detector. Overcoming problems of detector sensitivity inevitably results in preconcentration, elevations in matrix concentrations and a required increase in peak capacity for the separation system.

Two-dimensional gas chromatography plays an important role in such analysis, in combination with a wide array of sample preparation, and preconcentration techniques and injection devices (25–27). While preconcentration prior to the primary column separation is the first step in obtaining sufficient target compounds, 'on-column' subsequent refocusing at the midpoint between the dimensions can also be used as a method of preconcentration. This can be achieved through either multiple heart-cuts at the same primary column retention time or through zone compression, thus leading to a narrower eluting band (and hence greater mass/unit time) emerging from the second column into the detector.

Sample concentration, and hence enrichment, is certainly a key issue in this area of analysis, since complementary information obtained from NMR or IR spectroscopic detection is often desirable in conjunction with mass spectrometry. Detection methods such as these have far higher concentration thresholds than MS and obtaining adequate quantities of material for detection becomes a significant challenge.

It is worth noting that while a significant number of papers on flavour and fragrance analysis are published each year, this constitutes perhaps only a fraction of the amount of research time spent on this area, and it is likely that the commercial nature of such work has resulted in an under representation in the published literature. Much of the early two-dimensional GC analysis work on flavours and fragrances focused on detecting product deterioration rather than on identifying active fragrance components (28), while work by Nitz *et al.* (29) demonstrated the use of multidimensional separations in combination with human odour assessment. In this latter study, a two-dimensional GC technique was used to examine wheat grain samples, with pre-fractionation, followed by separation, organoleptic assessment and the final collection of fractions for further analysis. This final collection phase, described as 'micropreparative,' yielding sufficient material for either further GC analysis or spectroscopic measurement. The detection of a compound responsible for 'off-odour' in this type of product was found to be 2-methyllisoborneol, a trace constituent in what was a highly complex mixture. Its co-elution with higher-concentration species masked identification with single-dimension GC–MS and organoleptic assessment methods.

The analysis of 'more pleasant' odours associated with, for example, fruits, plants and extracted essential oils, is also an area that exploits the resolution possible with two-dimensional GC. The analysis of fruit extracts and products by using two-dimensional methods was first reported in the mid 1980s and much published work has followed since then (30, 31). Figure 3.6 demonstrates a two dimensional separation of orange oil extract (32), analysed on primary apolar and secondary Carbowax 20M columns. Two-dimensional GC in combination with organoleptic assessment indicated that co-elution of a minor concentration (but odour significant) compound with β-myrcene was occurring when the analysis was performed on a single column. The analysis made use of a non-polar primary column in combination with a carbowax secondary column, interfaced by using a Deans switch at the midpoint. The very short cut to the secondary column resulted in no requirement for an intermediate trap, and although the system was operated by using two ovens, the secondary oven was held isothermally.

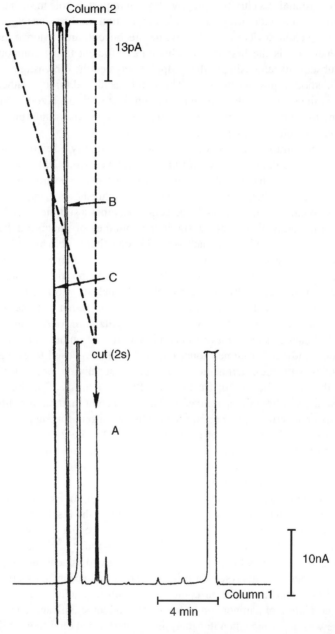

Figure 3.6 Two-dimensional gas chromatogram of an orange oil extract, in which a 2 s heart-cut has been made in the region A where β-mrycene has eluted on a non-polar column. Secondary analysis on a polar Carbowax 20M column indicated two compounds (marked B and C), both identified as 'odorous' by organoleptic assessment. Reproduced from P. A. Rodriguez and C. L. Eddy, *J. Chromatogr Sci.* 1986, **24**, 18 (32).

It is the determination of volatile organic compounds produced from natural products that requires separation techniques that allow isolation of stereoisomers. The most commonly determined groups are the terpene and sesquiterpene species present in essential oils, which are used as key indicators of biological factors such as the growth season, geographic location, climate, etc. These species are also released directly into the atmosphere by very many plants and trees, and make a substantial contribution to global biogeochemical cycles.

The determination of enantiomers is possible with GC–GC by combination of a nonpolar, basic fractionating primary column used in conjunction with a chiral selective secondary columns. The sophistication of chiral columns has increased significantly since their introduction in the late 1970s (33), with enantiomer-selective two-dimensional separations first being performed in 1983 on amino acids by using Deans switch technology (34). More recent examples related to the food and flavour industry include the work of Mondello et al. (35, 36)and Mosandl et al. (37). Recent analyses reported by Mondello operate without midpoint preconcentration and demonstrates that a high speed of transfer, without band broadening, may be achieved with modern valve interfacing.

The introduction of synthetic materials into natural products, often described as 'adulteration', is a common occurrence in food processing. The types of compounds introduced, however, are often chiral in nature, e.g. the addition of terpenes into fruit juices. The degree to which a synthetic terpene has been added to a natural product may be subsequently determined if chiral quantitation of the target species is enabled, since synthetic terpenes are manufactured as racemates. Two-dimensional GC has a long history as the methodology of choice for this particular aspect of organic analysis (38).

Figure 3.7 shows some early examples of this type of analysis (39), illustrating the GC determination of the stereoisomeric composition of lactones in (a) a fruit drink (where the ratio is racemic, and the lactone is added artificially) and (b) a yoghurt, where the non-racemic ratio indicates no adulteration. Technically, this separation was enabled on a short 10 m slightly polar primary column coupled to a chiral selective cyclodextrin secondary column. Both columns were independently temperature controlled and the transfer cut performed by using a Deans switch, with a backflush of the primary column following the heart-cut.

In any form of analysis it is important to determine the integrity of the system and confirm that artefacts are not produced as a by-product of the analytical procedure. This is particularly important in enantiomeric analysis, where problems such as the degradation of lactone and furanon species in transfer lines has been reported (40). As chromatography unions, injectors, splitters, etc. become more stable and greater degrees of deactivation are possible, problems of this kind will hopefully be reduced. Some species, however, such as methyl butenol generated from natural emissions, still remain a problem, undergoing dehydration to yield isoprene on some GC columns.

Isolated problems of racemization, rearrangement or dehydration should not overshadow the fact, however, that the range of species amenable to enantiomeric two-dimensional GC is very wide indeed, including not only terpenes and lactones,

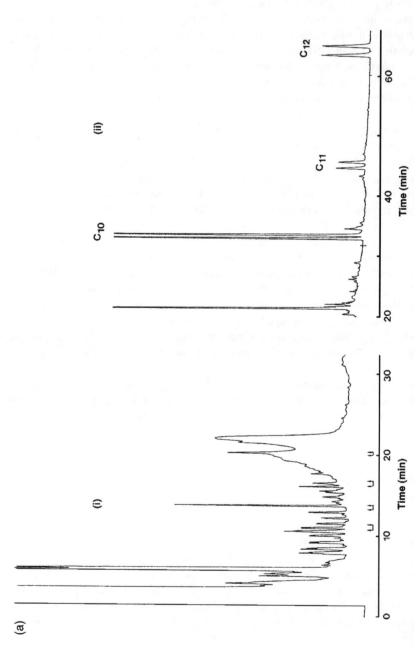

Figure 3.7 (a) Chromatograms of (i) the dichloromethane extract of a fruit drink analysed with an apolar primary column, with the heart-cut regions indicated, and (ii) a racemic mixture of γ-deca-(C_{10}), γ-undeca-(C_{11}) and γ-dodeca-(C_{12}) lactones isolated by heart-cut transfer, and separated by using a chiral selective modified cyclodextrin column.

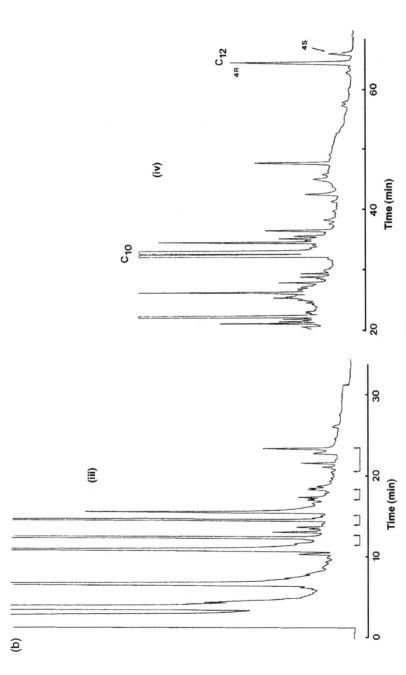

Figure 3.7 (*continued*) (b) Chromatograms of (iii) the dichloromethane extract of strawberry fruit yoghurt analysed with an apolar primary column, with the heart-cut regions indicated, and (iv) a non-racemic mixture of γ-deca-(C₁₀) and γ-dodeca-C₁₂ lactones isolated by heart-cut transfer, and separated by using a chiral selective modified cyclodextrin column. Reproduced from A. Mosandl, *et al.* J. High Resol. Chromatogr. 1989, **12**, 532 (39).

but also esters, alcohols and acids, all of which appear stable under two-dimensional GC conditions (41–44).

The study of biochemical natural products has also been aided through the application of two-dimensional GC. In many studies, it has been observed that volatile organic compounds from plants (for example, in fruits) show species-specific distributions in chiral abundances. Observations have shown that related species produce similar compounds, but at differing ratios, and the study of such distributions yields information on speciation and plant genetics. In particular, the determination of hydroxyl fatty acid adducts produced from bacterial processes has been a successful application. In the reported applications, enantiomeric determination of polyhydroxyl alkanoic acids extracted from intracellular regions has been enabled (45).

An obvious extension of enantiomeric two-dimensional GC of natural products is the application of carbon isotope mass spectrometry as the detection process. While the application of isotopic carbon abundance in pharmaceutical and food research has been commonplace for many years, its coupling to enantiomer-selective GC is still little explored. The ability to isotopically discriminate between the species evolved during biological processes can provide valuable checks on the authenticity of the analysis. More specifically, enantiomers from the same origin will have identical $^{12}C/^{13}C$ ratios. Since isotopic separations are generally not feasible, enantiomeric separation in combination with isotopic mass spectrometry offers a possible analysis route. At present, the sensitivity of $\Delta^{12/13}C$ analysers is significantly less than is normally achieved by using benchtop quadrupole or ion-trap configurations and so detector sensitivity is an important issue. When two-dimensional separations are applied to such a problem, the data obtained are more reliable than those obtained by single-column separations, since background interferences on the chromatogram are very greatly reduced (46, 47).

3.3.2 TYPICAL ENVIRONMENTAL CONTAMINANT ANALYSES

While a small number of fuel-derived organic emissions relevant to the environment have been described above, the main environmental application of two-dimensional GC has been in the analysis of halogenated persistent organic pollutants (POPs). This broad classification of pollutants contains species such as polychlorinated biphenyls (PCBs), dioxins, furans, toxaphenes and other persistent organochlorine molecules released via anthropogenic sources. The sources of species such as these are too numerous to list exhaustively but include, (famously) pesticides, and transformer insulator oil. It is their potency to detrimentally affect human health, coupled to long lifetimes in the environment, which has lead to such interest in their analysis. Atmospheric lifetimes are in excess of 100 years, with the major route for degradation being via microbial action in soils. This too, however, is achieved at only very slow rates. The analysis of PCBs is particularly challenging, given a combination of large diversity (over 209 individual biphenyl molecules have been produced in 150 commercial products (48)) and yet a very low abundance in the environment (e.g. pg–ng m^{-3} in air).

The analysis of persistent organic pollutants presents particular problems to the analyst. While instrumental methods exist for determining the bulk chlorinated content, the variability in toxicity effects means that these are of little use in health impact assessment. A fully speciated analysis of each individual congener is required, and the analysis is hampered not only by target species complexity, but the more concentrated and equally complex sample matrix. In both soils and atmospheric samples, the overwhelming organic background is that of toxicologically insignificant aliphatic species. The isolation of organochlorine compounds against this background requires considerable sample preparation prior to analysis, in combination with selective detection (often using electron-capture detectors (ECDs) or MS) and a high resolution separation. Ideally, a single column would be used for the universal separation of all congeners of interest; however, after many years of optimization this seems unlikely to emerge (49, 50). Extensive multilaboratory studies of retention behaviour on a variety of stationary phases (51) have highlighted that it is not possible to determine all species under a single set of single-column conditions. A number of approaches to deconvolution have been undertaken, including the use of (i) parallel columns to increase probability of isolation on at least one column, (ii) mass spectrometric deconvolution, (iii) serially coupled columns, and (iv) two-dimensional GC. While option (iv) requires significant apparatus and further development, it is the most reliable method on offer at the present time.

Work by Kinghorn *et al.* has demonstrated a two-dimensional separation of the PCB, Aroclor 1254. This separation used a non-polar primary column with selected cuts to a secondary chiral selective column (52). This system utilized a combination of a Deans switch transfer with a cryogenically cooled intermediate capillary column which was used to refocus the analytes prior to secondary column analysis. Poorly resolved single-column peaks were well resolved on application of a second separation in combination with the refocusing step, with the exception being congener 138, which may be resolved only through the use of very polar cyanopropyl or liquid crystal phase columns (53, 54). Chromatograms from this work are shown later in Chapter 13 (see Figure 13.1).

Of the 209 PCB congeners, 78 are known to exist in two chiral forms. Rather than chirality based around a central carbon atom, asymmetric substitution of both phenyl rings leads to axial chirality of all non-planar conformations. Many of these have low energies of transformation and are able to interconvert by rotation about the central $C-C$ bond and form racemic mixtures. There are however 10 chiral tri- or tetraortho-substituted PCBs which have rotational energy barriers sufficiently high to be conformationally stable and thus will not undergo racemization–these are known as atropisomers. There has been recent evidence to show that under certain conditions the biodegradation of PCBs favours one enantiomer and the relative ratio of the respective enantiomers may be used to study this phenomenon (since the physico-chemical and transport properties will not affect the enantiomeric ratios).

It is in the study of this phenomenon where two-dimensional GC offers by far the most superior method of analysis. The use of chiral selector stationary phases, in particular modified cyclodextrin types, allows apolar primary and atropisomer selective secondary separation. Reported two-dimensional methods have been successful in isolating several important pairs of atropisomers (55), and quantitation has been possible since peaks have been baseline-separated by using simple well character-ized detectors such as ECDs. The examples of quantitative two-dimensional separations which have been used to identify non-racemic distributions of PCBs in river and marine samples have been reported recently (56).

Chlorinated dioxins and benzofurans are perhaps the most toxic of all persistent organic compounds found in the environment. Their toxicity is by no means univer-sal, and variations between apparently similar molecules can vary by several orders of magnitude. Because of this specificity, analysis must be similarly species-specific. This group of compounds suffer from the common problem of being present in a matrix of hydrocarbon species at a far higher concentration. In order to simplify the separation, a very extensive sample clean-up is often required, and this can aid in gaining reasonably high resolution even on a single column. The use of two-dimen-sional techniques reduces the degree to which interfering compounds must be removed at the off-line clean-up stage, thus increasing reproducibly and reducing sample handling. Work carried out in the mid 1980s demonstrated the two-dimen-sional separation of several polychlorinated dibenzo-*p*-dioxins (PCDDs) congeners, and also illustrated how the period of heart-cut transfer was critical in obtaining a baseline-resolved secondary chromatogram (57). In addition, for PCDDs and PCBs, the use of ECDs is widespread. This is for a number of reasons, related to cost, ease of quantitation and sensitivity, which all favour the use of ECDs as opposed to mass spectrometric detectors. It is important, however, since identification is based on retention index alone, that the timing of heart-cuts in the primary dimension are extremely precise. Recent developments in electronic pressure and flow rate control has greatly improved this reliability.

Chlorinated camphene and bornane compounds (referred to generally as toxaphenes) are a further group of anthropogenically produced persistent organic pollutants (58). While most of the 200 reported derivatives are chiral in nature, they are manufactured and subsequently occur in the environment as racemates. The degradation of such species within organisms, however, occurs non-racemically and this phenomenon may be used to study the exact metabolic processes involved in their degradation. In a similar way to PCBs, this places an analytical requirement that any determination is both isomeric- and enantiomeric-specific. Two-dimen-sional GC had been applied to the study of toxaphenes in marine mammal samples (59), by using an apolar primary column and a chiral secondary dimension. The chromatogram shown in Figure 3.8 illustrates several heart-cuts from the primary column, demonstrating that in many instances a non-racemic mixture of enantiomers is obtained, as a result of metabolic degradation. The separation was enabled by using non-polar and chiral columns, with independent temperature programming of both of these. While a large number of target compounds with potentially differing enantiomeric ratios were identified, each pair required a heart cut and subsequent analysis. The authors noted that although the separations were baseline, and thus only possible through this analytical route, the number and length of analyses required to characterize large number of species in a complex mixture is still a major limitation of GC–GC.

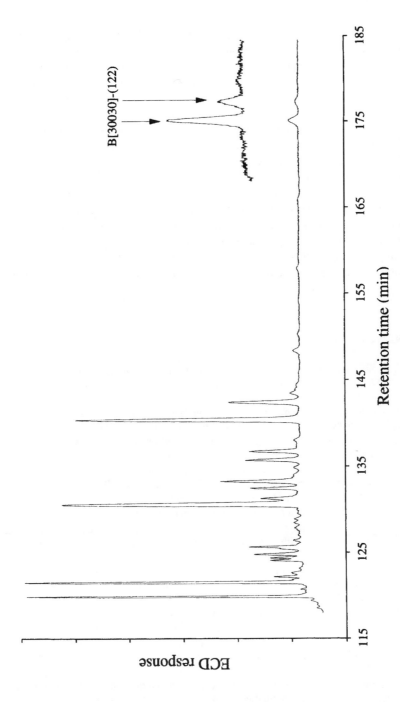

Figure 3.8 Second-dimension chiral cyclodextrin capillary column separation of a non-racemic pair of nonachlorobornane compounds extracted from dolphin blubber, shown with expanded attenuation in the inset. The primary separation (not shown) was performed on an apolar primary capillary column. Reproduced from H.-J. de Geus *et al. J. High Resol. Chromatogr.* 1998, **21**, 39 (59).

3.4 CONCLUSIONS

While far less widely reported in the literature than GC–MS coupled techniques, heart-cut two-dimensional gas chromatography offers very many solutions to the analysis of trace level components in complex mixtures. The area of petrochemical analysis is likely to continue to be a source of new innovative methods for the analysis of aromatic and polynuclear aromatic compounds, as well as pioneering the isolation of heteroatom species. It is the analysis of enantiomeric species, however, where two-dimensional GC currently excels. The ability to study chiral compounds released as a result of natural processes may hold the key to a better understanding of many biological and biochemical processes, as well as giving essential insights into food, flavour and fragrance science. The ability to differentiate between atropisomers related to persistent chlorinated organic pollutants will also have a major impact on the way that we study the remediation and cycling of such species in the environment. Since two-dimensional separations already allow the isolation of many trace species at both isomeric and enantiomeric levels, it is likely that future developments and coupling to isotope-ratio mass spectrometry will also offer new possibilities in tracing the formation and fate of organic compounds in both industrial and natural processes.

REFERENCES

1. L. A. Luke and J. V. Brunnock, 'Separation of naphthenic and paraffinic hydrocarbons up to C_{11} from hydrocarbon mixtures by gas chromatography on faujasite molecular sieves', *Ger. Offen. 1 908 418* (1968).
2. J. C. Giddings, 'Maximum number of components resolvable by gel filtration and other elution chromatographic methods', *Anal. Chem.* **39**: 1027–1028 (1967).
3. T. A. Berger, 'Separation of a gasoline on an open tubular column with 1.3 million effective plates', *Chromatographia* **42**: 63–71 (1996).
4. Z. Liu and J. B. Phillips, 'Comprehensive two-dimensional gas chromatography using an on-column thermal modulator interface', *J. Chromatogr. Sci.* **29**: 227–231 (1991).
5. K. Grob, *Split and Splitless Injection in Capillary GC*, W. Bertsch, W. G. Jennings and P. Sandra (Series Eds), Hüthig, Heidelberg, Germany (1991).
6. E. Boselli, B. Grolimund, K. Grob, G. Lercker and R. Amadò, 'Solvent trapping during large volume injection with an early vapor exit. Part 1: description of the flooding process', *J. High Resolut. Chromatogr.* **21**: 355–362 (1998).
7. B. M. Gordon, C. E. Rix and M. F. Borgerding, 'Comparison of state-of-the-art column switching techniques in high resolution gas chromatography', *J. Chromatogr. Sci.* **23**: 1–10 (1985).
8. P. L. Mills and W. E. Guise, 'A multidimensional gas chromatographic method for analysis of *n*-butane oxidation reaction products', *J. Chromatogr. Sci.* **14**: 431–459 (1996).
9. S. T. Adam, 'Quality test of a mechanical switching valve for two-dimensional open tubular gas chromatography', *J. High Resolut. Chromatogr. Chromatogr. Commun.* **11**: 85–89 (1988).
10. K. Shiomi, 'Determination of acetaldehyde, acetal and other volatile congeners in alcoholic beverages using multidimensional capillary gas chromatography', *J. High Resolut. Chromatogr. Chromatogr. Commun.* **14**: 136–137 (1991).

11. R. R. Deans, 'A new technique for heart cutting in gas chromatography', *Chromatographia* **1**: 18–22 (1968).

12. W. Bertsch, 'Multidimensional gas chromatography', in *Multidimensional Chromatography. Techniques and applications*, H. J. Cortes (Ed.), Chromatographic Science Series, Vol. 50, Marcel Dekker, New York, pp. 74–144 (1990).

13. D. C. Fenimore, R. R. Freeman and P. R. Loy, 'Determination of Δ^9-tetrahydrocannabinol in blood by electron capture gas chromatography', *Anal. Chem.* **45**: 2331–2335 (1973).

14. H. J. Neumann, B. Paczynska-Lahme and D. Severin, *Geology of Petroleum,* Vol. 5, Composition and Properties of Petroleum, Ferdinand Enke, Stuttgart Germany (1981).

15. H. Boer and P. van Arkel, 'Automatic PNA (paraffin-naphthene-aromatic) analyzer for (heavy) naphtha', *Chromatographia* **4**: 300–308 (1971).

16. P. van Arkel, J. Beens, H. Spaans, D. Grutterink and R. Verbeek, 'Automated PNA analysis of naphthas and other hydrocarbon samples', *J. Chromatogr. Sci.* **25**: 141–148 (1988).

17. P. Coleman and L. S. Ettre, 'Analysis of gases containing inorganic and organic compounds using a combination of a thick-film capillary column and a packed adsorption column', *J. High Resolut. Chromatogr Chromatogr. Commun.* **8**: 112–118 (1985).

18. H. Tani and M. Furuno, 'Rapid analysis of hydrocarbons and inert gases by a multidimensional gas chromatograph', *J. High Resolut. Chromatogr. Chromatogr. Commun.* **9**: 712–716 (1985).

19. S. Wu, W. H. Chatham and S. O. Farwell, 'Multidimensional HRGC for sample components with a wide range of volatilities and polarities', *J. High Resolut. Chromatogr.* **13**: 229–233 (1990).

20. R. G. Schäfer and J. Höltkemeier, 'Determination of dimethylnaphthalenes in crude oils by means of two-dimensional capillary gas chromatography', *Anal. Chim. Acta* **260**: 107–112 (1992).

21. *J. High Resol. Chromatogr*, **17** (4 and 6) (1994).

22. J. J. Szakasits and R. E. Robinson, 'Hydrocarbon type determination of naphthas and catalytically reformed products by automated multidimensional gas chromatography', *Anal. Chem.* **63**: 114–120 (1991).

23. B. M. Gordon, M. S. Uhrig, M. F. Borgerding, H. L. Chung, W. M. Coleman, J. F. Elder, J. A. Giles, D. S. Moore, C. E. Rix and E. L. White, 'Analysis of flue-cured tobacco essential oil by hyphenated analytical techniques', *J. Chromatogr. Sci.* **26**: 174–180 (1988).

24. A. C. Lewis, K. D. Bartle and L. Rattner, 'High-speed isothermal analysis of atmospheric isoprene and DMS using online two-dimensional gas chromatography', *Environ. Sci. Technol.* **31**: 3209–3217 (1997).

25. S. Blomberg and J. Roeraade, 'Preparative capillary gas chromatography. II. Fraction collection on traps coated with a very thick-film of immobilized stationary phase', *J. Chromatogr.* **394**: 443–453 (1987).

26. J. P. E. M. Rijks and J. A. Rijks, 'Programmed cold sample introduction and multidimensional preparative capillary gas chromatography. Part I: introduction, design and operation of a new mass flow controlled multidimensional GC system', *J. High Resolut. Chromatogr.* **13**: 261–266 (1990).

27. O. Nishimura, 'Application of a thermal desorption cold trap injector to multidimensional GC and GC–MS', *J. High Resolut. Chromatogr.* **18**: 699–704(1995).

28. R. H. M. van Ingen and L. M. Nijssen, 'Determination of diethylene glycol monoethyl ether in flavors by two-dimensional capillary gas chromatography', *J. High Resolut. Chromatogr.* **12**: 484–485 (1989).

29. S. Nitz, F. Drawert, M. Albrecht and U. Gellert, 'Micropreparative system for enrichment of capillary GC effluents', *J. High Resolut. Chromtogr.* **11**: 322–327 (1988).

30. D. W. Wright, K. O. Mahler and L. B. Ballard, 'The application of an expanded multidimensional GC system to complex fragrance evaluations', *J. Chromatogr. Sci.* **24**: 60–65 (1986).

31. M. Herraiz, G. Reglero, T. Herraiz and E. Loyola, 'Analysis of wine distillates made from muscat grapes (Pisco) by multidimensional gas chromatography and mass spectrometry', *J. Agric. Food Chem.* **38**: 1540–1543 (1990).

32. P. A. Rodriguez and C. L. Eddy, 'Use of a two-dimensional gas chromatograph in the organoleptic evaluation of an orange extract', *J. Chromatogr. Sci.* **24**: 18–21 (1986).

33. Frank H, G. J. Nicholson and E. Bayer, 'Rapid gas chromatographic separation of amino acid enantiomers with a novel chiral stationary phase', *J. Chromatogr. Sci.* **15**: 174–176 (1977).

34. C. Wang, F. Hartmut, G. Wang, L. Zhou, E. Bayer and P. Lu, 'Determination of amino acid enantiomers by two-column gas chromatography with valveless column switching', *J. Chromatogr.* **262**: 352–359 (1983).

35. L. Mondello, M. Catalfamo, P. Dugo and G. Dugo, 'Multidimensional capillary GC–GC for the analysis of real complex samples. Part II. Enantiomeric distribution of monoterpene hydrocarbons and monoterpene alcohols of cold-pressed and distilled lime oils', *J. Microcolumn Sep.* **10**: 203–212 (1998).

36. L. Mondello, M. Catalfamo, A. R. Proteggente, I. Bonaccorsi and G. Dugo, 'Multidimensional capillary GC–GC for the analysis of real complex samples. 3. Enantiomeric distribution of monoterpene hydrocarbons and monoterpene alcohols of mandarin oils', *J. Agric. Food Chem.* **43**: 54–61 (1998).

37. A. Mosandl, K. Fischer, U. Hener, P. Kreis, K. Rettinger, V. Schubert and H.-G. Schmarr, 'Stereoisomeric flavor compounds. 48. Chirospecific analysis of natural flavors and essential oils using multidimensional gas chromatography', *J. Agric. Food Chem.* **39**: 1131–1134 (1991).

38. G. Full, P. Winterhalter, G. Schmidt, P. Herion and P. Schreier, 'MDGC–MS: a powerful tool for enantioselective flavor analysis', *J. High Resolut. Chromatogr.* **16**: 642–644 (1993).

39. A. Mosandl, U. Hener, U. Hagenauer-Hener and A. Kustermann, 'Direct enantiomer separation of chiral γ-lactones from food and beverages by multidimensional gas chromatography', *J. High Resolut. Chromatogr.* **12**: 532–536 (1989).

40. D. Häring, T. König, B. Withopf, M. Herderich and P. Schreier, 'Enantiodifferentiation of α-ketols in sherry by one- and two-dimensional HRGC techniques', *J. High Resolut. Chromatogr.* **20**: 351–354 (1997).

41. W. A. König, B. Gehrcke, D. Icheln, P. Evers, J. Dönnecke and W. Wang, 'New selectively substituted cyclodextrins as stationary phases for the analysis of chiral constituents of essential oils', *J. High Resolut. Chromatogr.* **15**: 367–372 (1992).

42. C. Bicchi and A. Pisciotta, 'Use of two-dimensional gas chromatography in the direct enantiomer separation of chiral essential oil components', *J. Chromatogr.* **508**: 341–348 (1990).

43. C. Askari, U. Hener, H-G. Schmarr, A. Rapp and A. Mosandl, 'Stereodifferentiation of some chiral monoterpenes using multidimensional gas chromatography', *Fresenius J. Anal. Chem.* **340**: 768–772 (1991).

44. V. Karl, H-G. Schmarr and A. Mosandl, 'Simultaneous stereoanalysis of 2-alkyl-branched acids, esters and alcohols using a selectivity-adjusted column system in multidimensional gas chromatography', *J. Chromatogr.* **587**: 347–350 (1991).

45. A. Kaunzinger, F. Podebrad, R. Liske, B. Maas, A. Dietrich and A. Mosandl, 'Stereochemical differentiation and simultaneous analysis of 3-,4- and 5-hydroxyalkanoic

acids from biopolyesters by multidimensional gas chromatography', *J. High Resolut. Chromatogr.* **18**: 49–53 (1995).

46. S. Nitz, B. Weinreich and F. Drawert, 'Multidimensional gas chromatography–isotope ratio mass spectrometry (MDGC–IRMS). Part A: system description and technical requirements', *J. High Resolut. Chromatogr.* **15**: 387–391 (1992).

47. D. Juchelka, T. Beck, U. Hener, F. Dettmar and A. Mosandl, 'Multidimensional gas chromatography coupled on-line with isotope ratio mass spectrometry (MDGC–IRMS): progress in the analytical authentication of genuine flavor components', *J. High Resolut. Chromatogr.* **21**: 145–151 (1998).

48. G. M. Frame, 'A collaborative study of 209 PCB congeners and 6 Aroclors on 20 different HRGC columns. Part 2. Semi-quantitative Aroclor congener distributions', *Anal. Chem.* **70**: 714–722 (1997).

49. B. R. Larsen, 'HRGC separation of PCB congeners', *J. High Resolut. Chromatogr.* **18**: 141–151 (1995).

50. P. Hess, J. de Boer, W. P. Cofino, P. E. G. Leonards and D. E. Wells, 'Critical review of the analysis of non- and mono-ortho-chlorobiphenyls', *J. Chromatogr.* **703**: 417–465 (1995).

51. G. M. Frame, 'A collaborative study of 209 PCB congeners and 6 Aroclors on 20 different HRGC columns. Part 1. Retention and coelution database', *Fresenius' J. Anal. Chem.* **357**: 701–713 (1997).

52. R. M. Kinghorn, P. J. Marriott and M. Cumbers, 'Multidimensional capillary gas chromatography of polychlorinated biphenyl marker compounds', *J. High Resolut. Chromatogr.* **19**: 622–626 (1996).

53. J. de Boer and Q. T. Dao, 'Analysis of seven chlorobiphenyl congeners by multidimensional gas chromatography', *J. High Resolut. Chromatogr.* **14**: 593–596 (1991).

54. J. de Boer and Q. T. Dao, 'Interferences in the determination of 2,4,5,2',5'-pentachlorobiphenyl (CB 101) in environmental and technical samples', *Int. J. Environ. Anal. Chem.* **43**: 245–251 (1991).

55. E. Benická, R. Novakovsky, J. Hrouzek, J. Krupčík, P. Sandra and J. de Zeeuw, 'Multidimensional gas chromatographic separation of selected PCB atropisomers in technical formulations and sediments', *J. High Resolut. Chromatogr.* **19**: 95–98 (1996).

56. G. P. Blanch, A. Glausch, V. Schurig, R. Serrano and M. J. Gonzalez, 'Quantification and determination of enantiomeric ratios of chiral PCB 95, PCB 132 and PCB 149 in shark liver samples (*C. coelolepis*) from the Atlantic ocean', *J. High Resolut. Chromatogr.* **19**: 392–396 (1996).

57. G. Schomburg, H. Husmann and E. Hübinger, 'Multidimensional separation of isomeric species of chlorinated hydrocarbons such as PCB, PCDD and PCDF', *J. High Resolut. Chromatogr. Chromatogr. Commun.* **8**: 395–400 (1985).

58. D. Hainzl, J. Burhenne and H. Parlar, 'HRGC–ECD and HRGC–NICI SIM quantification of toxaphene residues in selected marine organism by environmentally relevant chlorobornanes as standard', *Chemosphere* **28**: 237–243 (1994).

59. H.-J. de Geus, R. Baycan-Keller, M. Oehme, J. de Boer and U. A. Th Brinkman, 'Determination of enantiomer ratios of bornane congeners in biological samples using heart-cut multidimensional gas chromatography', *J. High Resolut. Chromatogr.* **21**: 39–46 (1998).

4 Orthogonal GC–GC

P. J. MARRIOTT

Royal Melbourne Institute of Technology, Melbourne, Australia

4.1 INTRODUCTION TO MULTIDIMENSIONAL GAS CHROMATOGRAPHY

A multidimensional gas chromatography (MDGC) separation involves two columns in the separation process. However, simply the joining together of two columns is not sufficient to produce the MDGC process. The direct coupling of two different columns – called multichromatography (1) – is able to improve separation, but is essentially the same as mixing stationary phases (in this case, the two phases on the coupled columns) to obtain an improved selectivity in the separation. In Figure 4.1, the interface or valve at the confluence of the two columns is simply a column join in this instance. There is no increase in capacity (total number of available separation plates) in the system, but merely a shifting of the peak relative retentions. This experiment has been improved in its implementation by locating a valve between the two columns, thus allowing the mid-point pressure to be altered (the stimulus for the valve in Figure 4.1 is variable pressure). This is termed a pressure-tuning experiment (2), and by varying the pressure the relative contribution of each column to the separation can be varied, and so a wide range of apparent retention factors can be obtained. By optimization of the pressure and other experimental settings (3, 4), the chromatographic separation can be consequently optimized according to the mixture being studied. This method still does not qualify to be called MDGC. The MDGC definition must be cast to allow differentiation between the accepted MDGC experiment and the above coupled-column approaches. It must incorporate the process of isolation of zones of effluent from column 1 (also called dimension 1 (1D)) and then passing them to column 2 (also called dimension 2 (2D)). One could believe that this must involve extra-column couplings, but that has been obviated by recent developments (see below). Thus, the simplest MDGC arrangement will be to use a valve switching system that can pass zones of effluent from 1D to 2D. This will be the subject of other chapters in this book and so will not be treated any further here. In the case of the system shown in Figure 4.1, the interface or valve will be a mechanical valve or a pressure switching valve that can be switched to pass effluent from 1D to 2D, while the stimulus will be the electronic valve drive or balancing pressure supply. Those zones not passed to 2D will be directed to the detector at the end of 1D.

Multidimensional Chromatography, edited by L. Mondello, A. C. Lewis and K. D. Bartle
©2002 John Wiley & Sons Ltd.

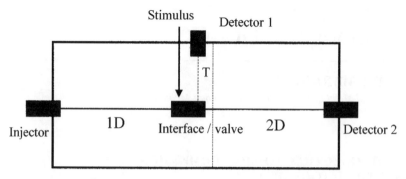

Figure 4.1   Schematic diagram of a coupled column system. The first column (1D) is connected to the second column (2D) through the interface or valve system. The interface can be a direct coupling, a live T-union, a complex multiport valve, or a thermal or cryogenic modulation system. The stimulus can be the switching of the valve, a balancing pressure to divert flow towards 2D, an added flow that is used in pressure tuning, or the drive mechanism for the modulator. The line to detector 1 will normally be a non-retaining section of column. In a two-oven system, 1D and 2D will be in different ovens; the dotted line indicates separately heated zones.

Figure 4.2(a) shows a schematic diagram of how the result might be viewed, with sections 1–4 transferred from 1D to 2D by what we call a heartcut operation. In the simplest mode, 2D can be a column which continues the chromatographic process immediately the zone is passed to it, or it can incorporate a cryotrap or cooled zone which focusses or traps all of the segments at the start of 2D before they are allowed to travel along the second column. The latter can be located in a second oven, such as is shown by the dotted line separating the two sections in Figure 4.1.

The separation zone on 1D has been expanded on 2D, and so if peaks are not resolved on 1D their separation on 2D will be achieved provided that the selectivity of the latter towards the specific components being transferred allows this. The total peak capacity is increased because we add the number of plates (or separable peaks) from 1D to that of 2D. In some cases, the components from the 1D heartcuts may overlap on 2D. This will reduce the total resolution. It is logical that such overlap should be minimal if we are to prevent solutes from different zones from overlapping on the subsequent 2D analysis. If the argument is extended to the limit, we get the situation shown in Figure 4.2(b) where every small zone of 1D is cut to 2D. The final result will almost be equivalent to just analysing the sample on 2D (or equivalent to the multichromatography method). Hence, MDGC with heartcuts has an inherent problem where many separate parts of the 1D analysis are of interest in respect of the analyst wanting to increase their resolution or separation.

The conventional MDGC experiment is essentially a limited multidimensional separation method, applying the advantage of MDGC to only limited portion(s) of a separation problem. Thus, we might conclude that MDGC can be applied to either relatively simple samples, or only to small zones of a complex sample. Hence, this could be construed as not achieving the overall goal of the multidimensional

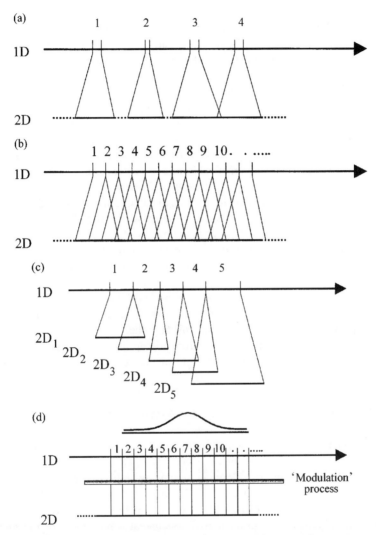

Figure 4.2 Examples of multidimensional gas chromatography separation arrangements: (a) small heartcut zones 1–4 are transferred to the second column and run together on this analytical column, where ideally greater selectivity sees the zones more spread-out on the latter column; (b) if contiguous zones from 1D are all simply passed to 2D, then we get severely overlapping zones, and as a result this will be similar to just analysing the sample on the second column.(c) the contiguous zones 1–5 are transferred to separate traps, and each can be independently analysed in the second dimension, thus giving rise to five 2D chromatograms; (d) if we implement the right modulation process on small contiguous zones from 1D, which is rapid with respect to the peak width times, between 1D and 2D, preferably with zone compression and fast GC conditions on 2D, then we can obtain the GC × GC technique.

analysis, unless the analyst was only specifically interested in increasing the separation of limited target zone(s).

This immediately also leads one to conclude that where a total sample should be analysed by MDGC – due to the less-than-satisfactory resolution of components achievable on a single column – or might be better analysed by MDGC applied over the whole analysis, conventional MDGC fails to deliver the separation power demanded. Wilkins and co-workers addressed this problem to some extent by using multiple traps (5, 6) where they could isolate contiguous fractions from a petroleum chromatographic separation on one column into different sorption traps, as shown by the approach indicated in Figure 4.2(c). The lower part of the latter shows that each 'trap' is analysed separately, and here gives five independent 2D chromatograms. Clearly, there is a physical and practical limitation to the number of traps that can be employed for this purpose. We should view Wilkins' work as an innovative transitional technology, recognizing the limitations of MDGC and trying to develop a system that offers a step towards maximizing the power of MDGC. Conceptually, we should wish to apply Wilkins' concept to the microcolumn environment, with minimum zone widths and on a rapid time-frame, but without the multitude of traps. We might thus seek to incorporate the idea of multiple-zone trapping into the column itself.

While it is generally acknowledged therefore that MDGC can increase resolution by transfering zones from one column to another, there is clearly a limited opportunity to improve the total analysis through this approach.

Conventional multidimensional gas chromatography operation procedures should now be reconsidered and redefined in light of the new method of comprehensive GC × GC technology, as discussed below.

4.2 INTRODUCTION TO GC × GC SEPARATION

The comprehensive GC × GC technique was introduced by Phillips and co-workers (7–9) and has been reviewed by de Geus *et al.* (10). GC × GC achieves the goal of applying multidimensional gas chromatography separation over the total analytical separation. The comprehensive GC × GC experiment is also defined as a system that allows all of the sample from the first column to be analysed on the second column. The key to the experiment is the technical achievement of the interface between the two dimensions, which is discussed in more detail below. Notwithstanding the above discussion, GC × GC is achieved precisely by the method shown in Figure 4.2(b), except that we have to modify the diagram slightly, and this is shown in Figure 4.2(d). The 2D analysis duration for the 'heartcut' zone is approximately the same as the zone duration taken or transferred from 1D. These zones are a small fraction of the elution time of an individual chromatographic peak, as shown in the figure. The transfer of solute from 1D to 2D involves, again, an innovative approach that leads to zone compression of the chromatographic band between the two dimensions. This is the 'modulator', and intervenes between the two dimensions to give two important results, and benefits, which arise from the modulation process. These are as follows:

(i)   the zone to be passed (or more correctly, pulsed) from 1D to 2D must be compressed in space;
(ii) the compressed zone must be delivered to 2D very rapidly, and as a sharp pulse;
(iii) 2D must be capable of giving fast GC results, achieved by a combination or all of the following, i.e. a short column, thin film thickness, narrow id column (giving high carrier linear velocity) and higher temperature (if a two-oven system is used);
(iv) the peaks produced at the detector will be increased in peak height response due to the above process;
(v)   all of the first column solute is transferred to the second column.

It is tempting to draw the analogy between GC × GC and 2D planar chromatography. On a TLC plate, the original single spot may be developed along one edge of the plate, which is then rotated 90° and placed in a second eluting solvent system to give rise to a different separation mechanism for separating compounds that were unresolved in the first step. The final plate could have solute spots distributed anywhere over the plate space that had been 'developed' by eluent. In a similar sense, the GC × GC experiment, if properly designed, could theoretically have peaks distributed over a space corresponding to the full range of possible distribution constants available to the mixture components on the columns used in the experiment. In TLC, however, zone visualization is conducted on the final plate (e.g. by densitometry), whereas in the GC case we must have a single detector recording the effluent from the second column. The GC × GC experiment also has all of the normal method advantages of the GC technique – sensitive analysis, hyphenation with mass spectrometry should be possible, readily automated methods, etc.

The ability of a GC column to theoretically separate a multitude of components is normally defined by the capacity of the column. Component boiling point will be an initial property that determines relative component retention. Superimposed on this primary consideration is then the phase selectivity, which allows solutes of similar boiling point or volatility to be differentiated. In GC × GC, capacity is now defined in terms of the separation space available (11). As shown below, this space is an area determined by (a) the time of the modulation period (defined further below), which corresponds to an elution property on the second column, and (b) the elution time on the first column. In the normal experiment, the fast elution on the second column is conducted almost instantaneously, so will be essentially carried out under isothermal conditions, although the oven is temperature programmed. Thus, compounds will have an approximately constant peak width in the first dimension, but their widths in the second dimension will depend on how long they take to elute on the second column (isothermal conditions mean that later-eluting peaks on 2D are broader). In addition, peaks will have a variance (distribution) in each dimension depending on the dispersion/diffusion processes in each column. For a discussion on two-dimensional separations, with zone formation, considerations of resolution, and expressions for dispersion in each dimension, the reader is directed to the treatment by Giddings (12). If single dimension capacity is calculated by how many peaks of basewidth x ($= 4\sigma$)s can be fitted into a total available elution time of y s, then approximately y/x peaks can be separated. For GC $\times$ GC, we can approximate the capacity by calculating how many peak 'areas' can be fitted into the total available area. Taking a peak as having a 15 s basewidth in dimension 1, and an average of 200 ms in the second dimension with a modulation 4 s, then for a 60 min (3600 s) analysis we should have a capacity of $(3600/15) \times (4/0.2) = 4800$ peaks. This statistically available capacity probably cannot all be used in an analysis, but more importantly it does increase dramatically the *potential* separation power, and many more compounds can be separated, hence achieving the goal of MDGC analysis. To properly implement the comprehensive GC $\times$ GC method, zone compression is used (an option not available in the TLC experiment or liquid mobile phase separations), which in combination with fast second-dimension analysis gives very much improved sensitivity (13, 14) with improvements of 50-fold or better being reported.

4.3   INTRODUCTION TO MODULATION TECHNOLOGY

There are essentially two procedures which can deliver the requirements of modulation between the two columns as defined above – these are the thermal sweeper/modulator of Phillips, and the cryogenic modulator of Marriott. A third option which has been reported is the valve/diaphragm modulator of Synovec and co-workers (15). The latter is capable of producing very narrow peak pulses to a second column, but it does this by a rapidly switching diphragm valve, which momentarily diverts a slice of the peak entering the valve to a second column. This may be as small as a 10 ms zone from each 1s portion of a peak, although the rest of the solute is vented. Whilst the chromatogram presentation seems to be analogous to that of the other comprehensive gas chromatography technologies, this diaphragm method does not increase the mass sensitivity of detection, and there is no solute compression. The first two options will be discussed further below.

4.3.1   THERMAL SWEEPER

The thermal sweeper is a commercial product licensed to Zoex Corporation, Lincoln, NA, USA (16). The sweeper incorporates a slotted heater (operated at about 100 °C above the oven temperature) which passes over the capillary column (normally an intermediate thicker film column is used in this region as an accumulator zone). Figure 4.3 is a schematic diagram of how the instrumental arrangement may be considered. The greater temperature of the rotating sweeper forces the solute which has been retained in the phase in the accumulator section to be volatilized out of the phase into the carrier gas stream, and then bunched up and brought forward

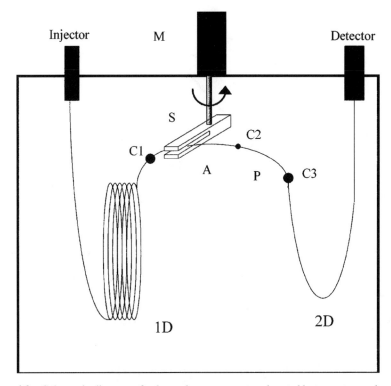

Figure 4.3 Schematic diagram of a thermal sweeper system located between two columns in the gas chromatograph; C1, C2 and C3 are column connections, with C2 being small enough to allow the slotted heater S to pass over the connection. The accumulator column A retards the travel of solutes through this section until the sweeper expels them out to the uncoated column (called a pigtail (P)), which then delivers them as a narrow band to the second column. The modulation drive, M, is external to the GC oven.

along the column as a focused band as the heater rotates over the column section. It is then delivered to 2D quickly, as required. The sweeper may have to pass over a column connection, and this had presented a robustness problem in the early development of the technology. Often the first column is connected to the accumulator column by using a short uncoated capillary, and another uncoated column is placed before 2D. This involves a total of four column connections. A more recent version of the commercial system incorporates a separately controllable heater block in which 2D is located in order to allow independent thermal control of the two columns. This may aid optimization of separation performance. Whether the heated block causes temperature control concerns in the main oven is unclear. In operation, the sweeper arm runs across the column section and sits over the accumulator column exit connection to ensure full transfer of solute out of that section. The sweeper then rotates to its home position and waits until the next sweep event is initiated. These steps are indicated on Figure 4.4.

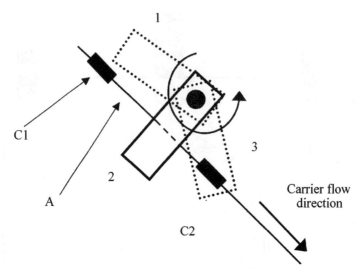

Figure 4.4   The slotted heater (see Figure 4.3) is held at its rest position 1 until it is moved to sweep across the accumulator column, as shown at position 2. When arriving at position 3, it rests again over the exit connection C2 to ensure solute is fully delivered to column 2. Then it rotates back to position 1, awaiting the next repetition of modulation.

4.3.2   LONGITUDINAL CRYOGENIC MODULATION

The cryogenic modulator (termed a longitudinally modulated cryogenic system (LMCS)), which was developed at The Royal Melbourne Institute of Technology, has a number of diverse applications mainly due to the universal value of cryogenic methods in GC, one of which is the comprehensive GC × GC technique. It offers some unique ways to manipulate chromatographic bands (17), which allows certain novel chromatographic results to be obtained. The LMCS uses a small cryogenic trap as a hollow sleeve about the capillary column to cool, and hence trap, any solute that enters the cold region (18). The key to this method is that the trap is moved along the capillary column to expose any trapped solute to the heat of the oven, which rapidly allows the solute to heat up and pass to 2D. Simply allowing the trap to heat up by turning off the CO_2 supply is not suitable (19); however, we have found that supplementary heating is also not required, with the prevailing heat of the oven being adequate to heat up the cold region of the column once the trap is slid away from the trapping region. Such a trap can perform a number of functions that conventional cryotraps are routinely used for. It is located inside the oven, and is used to cool a small length of fused silica column. Thus, it can focus volatile compounds at the injection end of the column system, or it can be used at the outlet end of the column inside the oven (just before the detector). These options are shown in Figure 4.5. The results of a definitive study of injection and detection applications of the LMCS have been reported (20), which suggest that the system should be useful

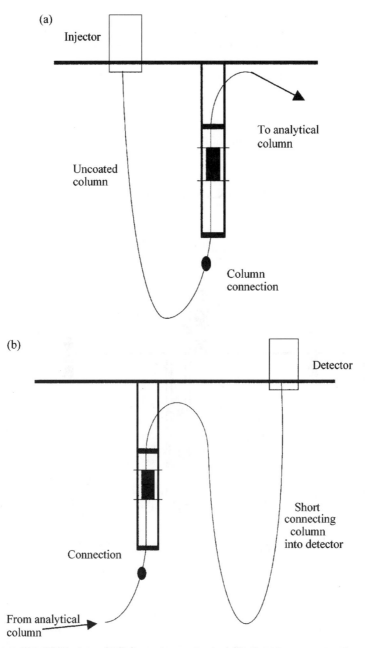

Figure 4.5   (a) By placing a short 'retention gap' column before the cryotrap, solute will be delivered to the collection zone, and when the trap is moved this will be effectively injected into the analytical column. (b) The LMCS can be used to focus solutes just before the detector. This gives very short, tall pulsed peaks which are many times taller than normal capillary GC peaks.

for headspace or thermal desorption experiments and, by extension, to solid-phase microextraction sample introduction, where the ability to focus the band at the head of the column is advantageous. When the LMCS is positioned between two columns in which we wish to perform some separation function, a range of multidimensional methods, some of which are novel approaches to solute manipulation, can be achieved (21). Figure 4.6 illustrates the mode where two separation columns are interfaced through the cryomodulator unit. Note that the capillary column is fed through the cryotrap, and so the solute is retained within the column (sorbed in or on the stationary phase on the walls of the column) which is surrounded by the cooled device.

4.3.3   MODES OF THE LMCS USING COUPLED SEPARATION COLUMNS

Once solute has been focused in space within the cryotrap, they are subsequently ejected out of the cold region (we can refer to this as being like a re-injection into the

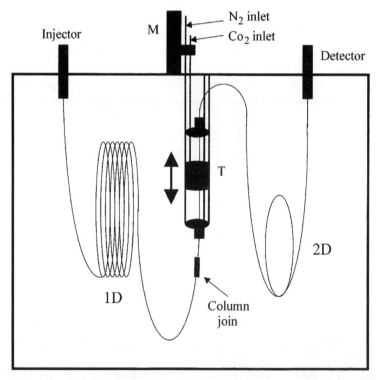

Figure 4.6   The LMCS system incorporates a moveable cryogenic trap T through which the capillary column is passed. The modulator (M) can be a pneumatic or motor driven device that moves T up and down as required, according to either preselected times (for a selective mode) or at a fast constant period for GC × GC.

downstream column) as a rapid pulse. The pulse width has been estimated to be as narrow as 10 ms or less, based on an experiment where a very short piece of column located between the cryotrap and detector gave peaks about 50 ms wide at their bases. Since the peak (or slice of a peak – see later) entering the cryotrap has a significantly broader bandwidth, the focusing process has the effect of concentrating the packet of solute – increasing its mass flow rate, which is translated into a peak height increase in the detection step. For instance, if a migrating peak is 40 cm wide, and is then focused to say 4 mm wide at the cryotrapping zone, we have a 100-fold increase in concentration. If the subsequent peak at the detector is 2 cm wide after it travels through a short (fast) column, then the peak will be about 20 times taller than that on a conventional capillary column operated normally. This argument can also be based on peak widths in units of s. If the conventional capillary peak is 5 s wide at its base, and the pulsed peak after the short second column is 200 ms wide, then a peak height increase of 25-fold is achieved. These improvements are easily attained in the LMCS experiment. Perhaps the more important effect of the focusing, however, is the potential for separation or resolution of solutes which are co-trapped in the cryotrap and collectively pulsed to column 2.

Accepting that the cryofocussing/remobilization process is both effective in the collection of discrete sections of the effluent from column 1, and very rapid in reinjection to column 2, we can now propose a number of ways of using the LMCS device in multidimensional gas chromatography modes.

4.3.3.1 Mode 1: Selective or Targeted Multidimensional Gas Chromatography

Many approaches have been described in the literature for achieving MDGC. Perhaps a common theme among these can be summarized as the need to isolate a small region of the first column separation and pass it to the second column, where we seek to attain greater separation. The analyst then only has to decide which choice of column will produce the desired result. Conventionally, however, the MDGC experiment has been much more involved and complex to implement than suggested by this simplified description of the process. The classical experiment will use flow switching valves or pressure-balanced systems that are not trivial to design, construct, operate and maintain. For these reasons, the number of analysts using MDGC on a routine basis is still very small, and often restricted to those laboratories where considerable expertise is available. With reference to Figure 4.1, the experiment will normally involve the following steps: (a) setting up the MDGC system; (b) running a normal GC analysis with the effluent directed to detector 1 through the non-retaining transfer line (T) (thus there is no time delay between peaks at the detector and their entry into the interface (I); (c) setting up the time sequence for heartcutting selected zones to 2D; (d) deciding if some trapping mechanism is required at the head of 2D; (e) running the experiment. The stimulus provided to the interface/value device (I/V) will be the event controlled by the multidimensional control unit that effects the heartcutting, and may be a valve switch/rotation, a valve closure, or a pressure change. Depending on the demands of the analysis, the target solutes might be only those transferred to the 2D, or might be peaks recorded at both detectors. There is no guarantee that both detectors respond to an equivalent degree. The use of modern electronic gas controls certainly adds to the precision of the MDGC experiment and should increase method robustness, but if there is a flow balancing problem between the two exit paths from the I/V device, the heartcutting event might be no longer reliable.

LMCS should provide an almost equivalent experimental result for the basic MDGC experiment described above. We have recently shown that within the single-capillary GC analytical run, a number of modes can be performed, including normal GC, targeted GC and comprehensive gas chromatography (22). The in-line cryotrap, which is capable of 'modulation' or pulsing of sections of effluent from column 1 to column 2, is conceptually the same as the heartcutting process in conventional MDGC, as described above. However, in order to create a workable system the second column must be operated under fast GC conditions (compared with the first column elution times) so that during the analysis of sequential sections of the first column zones, there is no overlap of adjacent pulsed zones. Instead of excising heartcut zones from column 1 to column 2, in this case we collect the desired zone in the cryotrap, and isolate it from the neighbouring zones by simply pulsing the collected zone to column 2. Any solute following this target zone in the first column is held back in the cryotrap while the previous zone is being analysed on column 2, and is retained in the cryotrap until itself is pulsed to column 2. If the column 2 analysis is fast (e.g. 0.2–0.5 min) and the column selectivity is sufficient, then many repeated pulsed zones can be analysed with target analytes resolved. Figure 4.7 shows how this concept can be viewed, with zones 1–9 collected and separately pulsed though to column 2. Dimension 2 is a fast separation and all peaks transferred to column 2 are eluted before the next pulse is delivered to column 2. Thus the three peaks in zone 3 will be resolved. This demands that these solutes have sufficient chemical difference to allow the phase on column 2 to differentiate them into individual peaks. Note that events 4 and 8 are blank (or dummy) events, and are inserted so that any trace components which may be collected since the previous zone are cleaned out of the cryotrap before the next peak is trapped. This could also be done to remove the buildup of column bleed in the cryotrap. Note that on the diagram, zone 1 is pulsed to dimension 2 when zone 2 is being collected, and so this explains why each peak or peak group is offset by one event step from the upper event sequence. The above procedure has been implemented in this laboratory for a number of applications such as semi-volatile aromatic hydrocarbons (23) and sterols (24).

It is again clear that the two benefits of increased sensitivity and better resolution are both achieved, where these arise from zone compression and phase selectivity, respectively. However, since this mode of analysis is relatively new, it has yet to be tested for a wide range of applications; such studies will be required to fully demonstrate its general utility. It is unclear whether this operational mode of selective MDGC constitutes a mode which is consistent with the definition of comprehensive

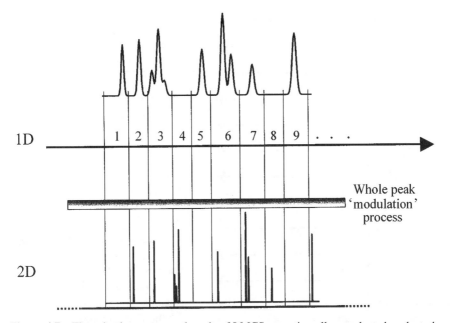

Figure 4.7 The selective or targeted mode of LMCS operation allows selected peaks to be collected sequentially in the cryotrap, and then pulsed rapidly to the second column. The resulting peaks are narrow and tall; provided that the second column phase selectivity and efficiency are adequate, they will also be resolved. The process is repeated as many times as required during the analysis. On this diagram, the lower trace response scale will be considerably less sensitive than on the upper trace.

gas chromatography as originally proposed, since this mode was not available until recently, and so neither its introduction nor its capability were anticipated by earlier chromatography researchers. However, it is probably better to only consider those modes which produce the more familiar GC × GC performance that should be termed comprehensive gas chromatography, as described below.

4.3.3.2 Mode 2: Comprehensive Gas Chromatography (GC × GC)

The process of GC × GC requires fast analysis of multiple slices of each peak eluting from column 1 to column 2. The cryogenic modulator was shown to produce this result (25) with the pulsed peak faithfully following the peak profile. Peak halfwidths of the order of 50 ms were reported, and a crude oil chromatogram presented. The process shown in Figure 4.8 illustrates the relative speed of modulation with respect to the peak width on the first column, which enters into the cryotrap. Thus about five or so slices are analysed successively on column 2. Figure 4.8 demonstrates that if two peaks are severely overlapping on the first column, and then subjected to the cryotrap-pulsing process, we can therefore obtain separate resolved peaks on column 2. By choosing conditions which are able to resolve the peaks on

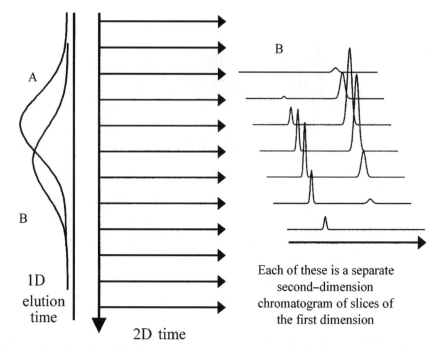

Figure 4.8 The GC × GC experiment can be considered to be a series of fast second chromatograms conducted about five times faster than the widths of the peaks on the first dimension. The 1D elution time is the total chromatographic run time, while the 2D time is the modulation period (e.g. 4–5 s). This figure shows two overlapping peaks A and B, with the zones of each peak collected together. When these slices are pulsed to the second column, they are resolved. Here, we show peak B eluting later on column 1, but earlier on column 2, with the 2D peak maxima tracing out a shape essentially the same as the original peak on 1D.

column 2 (e.g. using a more selective stationary phase) then each peak becomes a series of related second-column peaks. These secondary peaks will increase and then decrease according to how the primary peak distribution varies on the first column. Note that peak B has a longer retention time t_R on column 1, but is drawn as a shorter t_R on column 2. This would suggest that compound B has a smaller distribution coefficient (K) than compound A on column 2, and consequently a smaller retention factor, k, and thus a smaller retention time than A.

The peaks are revealed in the 2D separation space as oval-shaped peaks in a contour plot format, reproduced as a schematic diagram in Figure 4.9. The contour peaks are now completely separated in the 2D space, whereas they were severely overlapping on the first column.

This novel manifestation of the gas chromatographic separation demands that our fundamental understanding of the GC method – invariably of single-dimensional scope – is challenged as follows: concepts of column efficiency and separation are now supplanted by a need to compare the performances of two columns operating

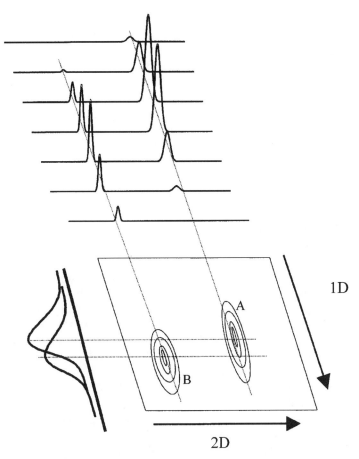

Figure 4.9   The peaks produced in the second dimension (see Figure 4.8) can be plotted as a contour shape in the retention or separation space, with characteristic retentions in each dimension. It can be seen that such peaks are now well resolved.

simultaneously; the usual post-run simple method reports are no longer simple since we have multiple peaks reported in the analytical results of a chromatogram for every single peak in a sample and recombining these back to the one component for purposes of quantitation are required; even retention is not an easy property to define because it will relate to the modulation period in addition to the first-column retention. These are, however, challenges which clever programming and familiarity with chromatogram interpretation will resolve in time. At present, the state-of-the-art of GC × GC appears to be in its qualitative capabilities. Selected highly complex samples have been reported in the GC × GC experiment, chiefly in the petroleum area – few other samples can be as complex as these – and the general consensus is that GC × GC has revealed complexity that has never before been realized, even though it may have been suspected. Not only have the samples been pulled apart in the 2D space, but even minor or trace components that would never have been exposed in the single column are often clearly placed as fully separated peaks.

The fact that we have peaks within a 2D space implies that where no peak is found represents a true detector baseline or electronic noise level. In a conventional petroleum sample, a complex unresolved mixture response causes an apparent detector baseline rise and fall throughout the GC trace. It is probably a fact that in this case the true electronic baseline is never obtained. We have instead a chemical baseline comprising small response to many overlapping components. This immediately suggests that we should have more confidence in peak area measurements in the GC × GC experiment.

4.3.4 CONSIDERATIONS OF RETENTION ON THE SECOND COLUMN – WRAPAROUND

For unresolved peaks eluting from the first column into the modulator, each pulsed solute packet delivers a group of components to column 2. It does this every modulation event, e.g. every 4 s. There is no guarantee that each solute will have a retention less than the modulation time, and so some peaks may not have reached the detector by the time the next packet of solute is launched into the short fast second column. The procedure for generation of the matrix format is illustrated in Figure 4.10. The column data are taken in blocks of data points equal to the number of points corresponding to the modulation time, and put into a column. This procedure is repeated until the full data are transformed. If a peak from one modulation event happens to be retained longer than the sampling period, then it will not be included in that period's transformed data block. Let us say the solute has a 2D retention time of 5 s. Since the data are transformed into matrix form for 2D presentation, using the 4 s modulation time as one matrix edge, then the 5 s retained solute will have an apparent transformed time of 1 ($= 5-4$) s, and its peak will appear to be centred on a time of 1 s in the 2D plot. Thus, while we normally use the labels 'first dimension retention' (which is correct) and 'second dimension retention' for the 2D space, the above solute does not have an absolute retention of 1 s on the second column. (It may be more appropriate to say 'apparent retention' or use a similar term to reflect that the time does not necessarily have an absolute meaning on the 2D axis.) Here, the solute has 'wrapped around' one set of matrix transformations of data and so appears in the subsequent matrix data line. It is possible to have solutes wraparound more than once. A useful way to decide if wraparound occurs is to study peak widths in the second dimension. Those peaks retained more will have increasingly wide peaks. In a recent study, for instance, of derivatized sterols (24), the high elution temperature meant that the sterols eluted during the isothermal hold period (at 280°C). Thus, the later peaks not only became increasingly broad in the first dimension, but because they were less volatile they were more retained on the second column. A modulation of 4 s was used, and the last sterol had a ^2t_R time of

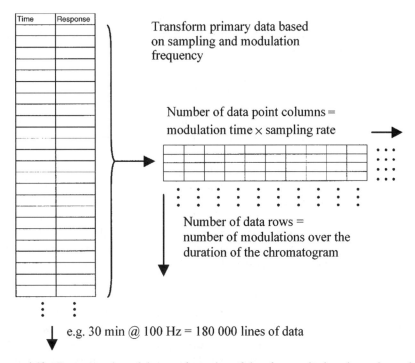

Figure 4.10 Representation of the transformation of data from a single-column data string to a matrix form, based on the sampling frequency and modulation time. The data points acquired for each modulation period are placed in a separate row of the matrix. The matrix data are then in a suitable format to read into an appropriate plotting package such as the 'Transform' program.

about 11 s, indicating that it was 'wrapped-around' twice. The sterol was located at 3 s in the 2D space, i.e. two wraparounds = 8 s, plus 3 s. The first eluting sterol had a ^2t_R time of about 3 s, while the others had ^2t_R times of about 4, 6 and 7 s as their retentions increased. The latter two peaks therefore appeared in the 2D space at about 2 s and 3 s. If the sterols were eluted during the temperature programme ramp, they would appear with a relatively constant ^2t_R time, since the decreasing volatility of the later sterols are compensated by the increased temperature at which they enter the second column.

In a novel experiment where a programmable temperature vaporizer (PTV) injector was used to continually supply dodecane sample into the first column, and presenting the retention on the second column for the dodecane during a temperature programmed analysis, it was possible to generate an isovolatility line for the second column which looked somewhat like a long tail extending through the separation space (26). Its retention decreased as the oven temperature increased, as was predicted. In some studies, in contrast, stationary phase bleed does not seem to trace out such a retention reduction. Presumably, since bleed comprizes highly volatile solutes, and only occurs at reasonably high oven temperature, it behaves almost as an unretained solute. In this case, the bleed retention profile may reflect the change in carrier gas flow in the second column as the temperature increases.

4.4 ORTHOGONALITY OF ANALYSIS

Multidimensional analysis methods rely on exploiting different properties or responses of solutes based on more than one physical/instrumental process, normally operated in series. Thus, a separation dimension coupled with a spectroscopic dimension constitutes a multidimensional analysis (27). The usual value of such a method is that if components are not uniquely identified in one dimension, then they will hopefully be identified in the second. Provided that the response bases of both dimensions are sufficiently different, then an orthogonal analysis results. If the mechanism of the two dimensions are similar, then we might propose that some degree of correlation exists, and this may reduce the identification power of the multidimensional analysis. For example, HPLC–diode array detection (DAD) can be regarded as orthogonal two-dimensional analysis because there is no correlation between the HPLC retention and the UV spectra of the components. Cortes has discussed a variety of coupled separation dimensions involving different chromatographic modes (28). An HPLC–high resolution GC system will likewise be orthogonal (29), and has been applied to such diverse applications such as oil fractions (30) and food and water analyses (31). The HPLC–capillary electrophoresis experiment involves two separation dimensions, but solutes respond in each dimension differently–HPLC according to the distribution constant of solutes between the mobile and stationary phases, and CE according to electrophoretic mobility (arising from solute size-to-charge properties). We might then conjecture that a multidimensional experiment employing reversed-phase HPLC and capillary electrochromatography on a non-polar phase material may possess a degree of correlation, since there will exist retention parameters strongly dependent upon partitioning processes into the reversed-phase stationary material which is common to both dimensions. In contrast, for the HPLC–DAD analysis the specificity of analysis then relies upon the components having uniquely identifiable or assignable spectra.

The most valuable tool for routine orthogonal two-dimensional analysis in GC is the GC–MS technique, with the dimensionality being shown schematically in Figure 4.11(a) for a GC–MS system where compound identity is defined by its time on the column (1D) and mass spectra (2D). Choosing unique or characteristic ions for overlapping components can produce reliable quantitative estimations of each component. How do we extend this argument to gas chromatography? In GC, MDGC is a two-dimension separation method. We might initially suggest that since we have two GC dimensions, then they must be correlated. However, the two dimensions will not be correlated provided that the mechanism of interaction between the components of a mixture and each column is different. Clearly, two columns of the same phase type operated at the same temperature will be correlated, and two components overlapping on the first column will not be expected to be separated on the

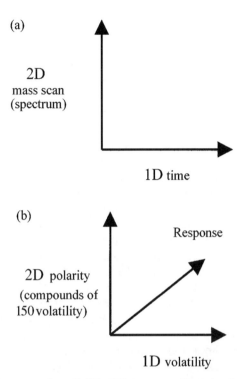

Figure 4.11 (a) Representation of GC–MS as a two-dimensional analysis method. (b) Representation of GC × GC as a two-dimensional separation, with separation mechanisms based of different chemical properties in each dimension.

second. We gain no extra separation or identification information from this experiment; there is no orthogonality. Thus, we ensure orthogonality of an analysis by employing techniques which exploit the chemical properties of the solutes to be studied, in a manner that gives independent responses in the two dimensions. In the case of two separation dimensions, we can force orthogonality by varying the analysis conditions of the second separation dimension as the first dimension proceeds. This can be referred to as tuning of the conditions, and is essentially achieved by using a second temperature zone for the second column.

4.4.1 HOW CAN WE ENSURE THAT THE TWO COLUMNS ARE ORTHOGONAL?

The elution of compounds on GC columns is a complex process related to the volatility of the compound, which results from its boiling point, and the chemical interactions between the compound and the stationary phase. These interactions are typically those which arise from polar–polar interactions, dispersion forces, dipole–dipole interactions, and so forth. Collectively, they are described by the term the chemical potential, $\Delta\mu°$, which derives from the potential for the compound to be transferred from the mobile to the stationary phase. This then alters the distribution constant and also the retention factor (i.e. retention time) of the compound. If two compounds have the same chemical potential on one column, then in order to separate them in a multidimensional experiment we have to alter their chemical potentials. This can be done by using a different temperature on the two dimensions, although a more useful approach is readily achieved by choosing different stationary phase types. Figure 4.11(b) shows that the first dimension separates by the property of volatility, and the second by polarity. The response axis indicates the detector response to the solute. Temperature variation on the two columns is a less practical solution, since we require two independently controllable temperature regions – such as a two-oven system. Most studies of multidimensional gas chromatography employ different column phases, and as a typical example we can consider a hypothetical experiment of a processed petroleum (e.g. kerosene) sample separation.

To a first approximation, the analysis of the petroleum sample on one stationary phase column versus another column will to a large extent appear very similar. The chromatogram will be dominated by the saturated alkanes, with the normal hydrocarbon suite providing the recognizable dome shape. Between the major components will be a range of branched and aromatic compounds. We cannot distinguish these minor components due to the large number of overlapping components, and one column is unlikely to be very much better than another. Admittedly, with mass spectral detection, we might be able to say that one column gives a better result than the other, but with flame-ionization detection the complexity is overwhelming. Different phases will shift the peak relative retentions, but for the kerosene trace components this is a little like changing one scrambled chromatogram for another. Each column separates or retains (primarily) by the component boiling point, and then imposes a selectivity or relative peak position adjustment based on what we might call polarity, but is better referred to as specific solute – stationary phase interactions.

With comprehensive GC, we can now choose a rational set of columns that should be able to 'tune' the separation. If we accept that each column has an approximate isovolatility property at the time when solutes are transferred from one column to the other, then separation on the second column will largely arise due to the selective phase interactions. We need only then select a second column that is able to resolve the compound classes of interest, such as a phase that separates aromatic from aliphatic compounds. If it can also separate normal and isoalkanes from cyclic alkanes, then we should be able to achieve second-dimension resolution of all major classes of compounds in petroleum samples. A useful column set is a low polarity 5 % phenyl polysiloxane first column, coupled to a higher phenyl-substituted polysiloxane, such as a 50 % phenyl-type phase. The latter column has the ability to selectively retain aromatic components.

The concept of tuning a separation was succinctly summarized by Venkatramani *et al.* who stated (32):

A properly tuned comprehensive 2-dimensional gas chromatograph distributes substances in the first dimension according to the strength of their dispersive interactions ... and in the second dimension according to their specific non-dispersive interactions . . . independent of any dispersive interactions the two stationary phases may have in common.

This statement refers to a non-polar first column and a more polar second column.

The ordering of classes of compounds within the separation space was summarized by Ledford *et al.* (33), who presented an analogy to the separation by using a mixture of objects of varied shapes, colours and sizes. The experimental dimensions could separate objects based on mechanisms which were *sensitive* to shape, size or colour, and the choice of two of these for the two-dimensional separation was illustrated. Applications showed a variety of petroleum products on different column sets, as well as a perfume sample.

Figure 4.12 is an example of such a separation obtained from work in our laboratory. Figure 4.12(a) shows a surface response plot which is reproduced in Figure 4.12(b) as a contour plot. The sample used is a light cycle oil, which has had certain components (presumably cyclic alkanes) removed through reforming the oil. Thus, there is a vacant region in the 2D space between the linear and branched alkane peaks (located at a 2D time of ~1.5–2.0 s), and aromatic solutes (benzenes, naphthalenes, etc., located from 2.5–6.0 s). The aromatic solutes are more polar and interact more strongly with the BPX50 stationary phase so are retained longer than the saturated compounds on the second column. The inset region of Figure 4.12(b) is expanded and shown in Figure 4.12(c). The linear alkanes are identified as peaks (a)–(e), with some selected other peaks (f), (g) and the groups (h) and (i) also being shown on this trace. Figure 4.12(d) shows the data obtained for a light gas oil, which has not been treated to remove the components noted as being absent in Figure 4.12(c). The same analysis conditions were used, and the same inset expansion is used for both of these figures. The same solutes (a)–(i) are identified in Figure 4.12(d), and the almost exact correspondence of these peaks is evident.

While samples such as these have obviously been the focus for much GC × GC work in the past, the technology still remains to be demonstrated for many other sample types. It is likely that in the near future, as many more applications are studied, a general theory–or at least a guide to column selection for GC × GC applications–will reveal a logical approach to selection of phases that embodies the principles of orthogonality of separation.

As a further example, we have recently completed one of the first studies of complex essential oil analysis by using GC × GC methods (34). While for petrochemical (35) and semivolatile aromatic compound analysis (36), we found the column set BPX5–BPX50 to be excellent–with the latter column able to separate saturated compounds from aromatics that coeluted on the first column–the nature of essential oil composition is such that we do not require the tuning of an aliphatic/aromatic separation, but rather we require a non-oxygenated/oxygenated component separation (with some degree of unsaturation also useable for separation). The obvious choice is a polar poly(ethylene glycol) column as the second dimension. Figures 4.13 and 4.14 show some typical results obtained for vetiver oil on BPX5–BPX50 and BPX5–BP20 column sets, respectively. Note that the

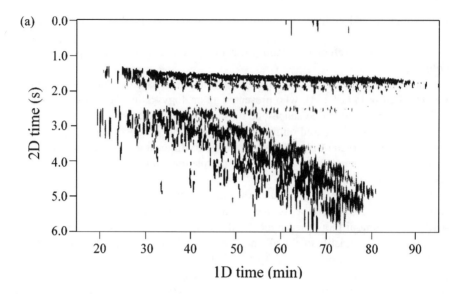

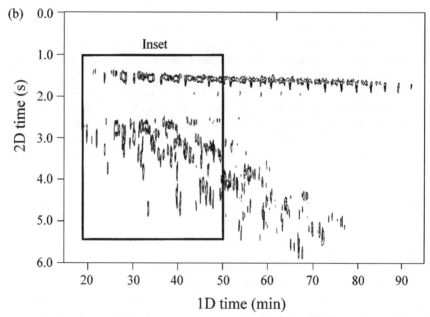

Figure 4.12   (a) Surface response plot of a light cycle oil analysed by GC × GC. The column set was a BPX5–BPX50 combination, with the second column being 0.8 m in length. A 6 s modulation time was used, in conjunction with a 1°C/min temperature programme rate. The cycle oil has been reformed to remove the cyclic alkanes. (b) The same data as in (a), but presented as a contour plot in this case.

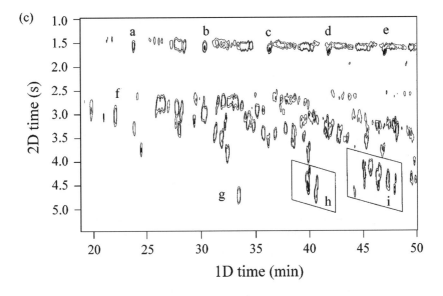

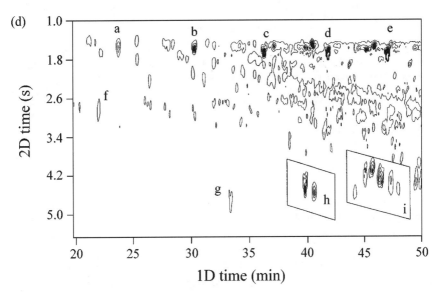

Figure 4.12 (*continued*) (c) An expansion of the inset region from (b), with the normal alkanes shown as (a–e). Other unidentified components (f–i) are presented to locate specific peaks for comparison purposes. (d) A light gas oil analysed under the same conditions as for the cycle oil, showing the same expanded region. In this case, the oil has not been treated in the same manner as the cycle oil, so it retains the components that were absent from the cycle oil. Peaks (a–i) are the same as those seen in (c).

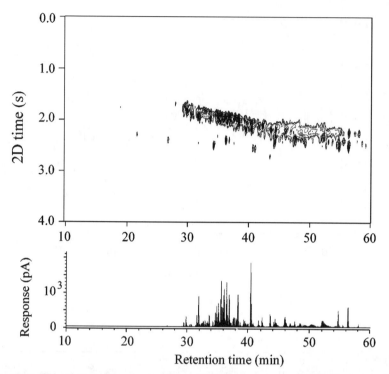

Figure 4.13 GC × GC analysis of vetiver essential oil: column 1, BPX5; column 2, BPX50 (0.8 m in length). The lower trace presents the pulsed peaks obtained from the modulation process, and shows such peaks in a manner that represents the normal chromatographic result presentation. This trace is many times more sensitive than a normal GC trace. In the upper plot, the 2D separation space shows that the BPX50 column is not very effective in separating components of the oils based on polarity, since all the components are bunched up along the same region of 2D time.

lower traces, which are the modulated data produced by 4 s cryogenic modulation, look very much alike for both column sets. In this presentation format, there is little difference in the effect that the short second column makes. The second column shows its effect when the data are transformed into the 2D separation format. The advantage of the BPX5–BP20 set is clear, with the more polar (oxygenated) components spread throughout the separation space. The BPX50 column does not appear to have a great ability to differentiate between the non-oxygenated and oxygenated components. In some instances in Figure 4.14, we can detect as many as eight peaks which co-eluted on the first column. At the present, very few individual peaks have been identified, but this should be a fertile area of research in future. More well-characterized samples should be an initial priority for these studies.

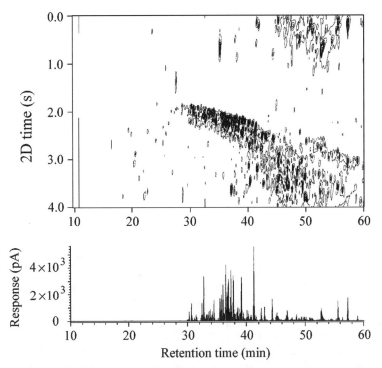

Figure 4.14   The same sample as that shown in Figure 4.13, but obtained in this case by using a BPX5–BP20 (1.0 m in length) column combination. The lower GC trace looks essentially the same as the equivalent one shown in Figure 4.13. The short second column apparently does not alter the total chromatogram greatly–most of the separation here is determined by the long first column. However, when the data are transformed to the 2D space, we can now obtain the full resolution produced by the polar BP20 column. This result shows that the selection of the second dimension column is critical to the success of the separation. We have 'tuned' the separation based on the need to separate oxygenated from non-oxygenated products, and different compound classes within oxygenated products, e.g. alcohols and ketones.

4.5   QUANTITATIVE ASPECTS

Applications of GC × GC can be grouped into the broad categories of qualitative and quantitative studies. By far the more numerous are qualitative studies, and in this group will be included those studies that also pay attention to the retention aspects of GC × GC. By quantitative studies, we take to mean those that involve measurement of peak responses in terms of area or height, and therefore offer the analysis of the relative proportions of the various components. Frysinger and co-workers (37, 38) have primarily studied the use of GC × GC to investigate petrochemical pollution at or around military (naval) bases. These authors have studied marker components and also absolute peak amounts that allow the comparison of spills with their likely sources. The goals of some studies were to provide the separation of target groups of compounds from gasoline (e.g. benzene, toluene, xylenes, C_9, C_{10} and C_{11} benzenes, naphthalene and C_{11} and C_{12} naphthalenes), quantitate these compounds by using internal standards, and then establish the feasibility of such studies by using GC × GC. They concluded that the GC × GC approach produced results directly comparable with ASTM methods. The ranges of individual compounds were of the order of 0.1–24% for calibration data, while the RSD% values were from 0.6–5% for individual components, and up to 14% for grouped components such as C_{12} naphthalenes. The ability of the GC × GC method to give the required data in a single run, using a new technique, with a result that was not too dissimilar to the ASTM result, was laudable. The GC × GC method was also able to produce data of sufficient reliability to analyse quantitatively the levels of target compounds in oil spills, for comparison with possible discharge sources.

Again in the petroleum realm, Beens et al. (39) compared the analytical results obtained by the GC × GC method against both single-column capillary GC and an LC–GC hyphenated system. A test mix of a variety of typical analytes gave essentially equivalent results in the first comparison, and the analysis of heavy gas oil by LC–GC and GC × GC for grouped classes of components (e.g. saturates and mono- di- and tri-aromatics) again appeared acceptable. These authors concluded that GC × GC was the technique of choice, although they acknowledged that the thermal sweeper had a restricted useable upper temperature limit. This work also discussed the data handling program 'Tweedee' that had been developed to provide some automated reporting capability of peak areas.

There are reported to be a number of important characteristics of GC × GC that permit more reliable peak response quantitation over single-column GC analysis. These are as follows:

(i)   There is less chance of peak overlap, which means that peak areas/heights will more reliably give these parameters of a pure peak rather than have contributions from minor unresolved constituents.

(ii) Related to this is the fact that the regions of the 2D space that do not contain any peaks will represent the true detector baseline rather than a chemical baseline comprising unresolved components.

(iii) GC × GC gives sharp peak pulses at the detector and so these peaks should be more reliably measured than broader peaks where the baseline construction may be less precise.

(iv) Quantitation requires calibration (response versus amount), and detection limit information should be available. Zone compression gives sensitivity improvement, and hence lower detectable amounts.

(v)   Faster data acquisition (e.g. 50/100 Hz versus 5 Hz for normal capillary GC) does lead to greater detection noise (by 3–4 times, respectively), and this means that detection limit increases are not quite as good as might be expected just on the basis of peak signal increases.

There remains one key concern for the routine use of GC × GC for quantitative analysis, and that is the need for a real-time, 2D data presentation package with the ability to present a comprehensive data report providing a list of all peaks found within the 2D space, along with their heights and areas. Ideally, further interpretive data might be provided, such as peak symmetry, resolution of neighbouring peaks, etc. Again, we await the general availability of such software, and perhaps this is the single most important issue to resolve before the methods outlined here become more widely accepted. However, the large data file sizes obtained with fast data acquisition should not be a problem.

4.5.1 COMPARISON OF FIGURES OF MERIT

Chromatographically, any procedure will only be of value if the analytical data or figures of merit are acceptable to the analyst. It is necessary to demonstrate the reliability of data when using the modulation process. The above points state in very general terms why modulation gives advantages, but this still requires demonstration. In a recent comprehensive study of analytical figures of merit for the analysis of derivatized sterols, we compared the three modes of operation of the LMCS, namely normal capillary GC, comprehensive GC × GC and targeted MDGC (24). This work revealed that improved precision for raw peak areas is achieved with the selective LMCS mode (RSD of ±2 %; $n = 5$), over that of normal GC and comprehensive gas chromatography analysis (both with RSD ca. ±4 %; $n = 5$). Peaks were generally 25–40 times taller for the selective mode when compared with normal GC peaks, based on peak height data, with detection limits of approximately 0.004 mg/l compared with 0.1 mg/l for normal capillary gas chromatography and 0.02 mg/l for the comprehensive mode. Note that peak areas are not increased when using the cryogenic system, although we can measure peak areas down to lower levels. The linearity of calibration was best for targeted MDGC (e.g. for the results as obtained by the procedure shown in the lower trace in Figure 4.7), presumably because there is less uncertainty in the peak measurement. Calibration curve linearity for comprehensive GC was similar to normal GC. The better height and area precision for the five replicate analyses for selective MDGC analysis may be due to the better confidence of measuring peak response when the peak is much higher than the level of detector noise. The same injection procedure (splitless) was used for all analyses, i.e. with an autoinjector, and hence the reproducibility of injection should have been similar for all of the operational modes.

The reason for the differences in comparative figures for the two cryogenic modes may be related to the fact that comprehensive GC gives a series of pulses for each peak, and these must be summed to get the total response, while targeted MDGC gives a single peak for the total component. Normal GC is improved upon by both the other modes. These results indicate that there is every reason to believe that the modulated peak response methods provide reliable and accurate data, at least as good as normal GC, for the analysis of chromatographic peaks.

4.6 FUTURE OPPORTUNITIES AND CHALLENGES OF GC × GC TECHNOLOGY

4.6.1 DIRECTIONS OF GC × GC APPLICATIONS AND RESEARCH

It is rarely wise to predict how a new technology will be accepted in the 'market-place' – in this case, in analytical laboratories where reliable, simple, informative and automated methods are demanded. However, given the possibilities that now present themselves here, such speculation is warranted. The reported applications of GC × GC are not many, and as stated earlier are predominantly in the petroleum area. However, the advantages that derive from GC × GC are equally desirable for, and of general relevance to, perhaps most separation studies. To say that all applications will be better conducted by employing GC × GC is maybe equivalent to saying that capillary GC is always better than packed column GC. There will always be an exception. In order to popularize GC × GC, systems must be widely available. Furthermore, to achieve system placements, users must be convinced that they can perform their analysis better by using GC × GC. This requires more literature stud-ies to demonstrate that these outcomes are likely. Thus, the future rests largely on (i) user-friendly GC × GC technology which is dependable, and readily available and implemented, (ii) a strong literature base of analytical studies, (iii) an interpretation protocol that assists users to assimilate the results of the 2D separation, and (iv) a data system that presents chromatographic reports in a manner that is familiar to users. The first feature has not yet been achieved due to limited market acceptance of the new approach, while the second is slowly being established, although clearly all of the major interest areas have not yet been tackled. The third and forth features are major challenges. In our laboratory, we can collect the primary GC × GC data, convert, transform and display the results reasonably quickly, but the fundamental interpretation that allows the less experienced user to understand the results must still be developed. The data system requires considerably more attention, and it will not just be a matter of smart programming to resolve this question. The challenge of the data system may in fact be the greatest opportunity for GC × GC to make its mark, since the information-rich separations will allow aspects of pattern recogni-tion and many other 'chemometric' tools to be investigated. However, until the rou-tine laboratory can be assured that at the end of the run, they will get a quantitative result printed on the post-run report, they are unlikely to rush into this technology. Nevertheless, it is much more likely that research laboratories will become increas-ingly attracted to the advantages of GC × GC, if only to evaluate the 'true' chemical nature of their materials, or if they are intrigued by the possibilities that the sensitiv-ity and separation power offers.

We can list the following areas as prime targets: essential oil and natural product analysis, chiral analysis (e.g. of fragrances), trace multi-residue analysis, pesticide monitoring, and further petroleum products applications, in fact any separation where simply greater resolution and sensitivity is demanded – which means probably *almost all* samples. One of the few pesticide samples analysed by comprehensive gas chromatography, using an early design system, indicated the promise that GC × GC held, but few more recent follow-up studies have been reported (40). The alternative capabilities of the cryogenic modulation system, with methods such as targeted MDGC, will hopefully be seen as another reason to investigate this new GC direction.

4.6.2 MASS SPECTROMETRY DETECTION

There is much excitement, but at the moment few results, on the use of mass spectrometry (MS) combined with GC × GC analysis. Such a system represents a three-dimensional analytical procedure, since it provides three orthogonal dimensions with each contributing to the identification of components. The MS final stage enables identification of the 2D separated peaks, and hence brings additional understanding to the unique structured chromatograms. The main criterion that decides if MS can be used with fast GC peaks is the scanning speed of the spectrometer, i.e. the number of scans per second. Since GC × GC peaks may be as narrow as 100 ms or less, and since reliable spectra or peak reproduction requires at least five scans (preferably ten) per peak, then a scan rate of 50–100 scans/s are needed. Time-of-flight (TOF) MS is the only practical technology that can deliver this speed, and since there are few such instruments for GC available, and even fewer in those laboratories with experience in GC × GC, therefore little has been published on this hyphenated technique. One approach that has been taken, however, for coupling quadrupole MS with GC × GC analysis, is to use a flow splitter before the modulator, so that at least some indication of the possible components which constitute the second-dimension result might be known (41). Another approach is to slow down the second dimension so that maybe two or three spectra can be acquired, and then hope that one suitable spectrum for library matching might be obtained (42). This study employed a long second-dimension column (14 m in length) which gave peaks about 1 s wide. The total GC analysis was developed over 440 min. This is unlikely to be very useful for routine analysis! More recently, the Centres for Disease Control have used a GC–TOFMS system for GC × GC, and the results of this work are eagerly anticipated.

Even though it appears that the technology has not been adopted yet, it is expected that TOF MS will be useful to validate the power of the GC × GC separation experiment by proving the separate identities of the vast number of resolved peaks and so show that the analyst who does not use GC × GC is missing valuable chemical compositional information on their samples. In addition, it is just as significant to TOFMS that GC × GC becomes a widespread separation tool, since this will then provide a demand for the powerful capabilities of TOFMS for identification. The GC community must wait for this to be demonstrated, and those who are working in GC × GC development are convinced that the wait will be worth it!

ACKNOWLEDGEMENTS

The author wishes to acknowledge the efforts of his various research students from whose work some of the applications presented here have been drawn. In particular, I have enjoyed the excellent partnership in the pioneering PhD research of Mr (now Dr) Russell Kinghorn. In addition, the provision of a gas chromatograph by Agilent Technologies and columns from SGE International have enabled our work to progress.

REFERENCES

1. J. V. Hinshaw, Jr and L. S. Ettre, 'Selectivity tuning of serially connected open-tubular (capillary) columns in gas chromatography. Part 1: fundamental relationships', *Chromatographia* **21**: 561–572 (1986).
2. M. Akard and R. Sacks, 'Pressure-tunable selectivity for high-speed gas chromatography', *Anal. Chem.* **66**: 3036–3041 (1994).
3. H. Smith and R. Sacks, 'Pressure-tunable GC columns with electronic pressure control', *Anal. Chem.* **69**: 5159–5164 (1997).
4. B. Lorentzeas, P. J. Marriott and J. Hughes, unpublished results.
5. C. L. Wilkins, 'Multidimensional GC for qualitative IR and MS of mixtures', *Anal. Chem.* **66**: 295A–301A (1994).
6. K. A. Krock, N. Ragunathan and C. L. Wilkins, 'Parallel cryogenic trapping multidimensional gas chromatography with directly linked infrared and mass spectral detection', *J. Chromatogr.* **645**: 153–159 (1993).
7. J. B. Phillips and J. Xu, 'Comprehensive multidimensional gas chromatography', *J. Chromatogr.* **703**: 327–334 (1995).
8. J. B. Phillips and E. B. Ledford, 'Thermal modulation: a chemical instrumentation component of potential value in improving portability', *Field Anal. Chem. Technol.* **1**: 23–29 (1996).
9. Z. Liu and J. B. Phillips, 'Comprehensive two-dimensional gas chromatography using an on-column thermal modulator interface', *J. Chromatogr. Sci.* **29**: 227–231 (1991).
10. H.-J. de Geus, J. de Boer and U. A. Th. Brinkman, 'Multidimensionality in gas chromatography', *Trends Anal. Chem.* **15**: 398–408 (1996).
11. J. C. Giddings, 'Sample dimensionality: a predictor of order–disorder in component peak distribution in multidimensional separation', *J. Chromatogr.* **703**: 3–15 (1995).
12. J. C. Giddings, 'Steady-state, two-dimensional and overlapping zones', in *Unified Separation Science,* John Wiley & Sons, New York, Ch. 6, pp. 112–140 (1991).
13. H.-J. de Geus, J. de Boer, J. B. Phillips, E. B. Ledford and U. A. Th Brinkman, 'Increased signal amplitude due to mass conservation in a thermal desorption modulator', *J. High Resolut. Chromatogr.* **21**: 411–413 (1998).
14. R. M. Kinghorn and P. J. Marriott, 'Enhancement of signal-to-noise ratios in capillary gas chromatography by using a longitudinally modulated cryogenic system', *J. High Resolut. Chromatogr.* **21**: 32–38 (1998).
15. C. A. Bruckner, B. J. Prazen and R. E. Synovec, 'Comprehensive two-dimensional high-speed gas chromatography with chemometric analysis', *Anal. Chem.* **70**: 2796–2804 (1998).
16. J. B. Phillips, R. B. Gaines, J. Blomberg, F. W. M. van der Wielen, J.-M. Dimandja, V. Green, J. Granger, D. Patterson, L. Racovalis, H.-J. de Geus, J. de Boer, P. Haglund, J. Lipsky, V. Sinha and E. B. Ledford-Jr, 'A robust thermal modulator for comprehensive two-dimensional gas chromatography', *J. High Resolut. Chromatogr.* **22**: 3–10 (1999).

17. P. J. Marriott and R. M. Kinghorn, 'Modulation and manipulation of gas chromatographic bands by using novel thermal means', *Anal. Sci.* **14**: 651–659 (1998).

18. P. J. Marriott and R. M. Kinghorn, 'Longitudinally modulated cryogenic system. A generally applicable approach to solute trapping and mobilization in gas chromatography', *Anal. Chem.* **69**: 2582–2588 (1997).

19. P. J. Marriott and R. M. Kinghorn, 'Studies on cryogenic trapping of solutes during chromatographic elution in capillary chromatography', *J. High Resolut. Chromatogr.* **19**: 403–408 (1996).

20. R. M. Kinghorn, P. J. Marriott and P. A. Dawes, 'Longitudinal modulation studies for augmentation of injection and detection in capillary gas chromatography', *J. Microcolumn Sep.* **10**: 611–616 (1998).

21. R. M. Kinghorn and P. J. Marriott, 'Comprehensive two-dimensional gas chromatography using a modulating cryogenic trap', *J. High Resolut. Chromatogr.* **21**: 620–622 (1998).

22. P. J. Marriott and R. M. Kinghorn, 'New operational modes for multidimensional and comprehensive gas chromatography by using cryogenic modulation', *J. Chromatogr. A* **866**: 203–212 (2000).

23. P. J. Marriott, R. C. Y. Ong, R. M. Kinghorn and P. D. Morrison 'Time-resolved cryogenic modulation for targeted multidimensional capillary gas chromatography analysis', *J. Chromatogr. A* **892**: 15–28 (2000).

24. T. Truong, P. J. Marriott and N. A. Porter, 'Analytical study of comprehensive and targeted multidimensional gas chromatography incorporating modulated cryogenic trapping', *J. AOAC Int.* submitted (2000).

25. R. M. Kinghorn and P. J. Marriott, 'High speed cryogenic modulation – A technology enabling comprehensive multidimensional gas chromatography', *J. High Resolut. Chromatogr.* **22**: 235–238 (1999).

26. J. Beens, R. Tijssen and J. Blomberg, 'Comprehensive two-dimensional gas chromatography (GC × GC) as a diagnostic tool', *J. High Resolut. Chromatogr.* **21**: 63–64 (1998).

27. M. Careri and A. Mangia, 'Multidimensional detection methods for separations and their application in food analysis', *Trends Anal. Chem.* **15**: 538–550 (1996).

28. H. J. Cortes, 'Developments in multidimensional separation systems', *J. Chromatogr.* **626**: 3–23 (1992).

29. L. Mondello, G. Dugo and K. D. Bartle, 'A multidimensional HPLC–HRGC system for the analysis of real samples', *LC–GC Int.* **11**: 26–31 (1998).

30. J. Beens and R. Tijssen, 'An on-line coupled HPLC–HRGC system for the quantitative characterization of oil fractions in the middle distillate range', *J. Microcolumn Sep.* **7**: 345–354 (1995).

31. L. Mondello, G. Dugo and K. D. Bartle, 'On-line microbore high performance liquid chromatography–capillary gas chromatography for food and water analyses. A review', *J. Microcolumn Sep.* **8**: 275–310 (1996).

32. C. J. Venkatramani, J. Xu and J. B. Phillips, 'Separation orthogonality in temperature-programmed comprehensive two-dimensional gas chromatography', *Anal. Chem.* **68**: 1486–1492 (1996).

33. E. B. Ledford, J. B. Phillips, J. Xu, R. B. Gaines and J. Blomberg, 'Ordered chromatograms: a powerful methodology in gas chromatography', *Am. Lab.* **28**: 22–25 (1996).

34. P. J. Marriott, R. Shellie, J. Fergeus, R. C. Y. Ong and P. D. Morrison, 'High resolution essential oil analysis by using comprehensive gas chromatographic methodology', *Flav. Fragr. J.* **15**: 225–239 (2000).

35. R. M. Kinghorn and P. J. Marriott, unpublished data.

36. P. J. Marriott, R. M. Kinghorn, R. Ong, P. Morrison, P. Haglund and M. Harju, 'Comparison of thermal sweeper and cryogenic modulator technology for comprehensive gas chromatography', *J. High Resolut. Chromatogr.* **23**: 253–258 (2000).

37. G. S. Frysinger, R. B. Gaines and E. B. Ledford-Jr, 'Quantitative determination of BTEX and total aromatic compounds in gasoline by comprehensive two-dimensional gas chromatography (GC × GC)', *J. High Resolut. Chromatogr.* **22**: 195–200 (1999).

38. R. B. Gaines, G. S. Frysinger, M. S. Hendrick-Smith and J. D. Stuart, 'Oil spill source identification using comprehensive two-dimensional gas chromatography', *Environ. Sci. Technol.* **33**: 2106–12 (1999).

39. J. Beens, H. Boelens, R. Tijssen and J. Blomberg, 'Quantitative aspects of comprehensive two-dimensional gas chromatography (GC × GC)', *J. High Resolut. Chromatogr.* **21**: 47–54 (1998).

40. Z. Liu, S. R. Sirimanne, D. G. Patterson-Jr, L. L. Needham and J. B. Phillips, 'Comprehensive two-dimensional gas chromatography for the fast separation and determination of pesticides extracted from human serum', *Anal. Chem.* **66**: 3086–3092 (1994).

41. R. M. Kinghorn and P. J. Marriott, unpublished results.

42. G. S. Frysinger and R. B. Gaines, 'Comprehensive two-dimensional gas chromatography with mass spectrometric detection (GC × GC/MS) applied to the analysis of petroleum', *J. High Resolut. Chromatogr.* **22**: 251–255 (1999).

5 Coupled-Column Liquid Chromatography

CLAUDIO CORRADINI

Institute of Chromatography, CNR Rome, Italy

5.1 INTRODUCTION

Today, the various high performance liquid chromatography (HPLC) techniques represent one of the major parts of modern analytical chemistry and are often at the forefront of new discoveries in chemistry, biochemistry, biology, pharmacy, clinical chemistry, food chemistry and environmental sciences. However, HPLC, which has had such an enormous impact on the practice of analytical chemistry, frequently approaches its limits when applied to complex mixtures. In order to overcome this difficulty, selectivity-enhancing steps, such as the use of more efficient columns or the employment of appropriate gradient elutions, may be successful. In addition, the derivatization of analytes with properly selected functional groups is usually carried out in order to enhance detectability and selectivity. Other important methods involve sample clean-up procedures, trace enrichment and matrix elimination. However, sample pretreatment frequently presents difficult practical problems and often takes up the majority of the total analysis time, as well as contributing significantly to the final cost of the analysis, both in terms of labour and the consumption of materials. In addition, the requisite manipulations may result in significant imprecision which can greatly outweigh any variables in the actual chromatographic process itself. Despite these complications, some type of sample treatment is frequently mandated in order to achieve satisfactory resolution and quantitation of the analyte of interest. Another closely related problem is column abuse, especially in trace analysis applications. In some instances, sample pretreatment can be quite effective; however, too often the capacity of the column has been exceeded by several orders of magnitude. Such excessive column fouling leads to poor peak shape and lower efficiencies, with a corresponding loss in resolution and shorter column lifetimes. The resolving power of HPLC may be enhanced significantly by the introduction of multidimensional liquid chromatographic techniques. These are chromatographic techniques aimed at the determination of the concentration of one or more specific analytes in a multicomponent complex matrix. When using such HPLC techniques, the sample is separated either by more than one column or chromatographic mode,

Multidimensional Chromatography, edited by L. Mondello, A. C. Lewis and K. D. Bartle
©2002 John Wiley & Sons Ltd.

or by various combinations of columns possessing complementary separation characteristics. This technique provides the optimum efficiency and selectivity for separations of the component of interest, while simultaneously minimizing the analysis
time by decreasing the time spent in separating those components of the sample that
are of no analytical interest. Multidimensional chromatography (also known as coupled-column chromatography or column switching) represents a powerful tool and
an alternative procedure to classical *one-dimensional* HPLC methods.

Multidimensional liquid chromatography can be performed either in an on-line or
off-line mode. With off-line operation, the fractions eluted from the primary column
are collected manually or by a fraction collector and then reinjected, either with or
without concentration, into a second column. This approach has the advantage of
being simple, does not need any switching valve, and the mobile phases used in each
column need not be mutually compatible. However, these procedures are labour-
intensive and time-consuming and the recovery of sample is often low. On-line techniques have the advantage of automation by using pneumatic or electronically
controlled valving, which switches the column effluent directly from the primary
column into the secondary column. Automation improves reliability and sample
throughput, and shortens the analysis time, as well as minimizing sample loss or
change since the analysis is performed in a closed-loop system. Obviously, on-line
techniques are preferred, although they are not always feasible from an operational
point of view. The main limitation is that the mobile phase system used in the coupled columns must be compatible in both miscibility and solvent strength. This
requirement arises since the eluent from the first column is the injection solvent for
the second column; consequently, not all column types are mutually compatible.
Furthermore, the use of two different separation principles may lead to an inversion
of the elution order on the two subsequent columns. Hence, the separation achieved
on the first column can be substantially reduced on the second column. A comparison of advantages and disadvantages of off- and on-line multidimensional liquid
chromatographic techniques is shown in Table 5.1.

Multidimensional LC separation has been defined as a technique which is mainly
characterized by two distinct criteria, as follows (1). The first criterion for a multidimensional system is that sample components must be displaced by two or more separation techniques involving orthogonal separation mechanisms (2), while the
second criterion is that components that are separated by any single separation
dimension must not be recombined in any further separation dimension.

Coupled-column liquid chromatography (LC–LC coupling) refers to the conventional two-dimensional mode of chromatography in which fractions from one column are selectively transferred to one secondary column for a further separation.
What characterizes LC–LC coupling when compared to conventional multistep
chromatography is the requirement that the whole chromatographic process be carried out on-line. The transferred volume of the mobile phase from the first column to
the second column can correspond to a group of peaks, a single peak or a fraction of
a peak, so that different parts of the sample may follow different paths through the
LC–LC configuration. A large number of factors play a role in the development of
an LC–LC procedure; these include the separating power of the chromatographic

Table 5.1 Comparison of off- and on-line multidimensional LC techniques

Off-line multidimentional chromatography	
Advantages	Disadvantages
• Easy to carry out by collection of column effluent	• Labour intensive
• Can concentrate trace solutes from large volumes	• More time consuming
• Can work with two LC modes that use incompatible solvents	• Sample loss or contamination during handling

On-line multidimentional chromatography	
Advantages	Disadvantages
• Easy to automate	• Incompatibility of different mobile phase system
• No loss or contamination	• Separation obtained on the first column can, at least partly, be reduced on the second column
• Decreased total analysis time	• Requires automated or semi-automated instrumentation
• More reproducible	

columns, mobile phase composition, the nature and number of analytes, and the type of matrix and its related interferences. In order to achieve this, simple valving circuits can be used with conventional LC apparatus.

5.2 THEORETICAL ASPECTS

In order to optimize separations on coupled-column liquid chromatographic systems under the conditions of solvent modulation, we need to consider the parameters which affect the resolution, as follows.

The basic measure of the efficacy of a single-column chromatographic system in separating two neighbouring peaks can be effected by the resolution (R_s), which is equal to the ratio between the two peak maxima, Δt_r (distance between the peak centres) and the average base width of the two peaks, as follows (3):

$$R_S = \frac{\Delta t_r}{w} = \frac{2(t_{r,2} - t_{r,1})}{(w_1 + w_2)} = \frac{\Delta t_r}{4\sigma_t} \tag{5.1}$$

where $t_{r,1}$ and $t_{r,2}$ are the retention times, w_1 and w_2 are the peak widths measured by the baseline intercept and σ_t is the mean standard deviation of a Gaussian peak.

For closely spaced peaks, the resolution may be expressed as the product of three factors (a,b,c), which are related to the adjustable variables of a chromatographic system as follows:

$$R_S = \underbrace{\left(\frac{a-1}{a}\right)}_{(a)} \underbrace{\left(\frac{k_2}{1+k_2}\right)}_{(b)} \underbrace{\left(\frac{N}{16}\right)^{1/2}}_{(c)} \tag{5.2}$$

where (α) is the selectivity, which is a useful measure of relative peak separation related to the discriminatory power of the chromatographic system, (b) is the retention, which expresses the retentive power of the chromatographic system, and (c) is the efficiency, which measures the peak broading that occurs in the chromatographic column (together with extra-column contributions, which in a well-designed system are small). The three terms of equation (5.2) are essentially independent, so that we can optimize first one, and then the others. The separation can be improved by varying the selectivity of the system (α) by increasing N, and also by increasing the retention factor by changing the solvent strength, until the term ($k'/(1 + k')$) reach a plateau. Resolution is seen to be proportional to the term ($k'/(1 + k')$) of equation (5.2), which corresponds to the fraction of the sample that is in the stationary phase. Small values of k' mean that the sample is largely in the mobile phase, and under these conditions a poor separation of the sample is achieved. At high k' values, the factor ($k'/(1 + k')$) $\rightarrow 1$ and it may therefore be thought that k' should be high. However, high k' values lead to very long retention times, with the concomitant elution of excessively broad peaks which can be undetectable with available detectors. It can be shown that k' should not be much greater than 5 if reasonable analysis times are to be obtained.

Therefore, the expression given in equation (5.2) suggests that when $\alpha > 1$, in order to optimize a given separation and to achieve a short time of analysis and good sensitivity, the first factor to be optimized is k', and it may be shown that values between 1.5 and 5 represent a reasonable compromise. Separation efficiency, as measured by N, can be varied by changing the column length L or the solvent velocity u. N is usually chosen to provide the maximum efficiency compatible with a reasonable analysis time. The operation of extended column lengths without increasing the flow rate of the eluent necessarily increases the analysis time, and hence decreases the throughput of samples. On the other hand, doubling the length of the column will require a twofold increase in the flow rate of the eluent and consequently a fourfold increase in inlet pressure, in order to maintain a constant retention time. Moreover, as the term (c) of equation (5.2) shows, the resolution is not proportional to N but increases with the square root of plate number, and thus the corresponding increase in resolution with increasing plate number is not so great.

Up to this point, we have looked only at the separation of two-component mixtures. The optimization of separation becomes more complicated for samples that contain many components of widely different k' values.

As described above, resolution can be improved by variations in plate number, selectivity or capacity factor. However, when considering the separation of a mixture which contains several components of different retention rates, the adjustment of the capacity factors has a limited influence on resolution. The retention times for the last eluted peaks can be excessive, and in some cases strongly retained sample components would not be eluted at all.

Improvement of column efficiency in terms of the number of theoretical plates realized by increasing column length often yields marginal increases in resolution, with a corresponding increase of analysis time to unacceptable levels. This behaviour, termed the *General Elution Problem* (4) is common to all forms of liquid chromatographic systems in which a mixture of various components, having a large spread of k' values, is eluted under isocratic conditions. A solution for solving this problem is to change the band migration rates during the course of separation by a gradient elution under precisely controlled conditions. A chromatographic separation can be considered complete when the column produces as many peaks as there are components in the analysed sample (5). In order to describe the effectiveness of most separation systems to resolve a multicomponent mixture, Giddings introduced the concept of peak capacity (6), which is defined as the maximum number of peaks, Φ, that can be fitted into the available separation space with a given resolution which satisfies the analytical purpose. Peak capacity can be expressed by the following equation (6):

$$\Phi = \left(1 + N^{1/2}/_r\right)\ln(1 + k'_i) \qquad (5.3)$$

where N is the number of theoretical plates, r is the number of standard deviations which equal the peak width ($r = 4$) when the resolution (R_s) $= 1$, and k'_i is the capacity factor of the last eluted peak in a series.

Theoretically, under gradient elution conditions, HPLC systems yield peak capacities which are calculated to be in the range $100-300$. These values would be adequate to resolve components in a mixture where the number of analytes is smaller than the peak capacity of the system. However, peak capacity is an 'ideal' number and expresses the maximum number of resolvable analytes which exceeds the real number by some factor determined by operational conditions, such as the allowable separation time (components in a complex mixture are usually not uniformly distributed and appear randomly, overlapping each other). In other words, often the information obtained from the chromatogram is not the true recognition of all individual analytes in complex multicomponents samples, but gives an indication of sample complexity based on the number of observed peaks (7). Davis and Giddings (8) developed a statistical model of component overlap in multicomponent chromatograms by which it was estimated that one never expects to observe more than 37% of the theoretically possible peaks with uniform spacing. This percentage, corresponding to the number of visible peaks, P, in a chromatogram can be estimated by the following equation:

$$P = m \ \exp\left(- m/\Phi\right) \qquad (5.4)$$

where m is the number of components in a multicomponent mixture and Φ is the peak capacity.

By assuming that α (selectivity of the chromatographic system) can be rewritten as follows:

$$\alpha = m/\Phi \qquad (5.5)$$

where this equation shows that α can be considered to be a kind of 'saturation' factor, expressing the ratio of components m to the hypothetical maximum number of separable compounds Φ, thus expressing the degree to which the separation space is saturated (8), and assuming that $m = \alpha\Phi$, we can replace the latter in equation (5.4), thus obtaining the follow equation:

$$P = \alpha\Phi \exp(-\alpha\Phi/\Phi) = \alpha\Phi \exp(-\alpha) \tag{5.6}$$

and consequently we can write the following:

$$P/\Phi = \alpha \exp(-\alpha) \tag{5.7}$$

The dimensionless ratio P/Φ corresponds to the ratio between the number of visible peaks, under the proposed chromatographic conditions, with the chromatographic column having a peak capacity Φ. Differentiation of equation 5.6 with respect to α gives the maximum possible value of the ratio P/Φ and shows this to occur at $\alpha = 1$; then, the maximum ratio P/Φ can be estimated by the following equation:

$$(P/\Phi)_{max} \; 5 \; \exp(2\;1) \; 5 \; 0.3679 \tag{5.8}$$

which reveals that, as postulated above, the maximum number of visible peaks will be equivalent to 37% of the capacity of the system peaks. Furthermore, the number of single-component peaks which can be expected is given by the following:

$$S = mP_1 \tag{5.9}$$

where m, as above in equation 5.4, is the number of components in a multicomponents mixture and P_1 is the probability that an analyte is eluted as a single-component peak, which can be expressed as follows (8):

$$P = \exp(-\alpha)\exp(-\alpha) = \exp(-2\alpha) \tag{5.10}$$

Consequently, the corresponding number of single-component peaks is given by the following:

$$S \; m \exp(-2\alpha) \tag{5.11}$$

As above, we can replace m by $\alpha\Phi$ and thus equation 5.11 can be rewritten as follows:

$$S = \alpha\Phi \exp(-2\alpha) \tag{5.12}$$

which can alternatively be presented as:

$$S/\Phi = \alpha \exp(-2\alpha) \tag{5.13}$$

Differentiation of equation (5.12), with respect to α, as above, shows the maximum to occur at $\alpha = 1/2$, and then the maximum S/Φ ratio can be estimated by the following equation:

$$(S/\Phi)_{max} = 0.5 \exp(-1) = 0.1839 \tag{5.14}$$

which leads to the fact that only about 18% of the analytes will emerge as single-component peaks.

The above theoretical analysis of the total number of resolvable components in a complex mixture has shown that in LC, relative to the maximum peak content or peak capacity for closely spaced peaks, a random chromatogram will never contain more than about 37% of its potential peaks and furthermore that only 18% of such components will emerge as single-component peaks having a minimum specified resolution with respect to the neighbouring peaks.

A practical method for enhancing the peak capacity, and thus the resolution of analytes in multicomponent complex mixtures, can be achieved by changing the mode of the separation during the chromatographic analysis, employing a column switching system in order to optimize a separation.

In LC–LC coupling (2D system), the peak capacity is the product of the peak capacities of its component one-dimensional (1D) processes (9). The power of the separation measured by the LC–LC peak capacity is given by the following:

$$\Phi_{LC-LC} = \Phi_{LC1}\Phi_{LC2} \tag{5.15}$$

By assuming that both LC modes have the same peak capacity, equation (5.15) becomes:

$$\Phi_{LC-LC} = \Phi^2 \tag{5.16}$$

More generally the peak capacity for a multidimensional system can be expressed by the following:

$$\Phi_1\Phi_2\Phi_3\cdots\Phi_n = \prod_{i}^{n}\Phi_i \tag{5.17}$$

and assuming that each LC mode has the same peak capacity, equation 5.17 can conveniently be expressed as follows:

$$\Phi = \Phi^n \tag{5.18}$$

On the other hand, supposing that we have n identical columns connected in series, the peak capacity is given, in analogy to equation 5.3, by the following expression (10):

$$\Phi_S \; 5 \; 1 \; 1 \; \frac{(nN)^{1/2}}{r} \ln \; (1 \; 1 \; k_n) \tag{5.19}$$

which by a combination of equations (5.19) and (5.3), can be approximated by the following:

$$\Phi_S > n^{1/2}\Phi \qquad (5.20)$$

where it is evident that the benefits implied by equation (5.20) for columns of corre-lated selectivity are small when compared to the exponential effect described by equation (5.18), regarding two columns of different selectivities.

Thus, if two identical column with a peak capacity of 35 are coupled in series, the resultant peak capacity calculated from equation (5.20) will be $\cong 49.5$, compared to a value of 1225 if the same columns are used in multidimensional mode. However, the exponential effect of equation (5.18) will be observed only if each column hav-ing a distinct selectivity will give elution sequences that are unrelated to those obtained by any of the other chromatographic mechanisms being used. If the columns are related (redundant selectivity), the system selectivity can be expressed by equation (5.20). For instance, in size exclusion chromatography, columns packed with porous particles having different exclusion limits are connected in series, thus allowing separation of sample components over a wide molecular-weight range.

5.3   LC–LC TECHNIQUES

As reported above, the coupling of individual separation techniques increases the total peak capacity of the chromatographic system, which is given by the product of the peak capacities of the individual dimensions (1). The full separation power of an LC–LC system can be better understood by considering a multicomponent mixture comprised of analytes having a wide range of distribution coefficients. In a single-dimension system, when employing a column of high selectivity, less retained ana-lytes will be eluted as well-resolved peaks. On the other hand, the more retained peaks may become excessively broad and detection may be difficult. In contrast, by selecting a column with lower selectivity, the retention time and width of the later peaks will become acceptable, but early peaks will be poorly resolved.

In LC–LC mode, two columns are linked together via a switching valve in such a manner that any component flowing through the first column can be directed to the detector, to waste or into the second column in which further resolution can occur before the sample passes into the detector cell. A typical LC–LC arrangement con-sists of two columns having the same packing material, but of different length, or two columns of similar length, but of different selectivity. In both systems there is a distinct difference in retention in the columns used. The first column is short or of lower selectivity and is employed to separate the most retained analytes of the multi-component mixture. The less retained analytes which are eluted from the first column are switched to the second column and remain there, while components which are most retained are selectively eluted on the first column. When these analytes have been separated, the first eluted peaks on the first column are then separated on the longer or more selective second column. When comparing the chromatogram obtained on eluting by the single-dimension system with the chromatographic profile achieved by LC–LC column switching, it is evident that the elution order of analytes will be reversed. Under the proposed conditions, the chromatographic profile shows first the most retained analytes and than the least retained components. Peaks are usually eluted in a smaller volume of the mobile phase with less dead space between each of them, and as sharper peaks they are easier to detect. Moreover, separation may be carried out isocratically, thus allowing detection with electrochemical (EC) and refractive index (RI) detectors, which are very sensitive to mobile phase changes and are best used under isocratic conditions.

An LC–LC coupling experiment system can be performed by employing a commercially available HPLC apparatus and involving various combinations of HPLC columns, eluents, additives, switching devices and detectors.

LC–LC coupling can be subdivided into both homomodal and heteromodal systems (11).

Homomodal LC–LC.   In this type of development, the chromatographic improvement occurs by switching columns of analogous selectivity. Mainly the goal is to optimize an already satisfactory separation, that is, to concentrate a dilute sample (sample enrichment) or to shorten an analysis time.

Heteromodal LC–LC.   Essentially, this type of development is achieved by varying the separation mechanism during the separation process; selectivity changes may be made by varying the nature of the stationary phase, which can posses complementary separation characteristics. The high power of heteromodal LC–LC is represented by equation (5.16) and mainly by equatiuon (5.18) for a multidimensional LC system. The term *LC–LC*, and more generally *Multidimensional*, is usually restricted to these LC systems, which involve separation modes which are as different as possible (orthogonal), and in which there is a distinct difference in retention mechanisms (2).

Trace enrichment and sample clean-up are probably the most important applications of LC–LC separation methods. The interest in these LC–LC techniques has increased rapidly in recent years, particularly in environmental analysis and clean-up and/or trace analysis in biological matrices which demands accurate determinations of compounds at very low concentration levels present in complex matrices (12–24). Both sample clean-up and trace enrichment are frequently employed in the same LC–LC scheme; of course, if the concentration of the analytes of interest are sufficient for detection then only the removal of interfering substances by sample clean-up is necessary for analysis.

Trace enrichment or preconcentration by LC–LC methods are based on the possibility that the analytes will be retained as a narrow zone on the top of the first column when a large volume of sample is pumped trough the column. Good reproducibility can be achieved when the column capacity is not exceeded and the column is not overloaded. Trace enrichment is usually performed when relatively non-polar components from aqueous solutions are injected on to a reversed-phase column. A similar outcome is achieved by adsorption chromatographic methods employing suitable solvent polarities. Subsequent elution with a stronger eluent will remove the retained analyte on the first column and will start the separation procedure on the analytical column (secondary column). A schematic drawing of a typical enrichment system assembled by employing a standard chromatograph, an additional LC pump and a six-port valve is presented in Figure 5.1. When the six-port valve is in position A (Figure 5.1 (a)), large volumes of sample can be injected into the enrichment column and flushed by using the mobile phase from pump A. In position B (Figure 5.1 (b)), the enrichment column switches to the reverse direction so that pump B back-flushes the cleaned-up and concentrated analytes on to the analytical column. When the analytes have been transferred on to the analytical column, the valve can be switched back to position A so that the enrichment column can be conditioned and the next sample can be injected into it. To explain the basic operation conditions of an LC–LC approach, we can expound a recently published paper where trace enrichment and sample clean-up were carried out in a single step (25). A schematic representation of such a system detailing each component is shown in Figure 5.2. The LC–LC network was developed by employing a precolumn packed with Bondopack C_{18} 37–53 μm particles and an analytical column which consisted of a silica-based reversed-phase column (Suplex pK_b–100, 5 μm, 250 × 4.6 mm id from Supelco).

In the proposed LC–LC configuration, pump 1 (MP1) was used to deliver the mobile phase 1, which consisted of a 1 : 3 dilution in water of mobile phase 2, employed to flush the precolumn (PC), when the switching valve connects the columns as depicted in Figure 5.2 (a). The Sample (2 ml) is injected and eluted through the precolumn. The mobile phase 1 separates the interfering analytes present in large quantities in the complex matrix (a mouse embryo homogenate). Retinoids, compounds which are structurally related to vitamin A, which are present in embryonic tissue as the trace compounds (26), were retained and concentrated on the precolumn. During the concentration step, the excess injection volume and the eluted analytes were carried to waste. After rotating the switching valve into the transfer position which connects the analytical column (AC), the components retained on the precolumn were back-flushed and separated on the analytical column (Figure 5.2 (b)) by isocratic elution, employing mobile phase 2 (MP2), consisting of acetonitrile/methanol/2% ammonium acetate/glacial acetic acid (79 : 2 : 16 : 3, vol/vol). Mobile phase 2, which had an higher elution power than the primary mobile phase, was able to remove those retinoids which had been strongly retained on the precolumn. Under the proposed conditions, sample clean-up, enrichment, separation and quantification of picogram amounts of retinoids in embryonic tissue were achieved in a single step. Very recently, an on-line trace enrichment method was developed for the rapid, sensitive and reproducible determination of microcystins from water samples without purification (27). The analysed microcystin-LR (containing the L-amino acid residues leucine and arginine in positions 2 and 4, respectively), microcystin-RR (two L-arginine residues in positions 2 and 4), and microcystin-YR (L-tyrosine and L-methionine residues in positions 2 and 4, respectively), are

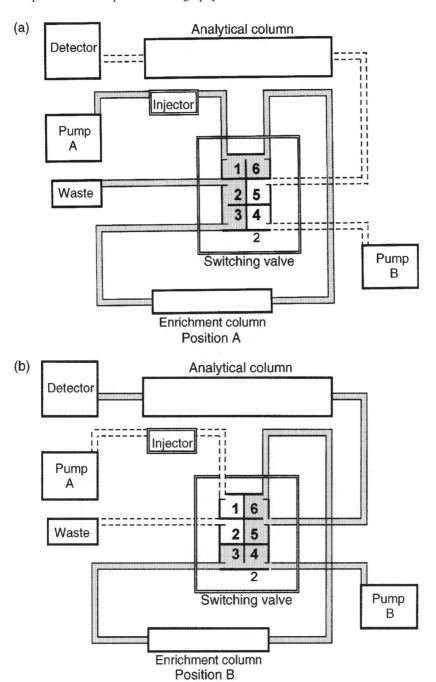

Figure 5.1 Schematic diagrams of a typical enrichment system: (a) forward-flush position; (b) back-flush position.

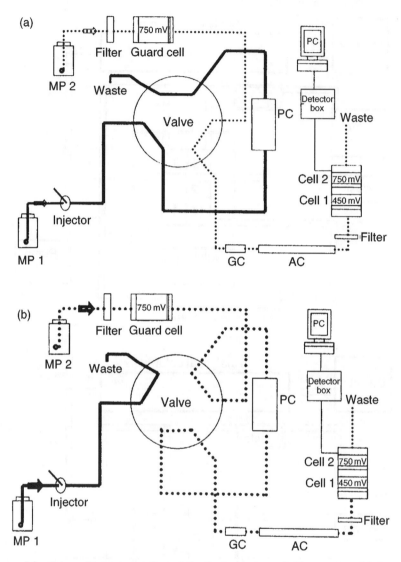

Figure 5.2 Schematic representation of the final column-switching system: (a) forward-flush position; (b) back-flush position (further details are given in the text). Reprinted from *Journal of Chromatography, A* **828**, A. K. Sakhi *et al.* 'Quantitative determination of endogenous retinoids in mouse embryos by high-performance liquid chromatography with on-line solid-phase extraction, column switching and electrochemical detection', pp. 451–460, copyright 1998, with permission from Elsevier Science.

strongly hepatotoxic cyclic heptapeptides produced by some species of freshwater cyanobacteria (blue–green algae) (28). These microcystins represent a health risk to humans through drinking water, since they have been found to act as tumor promoters (29). Several chromatographic analytical procedures for microcystins have been suggested and reversed-phase HPLC is frequently the technique of choice for these types of analyses (30). However, considering that these compounds may be present at low levels in natural and drinking water (≤ 1 μg/l), a preconcentration step before one-dimensional HPLC analysis is usually required (31–33). Furthermore, the proteins present in samples containing cyanobacteria are particularly troublesome since they tend to denature on reversed-phase packing and render the column useless (distorted peak shape, multiple peaks, etc.) (34). In the proposed LC–LC system, a precolumn consisting of a Zorbax CN cartridge was used for simultaneous enrichment and clean-up of the microcystins in water. The sample (100 ml) was passed through the cartridge on the enrichment side at a flow rate of 3 ml/min, and the microcystins were strongly retained in a narrow band on the top of the cartridge, due to their high hydrophobicity. At the same time, the analytical column was equilibrated with the starting mobile phase consisting of 10 mM phosphate buffer (pH, 2.5), containing 25% (vol/vol) acetonitrile. Desorption was performed by coupling the cartridge on-line with the analytical column and starting the gradient. Gradient elution was performed in the opposite direction to the sample preconcentration as follows: 0% of B at 0 min, 20% of B at 38 min, 60% of B at 42 min, and then isocratic under the same conditions until a period of 50 min. In this system, the mobile phase B was acetonitrile and the elution was carried out at a flow rate of 1 ml/min. Microcystins are peptides of differing hydrophobicity that can be readily chromatographed by using reversed-phase (RP)-HPLC. An example of the determination of microcystins (-LR, -RR and -YR) from a water sample by using the developed procedures is reported in Figure 5.3. Considering the two examples reported above regarding sample enrichment, the most important parameter is the sensitivity of the method (or minimal detectable concentration of analytes), determined by the sensitivity of the detector used, the adsorption capacity of the precolumn, the sample volume, the desorption and the chromatographic procedures. Two distinct processes are involved, i.e. (i) a frontal chromatography during the enrichment step, and (ii) a displacement chromatography during the desorption step.

A fundamental parameter characterizing the usefulness of a given precolumn for enrichment purposes is the breakthrough volume, V_B. This volume can be determined by monitoring continuously or discretely the detector signal at the outlet of the precolumn (35–37). The breakthrough volume can be defined by the following expression (37):

$$V_B = V_R - 2.3\,\sigma_V \qquad (5.21)$$

where σ_V is the standard deviation depending on the axial dispersion of analyte along the bed of particles in the precolumn. If the capacity factor, k'_S of the analyte eluted with a mobile phase that corresponds to the sample solvent, wash solvent or elution solvent can be predicted and if V_0, the dead volume of the precolumn, is determined, then V_R can be calculated by using the following expression:

$$V_R = V_0(1 + k'_S) \qquad (5.22)$$

while if the number of theoretical plates, N, of the precolumn is known, σ_V can be

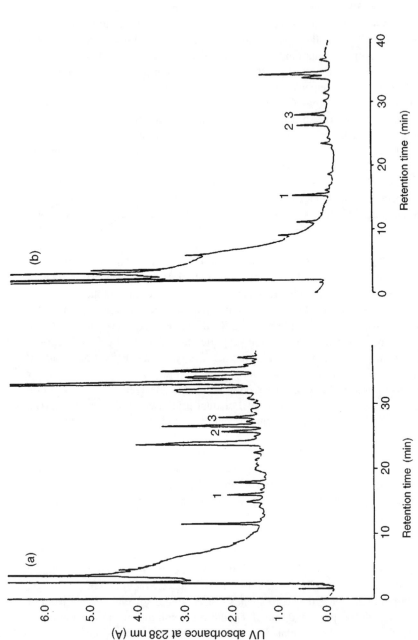

Figure 5.3 Analysis of 100 ml of (a) surface water and (b) drinking water sample spiked with 0.1 µg/ml of microcystins, using column-switching HPLC: 1, microcystin-RR; 2, microcystin-YR; 3, microcystin-LR. Reprinted from *Journal of Chromatography A*, **848**, H. S. Lee *et al.*, 'On-line trace enrichment for the simultaneous determination of microcystins in aqueous samples using high performance liquid chromatography with diode-array detection', pp 179–184, copyright 1999, with permission from Elsevier Science.

calculated from the following:

$$\sigma_v = \frac{V_0}{\sqrt{N}}(1 + k'_S) \tag{5.23}$$

By combining equations (5.22) and (5.23), equation (5.21) can be rewritten as follows:

$$V_B = (1 + k'_S)\left(1 - \frac{2.3}{\sqrt{N}}\right)V_0 \tag{5.24}$$

However, for very highly retained solutes direct measurement of the capacity factor k_S is not possible, and this parameter must be predicted on the basis of retention data determined with a stronger mobile phase. The determination of V_B is an essential step in the optimization of trace enrichment and clean-up procedures.

Loading of the analytes on the analytical column can be carried out either in the back-flush or forward-flush modes. The back-flush mode allows a better resharpening of the solute band than the forward-flush mode. However, such a flow reversal may lead to precolumn packing disturbances. In addition, back-flushing does not protect the analytical column as well as forward-flushing (10). In fact, other than analytes, a large number of contaminants can be simultaneously sorbed on, and then eluted from the precolumn. In the development of a new LC–LC enrichment method, we have to deal with the following: (i) the elaboration of a carefully designed gradient profile to achieve a more or less stepwise elution of the retained components from the precolumn; (ii) the choice of a more selective stationary phase for the trace-enrichment step; (iii) the use of a selective detection principle.

An LC–LC separation system may be used in either the profiling or targed mode (11). The purpose in the profiling mode is to separate all single components from a complex mixture. Every component from the first column (primary column) is fractionated and transferred in the second column (secondary column). In contrast, the purpose of LC–LC separation in the targeted mode is to isolate either a single or a few components of similar retention in a complex mixture containing components having a wide range of capacity factor values. Targeted component analysis is carried out by transferring a wide or narrow cut of the chromatographic effluent from the primary column to the secondary column by flow switching and the mobile phase is thereby diverted or reversed. The fraction of interest to be transferred on to the secondary column may be early-eluting analytes (first eluted zone, usually the named the 'front-cut'), or components eluted in the middle of the chromatographic effluent ('heart-cut') or at the end of the chromatogram ('end-cut') (38).

A schematic diagram of a 'heart-cut' LC–LC system is depicted in Figure 5.4. The column switching technique was developed by employing two high-pressure four-way pneumatic valves inserted before and after the precolumn (39). The 'front-cut' and the 'end-cut' of the sample eluted from the first column were vented to waste. The valves were manipulated to transfer only the 'heart-cut' of the analyte of interest to the analytical column. The detailed operational conditions for the four-step sequence of this system can be described as follows:

(Step 1) *Divert initial portion of chromatogram to waste.* The sample is injected with the valve A (left) closed; mobile phase flows through the precolumn to valve B (right), which is opened to waste.

Step (1) Early eluters to waste

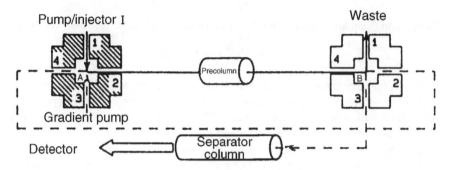

Step (2) "Heart-cut"

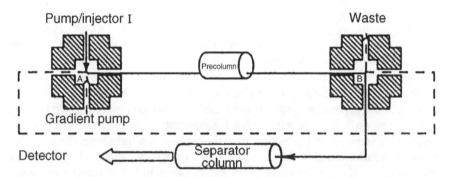

Step (3) By pass and complete analysis

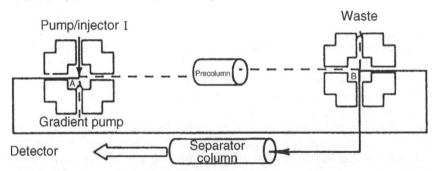

Figure 5.4 Schematic diagrams of a heart-cut valve configuration system. Reprinted from *Journal of Chromatography*, **602**, S. R. Villaseñor, 'Matrix elimination in ion chromatography by 'heart-cut' column switching techniques', pp. 155–161, copyright 1992, with permission from Elsevier Science.

(Step 2) *Introduce heart-cut to the analytical column and detector.* At the predetermined time interval, which was previously calculated by eluting analyte standards without the analytical column, i.e. the onset of the heart-cut, valve B is closed to divert the precolumn effluent to the analytical column.

(Step 3) *Bypass the precolumn and detection of the analyte of interest.* When all of the analytes of interest have been eluted from the precolumn, valve A is opened so that the eluent stream is diverted to valve B, which is immediately opened to allow eluent from valve A to flow into the analytical column, thus bypassing the precolumn.

(Step 4) *Precolumn clean-up not shown in Figure 5.4.* After the heart-cut analytes have been transferred to the analytical column, a step-gradient programme is used to flush the precolumn of the more strongly retained compounds. An additional pump configuration makes precolumn clean-up possible while the analysis is running.

The LC–LC configuration has been applied to sulfide analysis in a variety of different matrices, as well as for acetate and trifluoroacetate analysis in peptides (as shown in Figure 5.5).

A critical operation in target component analysis by LC–LC is the selection and transference of the eluent fraction from the primary to the secondary column. In complex samples, it is inevitable that a part of the interfering components will be transferred together with the analytes of interest. Selection is usually achieved by time-based valve switching, assuming that the analytes' retention times and peak widths are constant. This involves careful advance planning of the chromatographic conditions and imposes a standard of excellent retention time reproducibility for the analytes. Major drawbacks of the above method are that column ageing will change the retention time of the analyte, while peak width will increase due to column degradation. Under these conditions, the analyte will move either partially or totally out of the preselected time window for valve switching and quantitation will be compromised in the analytical column (12). Determination of the heart-cut timing parameters can be automated through repeated analysis with various retention-time windows of a sample containing a large amount of the analyte or a sample that has been spiked with the compound of interest (40).

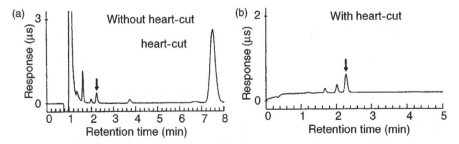

Figure 5.5 Trifluoroacetate determination in calcitonin acetate (a) without and (b) with heart-cut column switching. Reprinted from *Journal of Chromatography, 602*, S. R. Villaseñor, 'Matrix elimination in ion chromatography by 'heart-cut' column switching techniques', pp 155–161, copyright 1992, with permission from Elsevier Science.

As enunciated above, a high-resolving LC–LC system can be implemented by employing columns that operate by using a different separation mechanism (hetero-modal LC–LC). Several combinations of mechanisms with great dissimilarity are conceivable. These include the following: size exclusion–ion exchange; size exclusion–reversed phase; ion exchange–reversed phase; reversed phase (alkyl ligand)–reversed phase (ion-pairing eluent); reversed phase (alkyl ligand)–reversed phase (electron-pair acceptor or donator ligand); reversed phase–affinity (biospecific interactions); normal phase (plain silica)–normal phase (electron-pair acceptor or donator ligand (41). In addition, a significant number of applications describing the coupling of immunoaffinity chromatography and reversed-phase HPLC have been reported over the last ten years. A specific antibody is immobilized on a appropriate sorbent to form a so-called immunosorbent (IS) for packing into a HPLC precolumn. The antibodies are selected in order to involve antigen–antibody interactions, thus providing selective extraction methods based on molecular recognition. Samples or extracts from biological matrices are introduced on to this immunoaffinity system with little or no sample pretreatment. The analytes are then eluted from the immunoaffinity column and analysed directly by suitable on-line HPLC methods. Immunoaffinity columns can be packed with chemically activated sepharose beads, and antibodies will then covalently bind to these beads (42). However, Sepharose-based immunosorbents are not pressure resistant and therefore direct connection of the precolumn to the analytical column could not be achieved. When using these immunosorbents, analytes are usually desorbed at low pressure in a second precol-umn packed with C_{18}, which subsequently can be coupled on-line to the LC system (43–45). Antibodies have also been immobilized on silica-based sorbents. The par-ticular advantage of silica is its pressure resistance, which means that it can be used directly in on-line LC–LC systems (46–48). The on-line set-up using a silica-based immunosorbent precolumn is very simple and does not differ to any great extent from that which uses a single reversed-phase precolumn. Heteromodal LC–LC cou-pling has also been widely employed as a chiral separation technique, which usually involves sequential chromatography on a chiral and an achiral column. The consecu-tive order in which the columns are combined can be varied (e.g. first a chiral col-umn, then an achiral column or vice versa), depending on the problem to be solved and the main restrictions involved. Such restrictions may be a low sample amount, a low analyte concentration or a complex sample matrix, as well as a high degree of optical purity to be monitored (49–53). Enantiomeric separations can also be easily achieved by a two-dimensional HPLC system using achiral columns in both dimen-sions (54). Separation of unmodified amino acids in complex mixtures was achieved by employing two different separation methods. First, the amino acid separation was carried out by means of a cation-exchange column by elution with a lithium chlo-ride–lithium citrate buffer, and then each peak corresponding to an individual amino acid was switched to an achiral reversed-phase column where the chiral discrimina-tion was achieved by using a mobile phase containing a chiral copper (II) complex.

LC–LC coupling systems are also employed to perform separations requiring very large plate numbers. However, it has been demonstrated (see equation (5.20) that for coupled columns peak capacity increases linearly with the square root of n

(number of columns connected in series), although the column pressure drop increases linearly with the length. In practice, the operating conditions essentially limit the number of columns that may be coupled together and therefore restricts the total number of theoretical plates that can be generated in a reasonable time. An alternative way to improve the number of theoretical plates is the use of the same column in a recycling chromatographic system (55). This technique consists of a switching-column method which is accomplished by adding a switching valve to the fluid system to permit the eluent, or some designed portion thereof, to be directed from the column outlet back into the column inlet. The process can then be repeated, with each subsequent pass resulting in an increase of the separation between the peaks. In practice, recycling increases the effective length of the column by returning the eluted analyte to the head of the column for further separation. A schematic representation of a conventional chromatographic recycling system is depicted in Figure 5.6. The recycling technique offers, with respect to coupled columns, the possibility to increase the resolution by a large number of theoretical plates obtained by increasing the number of sample passes through the HPLC column and maintaining the column length and pressure constant. There are two limiting factors in recycling chromatographic methods, as follows: (i) the phenomenon of remixing of the separated analytes, which occurs when the fastest moving peak of the sample overtakes the slowest moving components after a certain number of cycles; (ii) the extracolumn band broading occurring in the pump, valve devices, connecting tubes, and detector at each cycle in the detector-to-injection transfer. The influence of column and extra-column effects on the maximum efficiency realizable in recycling

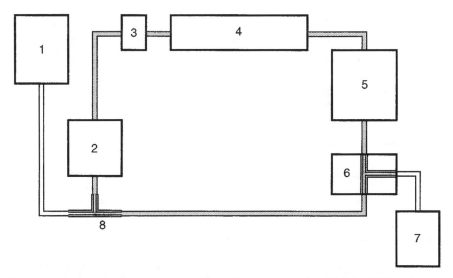

Figure 5.6 Schematic design for recycling chromatography in which the effluent is recycled through the pump: 1, eluent; 2, pump; 3, injector; 4, chromatographic column; 5, detector; 6, three way valve; 7, waste; 8, T-connection.

chromatography has been treated theoretically and mathematical models have been developed for the description of specific recycling techniques (56–60). Recycling is a chromatographic technique mainly used in preparative LC to increase the effective separation power (selectivity and efficiency), while avoiding the expense of purchasing and operating longer columns and additional column sections. The preparative recycling system can be improved by peak shaving (61). In this method, the recycling valve is used during each cycle and switched to direct unwanted components to waste, to collect portions of pure components of interest from peak fronts and tails, and to recycle the remainder of incompletely separated sample mixtures back through the column. An example of the advantages of a recycle chromatographic method with peak shaving to separate closely related compounds with an α value near to 1 is shown in Figure 5.7. In this figure, two polymethoxyflavones were separated to obtain pure compounds which were used as HPLC standards to construct calibration graphs, employing cumarin as the internal standard, for their quantitative evaluation in sweet orange and mandarin essential oils (62). Polymethoxyflavones were isolated by a recycling system combined with peak shaving, having a configuration similar to that shown in Figure 5.6. With the HPLC previously in the recycle mode, the essential oils were fractionated on a glass column (300 × 60 mm id.) filled with silica gel, using light petroleum/ethyl acetate (80:20 vol/vol) as the mobile phase at a flow rate of 1.0 ml/min (62). As illustrated in Figure 5.7, in order to

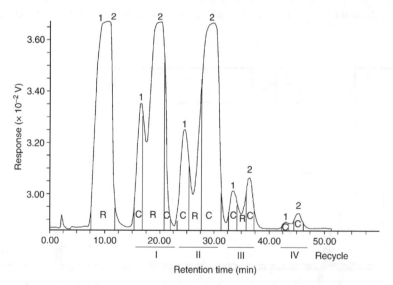

Figure 5.7 Separation of tangeretin and heptamethoxyflavon by recycle HPLC: R, recycle; C, collected; 1, tangeretin; 2, heptamethoxyflavon. Reprinted from *Essenze Derivati Agrumari*, **63**, L. Mondello *et al.*, 'Isolamento di polimetossiflavoni dagli olii essenziali di arancia dolce e di mandarino mediante cromatografia su colonna e HPLC semipreparativa con riciclo', pp. 395–406, 1993, with permission from Essenze Derivati Agrumari.

isolate tangeretin and heptamethoxyflavon, a large quantity of tangeretin was collected from the peak front and tail on the first pass. As the heavy sample load was reduced on each successive pass by shaving, the column's effective separation efficiency increased, thus reducing the number of passes necessary to separate and recover the two compounds, which were completely separated and collected by employing only four recycles. Under the same conditions, other oxygen heterocyclic compounds of citrus essential oils were isolated (63) and used to prepare a library of mass spectra to identify these compounds in samples of genuine cold-pressed citrus oils by HPLC hyphenated techniques (64). Recycling techniques included the simulated moving bed method (65, 66), mainly proposed for the large-scale chromatographic separation of enantiomers (67–73). Such methods are rather complex and require dedicated equipment. However, they usually require less solvent to separate a given quantity of enantiomers and the operating costs are therefore significantly lower than with batch chromatographic methods (74).

5.4 CONCLUSIONS

Today, the various chromatographic techniques represent the major parts of modern analytical chemistry. However, it is well known that the analysis of complex mixtures often requires more than one separation process in order to resolve all of the components present in a sample. This realization has generated a considerable interest in the area of two-dimensional separation techniques. The basics of LC–LC and its practical aspects have been covered in this chapter.

LC–LC systems can be divided into two different approaches; namely (i) where a small-sized column (or SPE cartridge) is mainly used as the first column for fast sample enrichment and/or clean-up, and (ii) an LC–LC coupling employing two full-sized separation columns operating in orthogonal mode, which provides, relative to one-dimensional or linear techniques, a greatly enhanced peak capacity. In many cases, the LC–LC coupling of conventional columns, as well as microbore columns, provides the optimum efficiency and selectivity for the separation of a wide range of compounds of interest in the biological field, as well as in environmental analysis.

Using LC–LC systems, a high degree of automation with a lower amount of sample and low detection limits is usually obtained. However, even with manual valve switching, these techniques are less time-consuming than most alternative HPLC methods. Furthermore, the combination of on-line coupling of LC–LC methods and a spectroscopic detection device which provides structural sample information, is a promising option for use in systems combining automated sample pretreatment and efficient and selective separations with high sensitivity detection. Reviews of combined LC–MS systems have been extensively published over the last few years (75–77) and the use in conjunction with hyphenated LC–LC methods has been proposed (18, 48, 78, 79), and its potential recently demonstrated (80).

REFERENCES

1. J. C. Giddings, 'Concepts and comparison in multidimensional separation', *J. High Resolut. Chromatogr. Chromatogr. Comm.* **10**: 319–323 (1987).

2. J. C. Giddings, 'Two-dimensional separation: concepts and promise', *Anal. Chem.* **56**: 1258A–1270A (1984).

3. Cs Horvàth and W. R. Melander, 'Theory of chromatography', in *Chromatography: Fundamentals and Applications of Chromatographic and Electrophoresis Methods. Part A: Fundamentals and Techniques*, Helftmann E. (Ed.), Journal of Chromatography Library, Vol. 22, Ch. 3, A28–A135 (1983).

4. B. L. Karger, L. R. Snyder and Cs Horvàth, *Introduction to Separation Science*, John Wiley & Sons, New York, pp. 146–155 (1973).

5. J. M. Davis and J. C. Giddings, 'Statistical method for estimation of number of components from single complex chromatograms: application to experimental chromatograms', *Anal. Chem.* **57**: 2178–2182 (1985).

6. J. C. Giddings, 'Maximum number of components resolvable by gel filtration and other elution chromatographic methods', *Anal. Chem.* **39**: 1027–1028 (1967).

7. J. C. Giddings, 'Use of multiple dimensions in analytical separations' in *Multidimensional Chromatography: Techniques and Applications,* H. J. Cortes (Ed.), Marcell Dekker, New York, Ch. 1–27 (1990).

8. J. M. Davis and J. C. Giddings, 'Statistical theory of component overlap in multicomponent chromatograms', *Anal. Chem.* **55**: 418–424 (1983).

9. J. C. Giddings, 'Future pathways for analytical separations', *Anal. Chem.* **53**: 945A–952A (1981).

10. D. F. Samain, 'Multidimensional chromatography in biotechnology', in *Advances in Chromatography-Biotechnological Applications and Methods,* J. C Giddings, E. Grushka and P. R. Brown (Eds), Marcel Dekker, New York, Basel, Ch. 2, pp. 77–132 (1989).

11. F. Regnier and G. Huang, 'Future potential of targeted component analysis by multidimensional liquid chromatography–mass spectrometry', *J. Chromatogr.* **750**: 3–10 (1996).

12. P. Campins-Falcó, R. Herráez-Hernández and A. Sevillano-Cabeza, 'Column-switching techniques for high performance liquid chromatography of drugs in biological samples', *J. Chromatogr.* **619**: 177–190 (1993).

13. E. A. Hogendoorn and P. van Zoonen, 'Coupled-column reversed-phase liquid chromatography in environmental analysis', *J. Chromatogr.* **703**: 149–166 (1995).

14. L. A. Holland and J. W. Jorgenson, 'Separation of nanoliter samples of biological amines by a comprehensive two-dimensional microcolumn liquid chromatography system', *Anal. Chem.* **67**: 3275–3283 (1995).

15. E. A. Hogendoorn, R. Hoogerbrugge, R. A. Baumann, H. D. Meiring, A. P. J. M. de Jong and P. van Zoonen, 'Screening and analysis of polar pesticides in environmental monitoring programmes by coupled-column liquid chromatography and gas chromatography–mass spectrometry', *J. Chromatogr.* **754**: 49–60 (1996).

16. J. A. Pascual, G. J. ten Hove and A. P. J. M. de Jong, 'Development of a precolumn capillary liquid chromatography switching system for coupling to mass spectrometry', *J. Microcolumn. Sep.* **8**: 383–387 (1996).

17. M. A. J. Bayliss, P. R. Baker and D. Wilkinson, 'Determination of the two major human metabolites of tipredane in human urine by high performance liquid chromatography with column switching', *J. Chromatogr.* **694**: 199–209 (1997).

18. Z. Yu and D. Westerlund, Characterization of the precolumn biotrap 500 C_{18} for direct injection of plasma samples in a column-switching system', *Chromatographia*, **47**: 299–304 (1998).

19. H. S. Lee, J. H. Kim, K. Kim and K. S. Do, 'Determination of myristicin in rat serum samples by using microbore high performance liquid chromatography with column-switching', *Chromatographia* **48**: 365–368 (1998).

20. J. G. Dorsey, W. T. Cooper, B. A. Siles, J. P. Foley and H. G. Barth, 'Liquid chromatography: theory and methodology', *Anal. Chem.* **70**: 591R–644R (1998).

21. Y. Huang, S. F. Mou and Y. Yan, 'Determination of bromate in drinking water at the μg/L level by column switching ion chromatography', *J. Liq. Chromatogr.* **22**: 2235–2245 (1999).

22. J. Slobodnik, H. Lingeman and U. A. Th Brinkman, 'Large-volume liquid chromatographic trace-enrichment system for environmental analysis', *Chromatographia* **50**: 141–149 (1999).

23. M. Cavalleri, W. Pollini and L. Colombo, 'Determination of ramoplanin in human urine by high performance liquid chromatography with automated column switching', *J. Chromatogr.* **846**: 185–192 (1999).

24. S. Nèlieu, M. Stobiecki and J. Einhorn, 'Tandem solid-phase extraction of atrazine ozonation products in water', *J. Chromatogr.* **866**: 195–201 (2000).

25. A. K. Sakhi, T. E. Gundersen, S. M. Ulven, R. Blomhoff and E. Lundanes, 'Quantitative determination of endogenous retinoids in mouse embryos by high-performance liquid chromatography with on-line solid-phase extraction, column switching and electrochemical detection', *J. Chromatogr.* **828**: 451–460 (1998).

26. H. L. Ang, L. Deltour, T. F. Hayamizu, M. Zgombic-Knight and G. Duester, 'Retinoic acid synthesis in mouse embryos during gastrulation and craniofacial development linked to class IV alcohol dehydrogenase gene expression', *J. Biol. Chem.* **271**: 9526–9534 (1996).

27. H. S. Lee, C. K. Jeong, H. M. Lee, S. J. Choi, K. S. Do, K. Kim and Y. H. Kim, 'On-line trace enrichment for the simultaneous determination of microcystins in aqueous samples using high performance liquid chromatography with diode-array detection', *J. Chromatogr.* **848**: 179–184 (1999).

28. G. A. Codd, 'Cyanobacterial toxins: occurrence, properties and biological significance', *Water Sci. Technol.* **32**: 149–156 (1995).

29. R. E. Honkanen, J. Zwiller, R. E. Moore, S. L. Daily, B. S. Khatra, M. Dukelow and A. L. Boynton, 'Characterization of microcystin-LR, a potent inhibitor of type 1 and type 2A protein phosphatases', *J. Biol. Chem.* **265**: 19401–19404 (1990).

30. J. Meriluoto, 'Chromatography of microcystins', *Anal. Chim. Acta.* **352**: 277–298 (1997).

31. D. Pyo and M. Lee, 'Chemical analysis of microcystins RR and LR in cyanobacterium using a prepacked cyano cartridge', *Chromatographia* **39**: 427–430 (1994).

32. R. W. MacKintosh, K. N. Dalby, D. G. Campbell, P. T. W. Cohen, P. Cohen and C. MacKintosh, 'The cyanobacterial toxin microcystin binds covalently to cysteine-273 on protein phosphatase 1', *FEBS Lett.* **371**: 236–240 (1995).

33. R. W. Moollan, B. Rae and A. Verbeek, 'Some comments on the determination of microcystin toxins in water by high performance liquid chromatography. *Analyst* **121**: 233–238 (1996).

34. K. Benedek, S. Dong and B. L. Karger, 'Kinetics of unfolding of proteins on hydrophobic surfaces in reversed-phase liquid chromatography', *J. Chromatogr.* **317**: 227–243 (1984).

35. W. Golkiewicz, C. E. Werkhoven-Goewie, U. A. Th Brinkman, R. W. Frei, H. Colin and G. Guiochon, 'Use of pyrocarbon sorbents for trace enrichment of polar compounds from aqueous samples with on-line HPLC analysis', *J. Chromatogr. Sci.* **21**: 27–33 (1981).

36. R. Ferrer, J. L. Beltràm and J. Guiteras, 'Mathematical procedure for the determination of the breakthrough volumes of polycyclic aromatic hydrocarbons', *Anal. Chim. Acta* **346**: 253–258 (1997).

37. M.-C. Hennion, 'Solid-phase extraction: method development, sorbents and coupling with liquid chromatography', *J. Chromatogr.* **856**: 3–54 (1999).

38. K. A. Ramsteiner, 'Systematic approach to column switching', *J. Chromatogr.* **456**: 3–20 (1988).

39. S. R. Villaseñor, 'Matrix elimination in ion chromatography by "heart-cut" column switching techniques', *J. Chromatogr.* **602**: 155–161 (1992).

40. S. R. Villaseñor, "Heart-cut" column switching techniques for the determination of an aliphatic amine in an organic matrix and for low levels of sulfate in an anion matrix', *J. Chromatogr.* **671**: 11–14 (1994).

41. H. J. Cortes and L. D. Rothman, 'Multidimensional high-performance liquid chromatography' in *Multidimensional Chromatography: Techniques and Applications*, H. J. Cortes (Ed.) Marcel Dekker, New York, Ch. 6, pp. 219–250 (1990).

42. A. Farjam, G. J. de Jong, R. W. Frei, U. A. Th Brinkman, W. Haasnoot, A. R. M. Hamers, R. Schilt and F. A. Huf, 'Immunoaffinity pre-column for selective on-line sample pre-treatment in high performance liquid chromatography determination of 19-nortestosterone', *J. Chromatogr.* **452**: 419–433 (1988).

43. W. Haasnoot, M. E. Ploum, R. J. A. Paulussen, R. Schilt and F. A. Huf, 'Rapid determination of clenbuterol residues in urine by high performance liquid chromatography with on-line automated sample processing using immunoaffinity chromatography', *J. Chromatogr.* **519**: 323–335 (1990).

44. A. Farjam, A. E. Brugman, A. Soldaat, P. Timmerman, H. Lingerman, G. J. de Jong, R. W. Frei and U. A. Th Brinkman, 'Immunoaffinity precolumn for selective sample pretreatment in column liquid chromatography: immunoselective desorption', *Chromatographia* **31**: 469–477 (1991).

45. G. S. Rule, A. V. Mordehai and J. Henion, 'Determination of carbofuran by on-line immunoaffinity chromatography with coupled-column liquid chromatography/mass spectrometry', *Anal. Chem.* **66**: 230–235 (1994).

46. V. Pichon, L. Chen and M.-C. Hennion, 'On-line preconcentration and liquid chromatographic analysis of phenylurea pesticides in environmental water using a silica-based immunosorbent', *Anal. Chim. Acta* **311**: 429–436 (1995).

47. V. Pichon, L. Chen, M.-C. Hennion, R. Daniel, A. Martel, F. Le Goffic, J. Abian and D. Barceló, 'Preparation and evaluation of immunosorbents for selective trace enrichment of phenylurea and triazine herbicides in environmental waters', *Anal. Chem.* **67**: 2451–2460 (1995).

48. I. Ferrer, V. Pichon, M-C. Hennion and D. Barceló, 'Automated sample preparation with extraction columns by means of anti-isoproturon immunosorbents for the determination of phenylurea herbicides in water followed by liquid chromatography–diode array detection and liquid chromatography–atmospheric pressure chemical ionization mass spectrometry', *J. Chromatogr.* **777**: 91–98 (1997).

49. N. C. van de Merbel, M. Stenberg, R. Öste, G. Marko-Varga, L. Gorton, H. Lingeman and U. A. Th Brinkman, 'Determination of D- and L-amino acids in biological samples by two-dimensional column liquid chromatography', *Chromatographia* **41**: 6–14 (1995).

50. M. Tanaka and H. Yamazaki, 'Direct determination of pantoprazole enantiomers in human serum by reversed-phase high performance liquid chromatography using a cellulose-based chiral stationary phase and column-switching system as a sample cleanup procedure', *Anal. Chem.* **68**: 1513–1516 (1996).

51. P. R. Baker, M. A. J. Bayliss and D. Wilkinson, 'Determination of a major metabolite of tipredane in rat urine by high performance liquid chromatography with column switching', *J. Chromatogr.* **694**: 193–198 (1997).

52. L. Liu, H. Cheng, J. J. Zhao and J. D. Rogers, 'Determination of montelukast (MK-0476) and its S-enantiomer in human plasma by stereoselective high performance liquid chromatography with column-switching', *J. Pharm. Biomed. Anal.* **15**: 631–638 (1997).

53. T. Fornstedt, A.-M. Hesselgren and M. Johansson, 'Chiral assay of atenolol present in microdialysis and plasma samples of rats using chiral CBH as stationary phase', *Chirality* **9**: 329–334 (1997).

54. A. Dossena, G. Galaverna, R. Corradini and R. Marchelli, 'Two-dimensional high performance liquid chromatographic system for the determination of enantiomeric excess in complex amino acid mixtures. Single amino acids analysis', *J. Chromatogr.* **653**: 229–234 (1993).

55. M. Martin, F. Verillon, C. Eon and G. Guiochon, 'Theoretical and experimental study of recycling in high performance liquid chromatography', *J. Chromatogr.* **125**: 17–41 (1976).

56. P. Kucera and G. Manius, 'Recycling liquid chromatography using microbore columns', *J. Chromatogr.* **219**: 1–12 (1981).

57. B. Coq, G. Cretier and J. L. Rocca, 'Recycling technique in preparative liquid chromatography', *J. Liq. Chromatogr.* **4**: 237–249 (1981).

58. F. Charton, M. Bailly and G. Guiochon, 'Recycling in preparative liquid chromatography', *J. Chromatogr.* **687**: 13–31 (1994).

59. H. Kalàsz, 'Peak and cycle capacity of recycling chromatography', *Chromatographia* **20**: 125–128 (1985).

60. J. Dingenen and J. N. Kinkel, 'Preparative chromatographic resolution of racemates on chiral stationary phases on laboratory and production scales by closed-loop recycling chromatography', *J. Chromatogr.* **666**: 627–650 (1994).

61. C. M. Grill, 'Closed-loop recycling with periodic intra-profile injection: a new binary preparative chromatographic technique', *J. Chromatogr.* **796**: 101–113 (1998).

62. L. Mondello, P. Dugo and I. Stagno-d'Alcontres, 'Isolamento di polimetossiflavoni dagli olii essenziali di arancia dolce e di mandarino mediante cromatografia su colonna e HPLC semipreparativa con riciclo', *Essenz. Deriv. Agrum.* **63**: 395–406 (1993).

63. P. Dugo, L. Mondello, E. Cogliandro, A. Cavazza and G. Dugo, 'On the genuineness of citrus essential oils. Part LIII. Determination of the composition of the oxygen heterocyclic fraction of lemon essential oils (*citrus limon* (L.) *burm. f.*) by normal-phase high performance liquid chromatography', *Flavour Frag. J.* **13**: 329–334 (1998).

64. P. Dugo, L. Mondello, E. Sebastiani, R. Ottanà, G. Errante and G. Dugo, 'Identification of minor oxygen heterocyclic compounds of citrus essential oils by liquid chromatography–atmospheric pressure chemical ionisation mass spectrometry', *J. Liq. Chromatogr.* **22**: 2991–3005 (1999).

65. F. Charton and R.-M. Nicoud, 'Complete design of a simulated moving bed', *J. Chromatogr.* **702**: 97–112 (1995).

66. A. Gentilini, C. Migliorini, M. Mazzotti and M. Morbidelli, 'Optimal operation of simulated moving-bed units for non-linear chromatographic separations. II. Bi-Langmuir isotherm', *J. Chromatogr.* **805**: 37–44 (1998).

67. R.-M. Nicoud, G. Fuchs, P. Adam, M. Bailly, E. Küsters, F. D. Antia, R. Reuille and E. Schmid, 'Preparative scale enantioseparation of a chiral epoxide: comparison of liquid chromatography and simulated moving bed adsorption technology', *Chirality* **5**: 267–271 (1993).

68. E. Küsters, G. Gerber and F. D. Antia, 'Enantioseparation of a chiral epoxide by simulated moving bed chromatography using chiralcel-OD', *Chromatographia* **40**: 387–393 (1995).

69. D. W. Guest, 'Evaluation of simulated moving bed chromatography for pharmaceutical process development', *J. Chromatogr.* **760**: 159–162 (1997).

70. B. Pynnonen, 'Simulated moving bed processing: escape from the high-cost box', *J. Chromatogr.* **827**: 143–160 (1998).

71. S. Nagamatsu, K. Murazumi and S. Makino, 'Chiral separation of a pharmaceutical intermediate by a simulated moving bed process', *J. Chromatogr.* **832**: 55–65 (1999).

72. L. Miller, C. Orihuela, R. Fronek, D. Honda and O. Dapremont, 'Chromatographic resolution of the enantiomers of a pharmaceutical intermediate from the milligram to the kilogram scale', *J. Chromatogr.* **849**: 309–317 (1999).

73. M. Juza, 'Development of a high performance liquid chromatographic simulated moving bed separation from an industrial perspective', *J. Chromatogr.* **865**: 35–49 (1999).

74. E. R. Francotte and P. Richert, 'Applications of simulated moving-bed chromatography to the separation of the enantiomers of chiral drugs', *J. Chromatogr.* **769**: 101–107 (1997).

75. J. F. Garcia and D. Barcelò, 'An overview of LC–MS interfacing systems with selected applications', *J. High Resolut. Chromatogr.* **16**: 633–641 (1993).

76. W. M. A. Niessen and A. P. Tinke, 'Liquid chromatography–mass spectrometry. General principles and instrumentation', *J. Chromatogr.* **703**: 37–57 (1995).

77. M. Careri, A. Mangia and M. Musci, 'Overview of the applications of liquid chromatography–mass spectrometry interfacing systems in food analysis: naturally occurring substances in food', *J. Chromatogr.* **794**: 263–297 (1998).

78. G. J. Opteck, K. C. Lewis, J. W. Jorgenson and R. J. Anderegg, 'Comprehensive on-line LC/LC/MS of proteins', *Anal. Chem.* **69**: 1518–1524 (1997).

79. J. Cai and J. Henion, 'Quantitative multi-residue determination of β-agonists in bovine urine using on-line immunoaffinity extraction-coupled column packed capillary liquid chromatography–tandem mass spectrometry', *J. Chromatogr.* **691**: 357–370 (1997).

80. C. S. Creaser, S. J. Feely, E. Houghton and M. Seymour, 'Immunoaffinity chromatography combined on-line with high performance liquid chromatography–mass spectrometry for the determination of corticosteroids', *J. Chromatogr.* **794**: 37–43 (1998).

6 Supercritical Fluid Techniques Coupled with Chromatographic Techniques

F. M. LANÇAS

University of São Paulo, São Carlos (SP), Brasil

6.1 INTRODUCTION

The first report on the occurrence of a supercritical fluid is usually attributed to Baron Cagniard de la Tour in 1822 (1). Working with a closed glass container he observed that for certain materials the gas–liquid boundary disappeared when the system was heated at a certain temperature. An extension of this observation was the discovery of the critical point of a chemical substance. Hannay and Hogarth (2, 3) carried out the next important step in the development of this area. These authors reported in 1879 (2) the results of their studies on the solubility of metal halides in ethanol under various experimental conditions. They found that the solubility increased by increasing the pressure and that decreasing the pressure then caused precipitation of the dissolved salts. This was the first practical demonstration of the solvating power of supercritical fluids, later confirmed by Buchner (4), among others.

There is no consensus about a rigorous definition of a supercritical fluid (SF) (5). In general, the concept of supercritical fluidity for a given substance is taken from its phase diagram (Figure 6.1). In this figure, the solid lines define the solid, liquid and gas states, as well as the possible transitions among them (sublimation, melting and boiling processes), while the points between the phases (along the lines) define the *equilibrium between the individual phases.*

The triple point (TP in Figure 6.1) is the point at which the substance coexists as a solid, liquid and gas. By increasing the temperature and pressure along the boiling line, the critical point (CP, Figure 6.1) is reached. The required pressure and temperature to reach the critical point (critical temperature (T_c); critical pressure (P_c)) varies from substance to substance. Any substance above its T_c and P_c is defined as a supercritical fluid (SF), while the region above the critical point, where a single phase exists which presents some properties of both a liquid and a gas, is termed the critical region. The critical temperature (T_c) is defined as the highest temperature at which a

Multidimensional Chromatography, edited by L. Mondello, A. C. Lewis and K. D. Bartle
©2002 John Wiley & Sons Ltd.

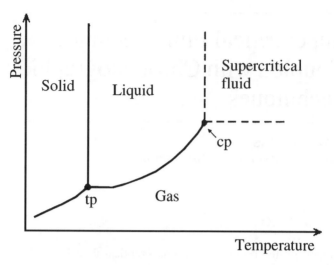

Figure 6.1   Phase diagram of a pure substance: cp, critical point; tp, triple point.

gas can be converted to a liquid by increasing its pressure, while the critical pressure (P_c) is the highest pressure at which a liquid can be converted to a gas by increasing its temperature (6–8).

Although several other early studies were conducted using Supercritical Fluids (SFs), also termed at that time as 'dense gases', this field did not receive the attention it deserved during the first half century after its discovery. Even then, the small number of investigators dedicated to further explore the scientific and technological potential of supercritical fluids concentrated their efforts on industrial rather than analytical applications (9, 10).

In spite of their tremendous potential, most of the analytical applications of SFs investigated after 1980 were concentrated in two areas, i.e. Supercritical Fluid Extraction (SFE) and Supercritical Fluid Chromatography (SFC). In the first case (SFE), the supercritical fluid replaces the use of organic solvents during the sample preparation step prior to further analytical determinations, which usually employ chromatographic techniques. In the second case, the supercritical fluid is used as the mobile phase in chromatographic analysis. The instrumentation used in both techniques presents several common features, with the major difference being that the analytical column used in SFC is replaced by an extraction cell in SFE, in most cases an open tube where the sample is placed for extraction. General schematic diagrams of the basic instrumentation for SFE and SFC are displayed in Figure 6.2. In both cases, a high-pressure system (usually a high-pressure pump) is used to deliver the fluid at the desired pressure, and an oven is then used to set and maintain the target temperature; a restrictor regulates the pressure inside the extraction cell (SFE) or inside the column (SFC), and the stream containing the supercritical fluid and the analytes is either vented (SFC) or collected (SFE).

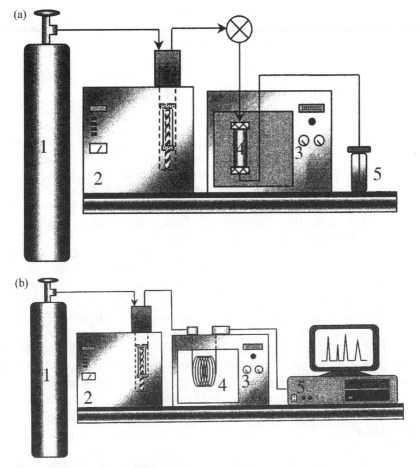

Figure 6.2 Schematic diagram showing the basic components of (a) SFE and (b) SFC instruments: 1, carbon dioxide; 2, high pressure pump; 3, oven; 4, extraction cell (SFE) or column (SFC); 5, collection vial (SFE) or data system (SFC).

Although SFE and SFC share several common features, including the use of a supercritical fluid as the solvent and similar instrumentation, their goals are quite distinct. While SFE is used mainly for the sample preparation step (extraction), SFC is employed to isolate (chromatography) individual compounds present in complex samples (11–15). Both techniques can be used in two different approaches: *off-line*, in which the analytes and the solvent are either vented after analysis (SFC) or collected (SFE), or *on-line* coupled with a second technique, thus providing a multidimensional approach. Off-line methods are slow and susceptible to solute losses and contamination; the on-line coupled system makes possible a decrease in the detection limits, with an improvement in quantification, while the use of valves for automation results in faster and more reproducible analyses (16). The off-line approach covering both techniques (SFE and SFC) has been recently covered in several books and reviews (17–21) and will not be discussed further in this present chapter, whose subject is the on-line approach.

6.2 ON-LINE COUPLING OF SFE WITH CHROMATOGRAPHIC TECHNIQUES

Hirschfeld, in his article on coupled techniques, defined, a hyphenated instrument as 'one in which both instruments are automated together as a single integrated unit via a hardware interface . . . whose function is to reconcile the often extremely contradictory output limitations of one instrument and the input limitation of the other' (22). Therefore, the key to combining SFE with chromatographic techniques is the interface which should allow the optimum and independent usage of each instrument while the couple still operate as an integrated unit (23).

Considering that in most SFE methods CO_2 is used as the extracting fluid, the on-line coupling of such a system with a chromatographic technique became easy due to the favourable properties of this fluid. For instance in coupling SFE with GC/flame-ionization detection (FID) there are no detection limitations since CO_2 does not present a signal with this detector. Therefore, the major requirement for a successful coupling is the quantitative transfer of the SF extract to the inlet of the chromatograph. SFE systems have been successfully coupled to GC (24, 25) SFC (26, 27), and LC (28, 29) systems. A review describing several applications where SFE has been coupled on-line to various chromatographic systems is available (30).

6.3 SFE–GC

On-line SFE–GC has been so far the most investigated coupling between SFE and chromatographic techniques. One of the reasons for this is that such a coupling is conceptually straightforward provided that the extracted analytes are quantitatively transferred to the GC inlet. Among the several publications on this subject, the simplest approach which has gained great popularity involves the use of either an on-column or a splitless injector as the interface. In this technique, the end of the restrictor from the SFE instrument is inserted inside the GC inlet and the extract is trapped inside the injector. After the extraction is accomplished, the injector is heated and the analytes are transferred to the analytical column to start the chromatographic run. The use of a cryogenic approach to trap the extracted analytes before the GC run has also been investigated (31, 32). The use of a programmed temperature vaporizer (PTV) injector as the interface for SFE-GC has been evaluated (33). Figure 6.3 shows a schematic drawing of a typical instrumental set-up used for on-line SFE–GC work.

The use of more selective detectors for SFE–GC, such as a thermo-energy analyzer for the detection of explosives (34) and a two-channel optical device for the

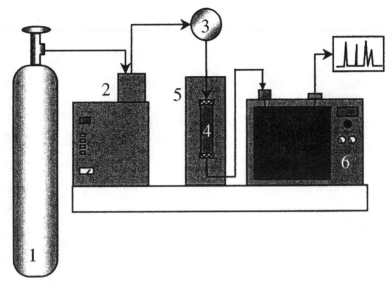

Figure 6.3 Schematic diagram of an on-line SFE–GC instrument: 1, carbon dioxide; 2, high-pressure syringe pump; 3, three-port valve; 4, extraction cell; 5, oven; 6, gas chromatograph.

selective detection of sulfur with a radiofrequency plasma detector (35), have been investigated. Review articles covering the instrumentation (24) and applications (30) of the on-line SFE–GC approach have also been published. Typical applications include those involving the analysis of organic compounds present in solid matrices that would require various extraction and fractionation procedures before the chromatographic analysis can take place. These are the slowest steps in a traditional analytical procedure and are also responsible for most of the errors introduced during the analysis (7). While an analysis using the traditional off-line extraction and clean-up methods can take days, the same analysis can be accomplished by an on-line coupled SFE–GC system in less than one hour and gives excellent results.

6.4 SPE–SFE–GC

Supercritical fluid extraction (SFE) and Solid Phase Extraction (SPE) are excellent alternatives to traditional extraction methods, with both being used independently for clean-up and/or analyte concentration prior to chromatographic analysis. While SFE has been demonstrated to be an excellent method for extracting organic compounds from solid matrices such as soil and food (36, 37), SPE has been mainly used for diluted liquid samples such as water, biological fluids and samples obtained after liquid–liquid extraction on solid matrices (38, 39). The coupling of these two techniques (SPE–SFE) turns out to be an interesting method for the quantitative transfer

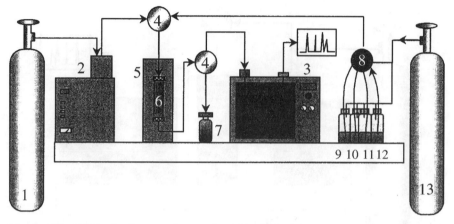

Figure 6.4 Schematic diagram of an on-line SPE–SFE–GC system (from ref. 40):1, carbon dioxide; 2, high-pressure syringe pump; 3, gas chromatograph; 4, three-port valve; 5, oven; 6, extraction cell; 7, waste; 8, ten-port valve; 9–11 conditioning and washing solvents; 12, sample; 13, nitrogen.

of analytes trapped on to a SPE sorbent to a GC column. As CO_2, the most widely used solvent for SFE, is a gas at room temperature, it therefore becomes the ideal solvent to quantitatively transport the analytes from the SPE cartridge to the GC column (5). Figure 6.4 shows a schematic drawing of a 'home-made' SPE–SFE–GC system (40).

The system is built by using three independent modules (SPE, SFE and GC) in such a way that it can be assembled to perform experiments in the on-line coupled mode (SPE–SFE, SFE–GC, SPE–GC and SPE–SFE–GC) or as independent units (GC, SPE, and SFE). This means that if we want to use the system for standard GC, there will be no problems, with the same applying for both SPE and SFE.

Operation of the SPE–SFE–GC system is very easy (see Figure 6.4). The SPE cartridge is conditioned (flasks 9 and 10) and washed (flask 11) depending upon the sorbent used. The first and second three-port valves (items 4) are then rotated to the SPE cartridge and waste positions, respectively. The sample contained in flask 12 is then loaded into the SPE cartridge (item 6) with the help of nitrogen (item 13). While the matrix and unwanted analytes goes to waste (item 7), the target compounds are trapped in the cartridge (item 6). After loading the sample into the extraction cell, the system is dried with nitrogen and the two valves are now switched to the SFE and GC feed positions, respectively, thus providing the transfer of the trapped analytes to the GC injector. A cryogenic system using liquid CO_2 is used to trap the analytes in the first section of the capillary column. Details of the interface used for coupling the SFE module to the GC injector is shown in Figure 6.5 (40). Figure 6.6 (b) displays an SPE–SFE–HRGC photo ionization detector (PID) chromatogram of a water sample contaminated with several aromatic compounds.

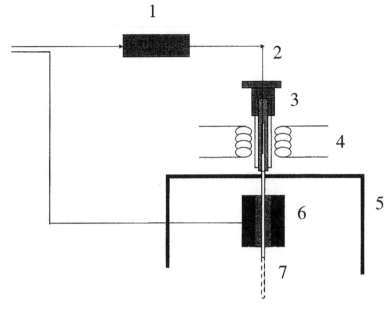

Figure 6.5 Schematic diagram of a typical interface used for on-line SFE–GC coupling (from ref. 40): 1, extraction cell; 2, restrictor; 3, on-column injector; 4, heater; 5, oven; 6, cryogenic module; 7, column.

6.5 SFE–SFC

The on-line coupling of supercritical fluid extraction with Supercritical Fluid Chromatography (SFE–SFC) is easier to achieve than those involving other chromatographic techniques. This is particularly the case when using neat CO_2 in both systems since it does not requires changes in the fluid composition or physical state. The SF extracts will contain the extracted analytes in CO_2 and these can be directly transported into the SFC system, which also uses CO_2 as the mobile phase. Different approaches have been used to trap and focus the analytes into the SFC column (26, 27), including a dual-trapping system to eliminate the modifier solvent (41).

Figure 6.7 shows a schematic diagram of an on–line SFE–SFC coupled system (42), with details of the interface being shown in Figure 6.8 (42).

6.6 SFE-LC

Unfortunately, not much experimental work has been carried out on the combination of supercritical fluid extraction and liquid chromatography systems (43, 44). One of the reasons for this arises from the difficulties in achieving compatibility between the extraction solvent and the LC mobile phase. Baseline perturbations have been

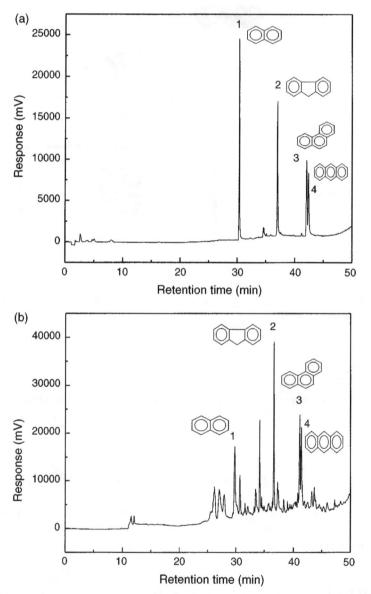

Figure 6.6 SPE–SFE–HRGC(PID) chromatograms obtained for (a) a mixture of analytical standards of selected aromatic compounds in water, and (b) the analysis of a water sample contaminated with various aromatic compounds.

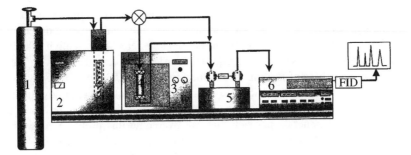

Figure 6.7 Schematic diagram of an on-line SFE–SFC system (from ref. 42):1, carbon dioxide; 2, pump; 3, oven; 4, extraction cell; 5, interface; 6, SFC unit.

observed with UV detection, attributed to the solubility of the CO_2 in the aqueous mobile phase, particularly at low flow rates (28).

6.7 ON-LINE COUPLING OF SUPERCRITICAL FLUID EXTRACTION WITH CAPILLARY ELECTRODRIVEN SEPARATION TECHNIQUES (SFE–CESTs)

Electrodriven Separation Techniques encompass a wide range of analytical procedures based on several distinct physical and chemical principles, usually acting together to perform the required separation. Example of electrophoretic-based techniques includes capillary zone electrophoresis (CZE), capillary isotachophoresis (CITP), and capillary gel electrophoresis (CGE) (45–47). Some other electrodriven separation techniques are based not only on electrophoretic principles but rather on chromatographic principles as well. Examples of the latter are micellar

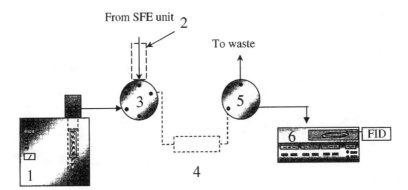

Figure 6.8 Schematic diagram of a typical interface used for on-line SFE–SFC coupling (from ref. 42):1, pump; 2, heated transfer line; 3, valve; 4, sample concentrator; 5, valve; 6, SFC unit.

electrokinetic chromatography (MEKC) and, more recently, capillary electrochromatography (CEC) (48–52). All capillary electrodriven separation techniques (CESTs) suffer from the same problem, i.e. the low detection sensitivity due to the use of narrow capillaries which are required in order to achieve the desired efficiency and prevent an excess of heating due to the Joule effect (53, 54). One way to overcome this limitation has been the use of membranes, cartridges, hollow fibers and similar approaches to concentrate the analytes prior to the electrophoretic separation (55–57). A more general approach that can overcome this problem, as well as to provide additional features, particularly for the analysis of complex real samples, involves the on-line coupling of supercritical fluid extraction and capillary electrodriven separation techniques (SFE–CESTs) (58–60). Figure 6.9 displays a schematic diagram of an on-line coupled SFE–CEST system (58). The developed system consists of three basic and independent parts, namely the SFE system, an interface to couple the two systems together and the capillary electrodriven apparatus. The sample is extracted in the SFE unit and is then directed to the interface in order to compatibilize the SFE conditions with those required by the electrodriven separation system. The interface developed for this instrument is displayed in Figure 6.10. By using this interface, the SFE–CEST system can be used for extraction, clean-up, sample concentration–enrichment and analysis all in just one step (58). The operational procedure is quite straightforward. Two different operational modes have been investigated with the developed interface, i.e. the concentration mode and the direct mode (see figures 6.11 and 6.12).

The **concentration mode** is used when the extract from the SFE module is not concentrated enough to be directly analyzed by the CE instrument, and thus requires a concentration step, which is carried out in the concentrator device. In this mode, during the concentration step the extract from the SFE cell enters the first valve (positioned in the concentration mode (Figure 6.11(a)), and is then directed to and adsorbed into the sample concentrator which contains an SPE cartridge. While the

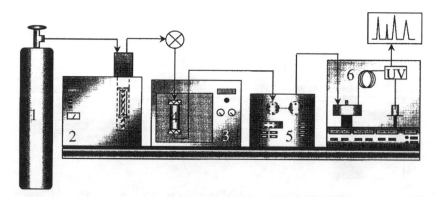

Figure 6.9 Schematic diagram of an on-line SFE–CEST system (from ref. 58):1, CO_2 cylinder; 2, syringe pump; 3, oven; 4, extraction cell; 5, interface; 6, CE instrument.

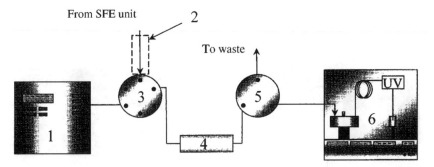

Figure 6.10 Schematic diagram of a typical interface used for on-line SFE–CEST coupling (from ref. 57): 1, micro-LC pump; 2, heated restrictor; 3, six-port valve; 4, sample concentrator; 5, three-port valve; 6, CE instrument.

analytes of interest are concentrated in the cartridge, the solvent is vented through the valve (item (5) in Figure 6.10). Upon completing the concentration step, the valves are then switched to the elution position (Figure 6.11(b), and a pump delivers the appropriate solvent to elute the sample from the SPE cartridge to the CE vial. After this step, the CE analysis is then carried out.

The **direct mode** is used when the concentration of the SFE extract is enough for direct analysis in the CE instrument without the need for a pre-concentration step. In this case, the sample concentrator is by-passed and the SFE extract goes directly to the CE instrument. The extract is collected in a CE vial containing an appropriate solvent and is thus ready for the CE analysis (Figure 6.12).

A practical application of SFE–CEST coupling is shown in Figure 6.13, which displays the electropherogram obtained for a tomato sample contaminated with a pesticide, i.e. carbaryl. The sample was placed in the SFE cell, extracted with CO_2

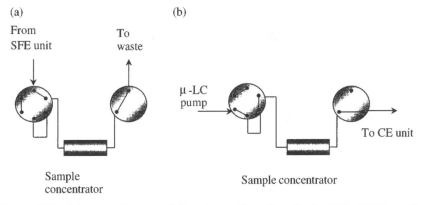

Figure 6.11 Schematic diagrams of the valve configurations for the SFE–CEST coupling in the concentration mode, shown for (a) the concentration step and (b) the elution step. (from ref. 58)

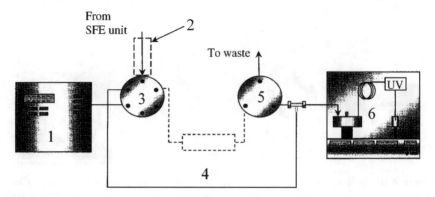

Figure 6.12 Schematic diagram of the interface used for direct SFE–CEST coupling without a sample pre-concentration step: 1, micro-LC pump; 2, heated restrictor; 3, six-port valve; 4, direct by-pass to the CE unit; 5, three-port valve; 6, CE instrument. (from ref. 58).

and the extract then concentrated in the concentrator device, while the solvent from the SFE step was led to waste. The concentrated extract was then desorbed from the concentrator to the CE unit where analysis was performed in the micellar electrokinetic chromatography (MEKC) mode. Therefore, this particular application describes the use of an on-line SFE–MEKC coupling system. Further details concerning this and other applications using such a system can be found elsewhere (58–60).

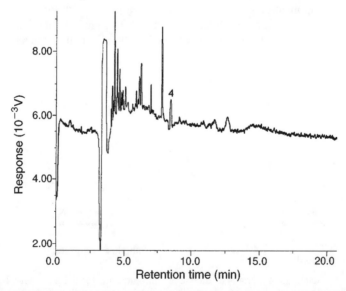

Figure 6.13 SFE–MEKC electropherogram of the pesticide carbaryl in a tomato sample; the peak assigned number (4) corresponds to the migration time of carbaryl. (from ref. 58).

6.8 FROM MULTIDIMENSIONAL TO UNIFIED CHROMATOGRAPHY PASSING THROUGH SUPERCRITICAL FLUIDS

In 1965, Giddings proposed that there are no theoretical boundaries between GC, LC and SFC, and that such distinctions are arbitrary, artificial and counter-productive (61). This concept was extended by Ishii *et al.* who proposed the concept of Unified Chromatography in 1988 (62). According to these authors, it is possible to demonstrate different-mode separations, i.e. liquid chromatography (LC), supercritical fluid chromatography (SFC) and gas chromatography (GC) by using a single chromatographic system (62). Still according to Ishii and co-workers, the separation mode could be selected by changing the pressure in the column and the column temperature (62). Since then a selected number of papers have appeared in the literature covering this subject (for a review of Unified Chromatography, the interested reader should consult references 63–65). Most of the work carried out so far is in fact related to multidimensional chromatography using different chromatographic modes, and not on Unified Chromatography as originally described by Ishii and co-workers. In this respect, several reports have dealt with techniques such as GC–GC, LC–LC, LC–GC, SFC–GC, and so on. Using such an approach, either the total or a fraction of the analytes exiting the first column is transferred (switched) to a second column for a second 'dimension' of analysis. The major conceptual difference between this approach and the Unified Chromatography concept is that in the latter system the sample is introduced into the first analytical unit (e.g GC) and part of it is then eluted from the column by using a 'proper' mobile phase. Following this and without changing the column, the other portion of the sample which is still retained on this column is then eluted in a different chromatographic mode (e.g. SFC), either with or without changing the mobile phase. Since both approaches, i.e. Multidimensional Chromatography and Unified Chromatography, use equivalent symbols to express the hyphenated character of both (e.g. GC–SFC), there is often confusion about their fundamental principles. In short, Multidimensional Chromatography uses a column-switching approach, with the analyte being transferred from one to the other column, while Unified Chromatography uses the same column for all of the chromatographic modes, without any analyte transfer. Since in both cases the techniques are coupled on-line they are both considered to be hyphenated systems. Supercritical fluid chromatography plays a very important role in the Unified Chromatography approach. Most of the (albeit, rather few) papers published on Unified Chromatography use SFC as a 'bridge' between the two limiting extremes of the mobile phase, represented by a gas (in GC) and a liquid (in LC). This subject will not be covered any further here since it is discussed in detail elsewhere in this present volume (see Chapter 7).

REFERENCES

1. C. Cagniard de la Tour, *Ann. Chim. Phys.* **21**: 27 (1822).
2. J. B. Hannay and J. Hogarth, On the solubility of solids in gases', *Proc. R. Soc. (London)* **29**: 324–326 (1879).

3. J. B. Hannay and J. Hogarth, '*Proc. R. Soc. (London)*', **30**: 178 (1880).

4. E. H. Buchner, *Z. Phys. Chem.*, **54**: 665 (1906).

5. T. L. Chester, J. D. Pinkston and D. E. Raynie, 'Supercritical fluid chromatography and extraction', *Anal. Chem.* **70**: 301R–319R (1998).

6. L. T. Taylor *Introduction to Supercritical Fluid Extraction,* Research and Development Magazine Publishers, USA (1995).

7. R. E. Majors, 'An overview of sample preparation', LC–GC **9**: 16–18 (1991).

8. S. S. H. Rizvi and A. L. Benado, 'Supercritical propane systems for separation of continuous oil mixtures', *Ind. Eng. Chem. Res.* **26**: 731–737 (1987).

9. R. E. Wilson, P. C. Keith and R. E. Haylett, 'Liquid propane. Use in dewaxing, deasphalting and refining heavy oils', *Ind. Eng. Chem.* **28**: 1065–1078 (1936).

10. N. L. Dickerson, J. M. Meyers, *J. Am. Oil Chem. Soc.* **29**: 235–239 (1952).

11. F. M. Lanças and S. R. Rissato, 'Influence of temperature, pressure, modifier and collection mode on supercritical CO_2 extraction efficiencies of diuron from sugar cane and orange samples', *J. Microcolumn Sep.* **10**: 473–478 (1998).

12. S. R. Sargenti and F. M. Lanças, 'Influence of the extraction mode and temperature in the supercritical fluid extraction of *Tangor murcote* (Blanco) for *Citrus sinensis* (Osbeck)', *J. Microcolumn Sep.* **10**: 213–223 (1998).

13. F. M. Lanças, S. R. Rissato and M. S. Galhiane, 'Off-line SFE–MEKC determination of diuron in sugar cane and orange samples', *J. High Resolut. Chromatogr.* **21**: 519–522 (1998).

14. M. C. H. Tavares, J. H. Vilegas and F. M. Lanças, 'Capillary supercritical fluid chromatography (c-SFC) analysis of underivatized triterpenic acids', *Phytochem. Anal.* **12**: 134–137 (2001).

15. M. C. H. Tavares, 'Project, construction and applications of a new capillary supercritical fluid chromatography', *Ph. D. Thesis,* University of São Paulo, Brazil. (1999)

16. K. Hartonen and M.-L. Riekkola, 'Detection of β-blockers in urine by solid-phase extraction–supercritical fluid extraction and gas chromatography–mass spectrometry', *J. Chromatogr.* **676**: 45–52 (1996).

17. L. T. Taylor, *Supercritical Fluid Extraction,* John Wiley & Sons, New York (1996).

18. S. A. Westwood (Ed.), *Supercritical Fluid Extraction and its Use in Chromatographic Sample Preparation,* CRC Press, Boca Raton, FL (1992).

19. K. Jinno (Ed.), *Hyphenated Techniques in Supercritical Fluid Chromatography and Extraction,* Elsevier, Amsterdam (1992).

20. B. Wenclawiak (Ed.), *Analysis with Supercritical Fluids: Extraction and Chromatography,* Springer-Verlag, Berlin (1992).

21. M. L. Lee and K. E. Markides, *Analytical Supercritical Fluid Chromatography and Extraction,* 'Chromatography Conferences', Provo, USA (1990).

22. T. Hirschfeld, 'Hyphenated methods', *Anal. Chem.* **52**: 297A–312A (1980).

23. I. L. Davies, M. W. Raynor, J. P. Kithinji, K. D. Bartle, P. T. Williams and G. E. Andrews, 'LC–GC, SFC–GC and SFE–GC interfacing', *Anal. Chem.* **60**: 683A–702A(1988).

24. M. D. Burford, K. D. Bartle and S. B. Hawthorne, 'Directly coupled (on-line) SFE–GC: instrumentation and applications', *Adv. Chromatogr.* **37**: 163–204 (1997).

25. P. Sandra, A. Kot, A. Medvedovici and F. David, 'Selected applications of the use of supercritical fluids in coupled systems', *J. Chromatogr.* **703**: 467–478 (1995).

26. H. Daimon and Y. Hirata, 'Trapping efficiency and solute focusing in on-line supercritical fluid extraction/capillary supercritical fluid chromatography', *J. Microcolumn Sep.* **5**: 531–535 (1993).

27. H. Daimon and Y. Hirata, 'Direct coupling of capillary supercritical fluid chromatography with supercritical fluid extraction using modified carbon dioxide', *J. High Resolut. Chromatogr.* **17**: 809–813 (1994).

28. M. Ashraf-Khorassani, M. Barzegar and Y. Yamini, 'On-line coupling of supercritical fluid extraction with high performance liquid chromatography', *J. High Resolut. Chromatogr.* **18**: 472–476 (1995).

29. D. C. Tilotta, D. L. Heglund and S. B. Hawthorne, 'Online SFE–FTIR spectrometry with a fiber optic transmission cell', *Am. Lab.* **28**: 36R–36T (1995).

30. T. Greibrokk, 'Applications of supercritical fluid extraction in multidimensional systems', *J. Chromatogr.* **703**: 523–536 (1995).

31. R. Fuoco, A. Ceccarini, M. Onor and S. Lottici, 'Supercritical fluid extraction combined online with cold-trap gas chromatography/mass spectrometry', *Anal. Chim. Acta* **346**: 81–86 (1997).

32. J. T. B. Strode and L. T. Taylor, 'Supercritical fluid extraction employing a variable restrictor coupled to gas chromatography via a sample pre-concentration trap', *J. High Resolut. Chromatogr.* **19**: 651–654 (1996).

33. X. Lou, H.-G. Janssen and C. A. Cramers, 'Investigation of parameters affecting the online combination of supercritical fluid extraction with capillary gas chromatography', *J. Chromatogr.* **750**: 215–226 (1996).

34. E. S. Francis, M. Wu, P. B. Farnsworth and M. L. Lee, 'Supercritical fluid extraction/gas chromatography with thermal desorption modulator interface and nitro-specific detection for the analysis of explosives', *J. Microcolumn Sep.* **7**: 23–28 (1995).

35. P. B. Farnsworth, M. Wu, M. Tacquard and M. L. Lee, 'Background correction device for enhanced element-selective gas chromatographic detection by atomic emission spectroscopy', *Appl. Spectr.* **48**: 742–746 (1994).

36. D. R. Gere, C. R. Knipe, P. Castelli, J. Hedrick, Randall Frank, L. G., H. Schulenberg-Schell, R. Schuster, L. Doherty, J. Orolin and H. B. Lee, 'Bridging the automation gap between sample preparation and analysis: an overview of SFE, GC, GC–MS and HPLC applied to environmental samples', *J. Chromatogr. Sci.* **31**: 246–258 (1993).

37. E. Anklam, H. Berg, L. Mathiasson, M. Sharman and F. Ulberth, 'Supercritical fluid extraction (SFE) in food analysis: a review', *Food Additive Contam.* **15**: 729–750 (1998).

38. V. Janda, M. Mikešová and J. Vejrosta, 'Direct supercritical fluid extraction of water-based matrices', *J. Chromatogr.* **733**: 35–40 (1996).

39. E. M. Thurman and M. S. Mills, *Solid-Phase Extraction: Principles and Practice,* John Wiley & Sons, New York (1998).

40. J. S. S. Pinto, E. Cappelaro and F. M. Lanças, 'Design and construction of an on-line SPE–SFE–HRGC system', *J. Braz. Chem. Soc.* **12**: 192–195 (2001).

41. U. Ullsten and K. E. Markides, 'Automated on-line solid phase adsorption/supercritical fluid extraction/supercritical fluid chromatography of analytes from polar solvents', *J. Microcolumn Sep.* **6**: 385–393 (1994).

42. M. C. Tavares and F. M. Lanças, "On-line coupling of supercritical fluid extraction with supercritical fluid chromatography", *J. Braz. Chem. Soc.* in press (2001).

43. C. Mougin and J. Dubroca, 'On-line supercritical fluid extraction and high performance liquid chromatography for determination of triazine compounds in soil, *J. High Resolut. Chromatogr* **19**: 700–702 (1996).

44. M. H. Liu, S. Kapila, K. S. Nam and A. A. Elseewi, 'Tandem supercritical fluid extraction and liquid chromatography system for determination of chlorinated phenols in solid matrices', *J. Chromatogr.* **639**: 151–157 (1993).

45. H. Engelhardt, W. Beck and T. Schmitt, *Capillary Electrophoresis: Methods and Potentials,* Vieweg, Wiesbaden, Germany (1996).

46. S. F. Y. Li, *Capillary Electrophoresis. Principles, Practice and Applications,* Elsevier, Amsterdam (1992).

47. B. Chankvetadze, *Capillary Electrophoresis in Chiral Analysis,* John Wiley & Sons, New York (1997).

48. M. G. Cikalo, K. D. Bartle, M. M. Robson, P. Myers and M. R. Euerby, 'Capillary electrochromatography', *Analyst* **123**: 87R–102R (1998).

49. C. Fujimoto, 'Packing materials and separation efficiencies in capillary electrochromatography', *Trends Anal. Chem.* **18**: 291–301 (1999).

50. K. D. Altria, 'Overview of capillary electrophoresis and capillary electrochromatography', *J. Chromatogr.* **856**: 443–463 (1999).

51. J. Viudevogel and P. Saundre, 'Introduction to unicellar electrokinetic chromatography', Hüthig Verlag GmbH Heidelberg, Germany (1992).

52. S. Terabe, 'Electrokinetic chromatography: an interface between electrophoresis and chromatography', *Trends Anal. Chem.* **8**: 129–134 (1989).

53. J. P. Landers (Ed.), *Handbook of Capillary Electrophoresis,* 2nd Edn, CRC Press, Boca Raton, FL (1996).

54. A. J. Tomlinson and S. Naylor, 'Enhanced performance membrane preconcentration–capillary electrophoresis–mass spectrometry (mPC–CE–MS) in conjunction with transient isotachophoresis for analysis of peptide mixtures, *J. High Resolut. Chromatogr.* **18**: 384–386 (1995).

55. A. J. Tomlinson and S. Naylor, 'Systematic development of on-line membrane preconcentration–capillary electrophoresis–mass spectrometry for analysis of peptide mixtures, *J. Capillary Electrophoresis* **2**: 225–233 (1995).

56. N. Guzman, On-line bioaffinity, molecular recognition and preconcentration in CE technology', LC–GC, **17**: 16 (1999).

57. N. M. Djordjevic and K. Ryan, 'An easy way to enhance absorbance detection on Waters Quanta-4000 capillary electrophoresis system', *J. Liq. Chromatogr.* **19**: 201–206 (1996).

58. F. M. Lanças and M. A. Ruggiero, 'On-line coupling of supercritical fluid extraction to capillary column electrodriven separation techniques', *J. Microcolumn Sep.* **12**: 61–67 (2000).

59. M. A. Ruggiero and F. M. Lanças, 'On-line supercritical fluid extraction–capillary electrophoresis (SFE–CE) analysis: development and applications', in *Proceedings of the 20th International Symposium on Capillary Chromatography*, Riva del Garda, Italy, May 26–29, Sandra P (Ed.), Hüthig, Hedelberg, pp. 22–23 (1998).

60. M. A. Ruggiero and F. M. Lanças, 'Approaching the ideal system for the complete automation in trace analysis by capillary electrophoresis', in *Proceedings of the 3rd Latin American Symposium on Capillary Electrophoresis*, Buenos Aires, Argentine, November 30–December 2. p. 1 (1997).

61. J. C. Giddings, *Dynamics of Chromatography. Part I: Principles and Theory,* Marcel Dekker, New York (1965).

62. D. Ishii, T. Niwa, K. Ohta and T. Takeuchi, 'Unified capillary chromatography', *J. High Resolut. Chromatogr.* **11**: 800–801 (1988).

63. K. D. Bartle, I. L. Davies, M. W. Raynor, A. A. Clifford and J. P. Kithinji, 'Unified multidimensional microcolumn chromatography', *J. Microcolumn Sep.* **1**: 63–70 (1989).

64. D. Tong, K. D. Bartle and A. A. Clifford, 'Principles and applications of unified chromatography', *J. Chromatogr.* **703**: 17–35 (1995).

65. F. M. Lanças GC × SFC × LC: Towards unified chromatography', in *Proceedings of the 6th Latin American Congress on Chromatography (COLACRO VI)*, Caracas, Venezuela Intevep Ed, 25–P1 (1996).

7 Unified Chromatography: Concepts and Considerations for Multidimensional Chromatography

T. L. CHESTER

The Procter & Gamble Company, Cincinnati, OH, USA

7.1 INTRODUCTION

Multidimensional chromatography brings together separations often based on different selectivity mechanisms. Although the forms of the mobile phase are not required to be different in the individual steps of a multidimensional separation, we usually strive to achieve orthogonal selectivity of these individual separation steps (1).

This is usually not a problem when bringing together two techniques with very different mobile phases, such as liquid chromatography (LC) and gas chromatography (GC), for example, to carry out a two-dimensional LC–GC separation. In GC, the only significant intermolecular forces are between the solutes and the stationary phase, and there are no attractive forces of any consequence involving the mobile phase. Rather, it is just an inert carrier that transports the gaseous fraction of the solute through the interparticle volume of the column whenever the solute is not associated with the stationary phase. Once a GC column is chosen, control of solute partitioning between the stationary and mobile phases is simply a function of temperature and nothing else.

However, in LC solutes are partitioned according to a more complicated balance among various attractive forces: solutes interact with both mobile-phase molecules and stationary-phase molecules (or stationary-phase pendant groups), the stationary-phase interacts with mobile-phase molecules, parts of the stationary phase may interact with each other, and mobile-phase molecules interact with each other. Cavity formation in the mobile phase, overcoming the attractive forces of the mobile-phase molecules for each other, is an important consideration in LC but not in GC. Therefore, even though LC and GC share a considerable amount of basic theory, the mechanisms are very different on a molecular level. This translates into conditions that are very different on a practical level; so different, in fact, that separate instruments are required in modern practice.

Multidimensional Chromatography, edited by L. Mondello, A. C. Lewis and K. D. Bartle
©2002 John Wiley & Sons Ltd.

Why are these techniques so different, and how did this division of practice arise? Helium and hydrogen are used in GC in preference to other mobile-phase gases because of diffusion. Since there are no interactions involving the mobile phase, gases with the fastest diffusion rates produce analyses in the shortest times. Tswett and the other pioneers of LC chose from among a few convenient liquids to use as mobile phases. These liquids had to have several important features. They could not be so viscous that they would not flow through packed columns by gravity or with the application of modest pressure. They could not be so volatile that they would evaporate before reaching the column outlet. In short, they had to be *well-behaved* liquids.

This thinking has carried through to the present day and is reflected in our choices of mobile-phase fluids in LC: water, acetonitrile, methanol, tetrahydrofuran, hexane, etc., are still among our popular choices. However, these particular materials are completely dependent on the conditions of column temperature and outlet pressure. Tswett's original conditions at his column outlet, actually the earth-bound defaults we call *ambient* temperature and pressure, determined his solvent choices and continue to dominate our thinking today.

If our ambient conditions were different from what they actually are, then we would surely have a much different list of our favorite liquid solvents. *Unified chromatography begins with the simple notion of rejecting ambient temperature and outlet pressure as being necessarily correct.* Instead, we simply include these parameters with the other chromatographic variables and find the best values of all the parameters, taken together, to accomplish our separation goals. This simple notion leads to a greatly expanded list of possible mobile phases. It also leads to new and different chromatographic performance characteristics that may be of great utility in trying to achieve orthogonality in multidimensional separations.

The history of unified thinking has been driven by a few pioneers. Martire was among the first to address theoretical aspects (2–7). His unified theory of chromatography spans a variety of mobile-phase conditions from a thermodynamics and statistical mechanics standpoint. Ishii and Takeuchi built instruments that could perform GC, supercritical fluid chromatography (SFC), and LC, usually in discontinuous steps involving a change in mobile phase (8, 9). From this came the concept Ishii called *Troika*, after the Russian carriage drawn by three horses (GC, SFC, LC). Bartle's group also contributed developments in instrumentation aimed primarily at performing multiple separation modes sequentially (10–15). Beginning in the 1960s, Giddings and co-workers did a great deal of work using dense-gas and supercritical-fluid mobile phases. Although Giddings' research focus turned to field flow fractionation later in his career, his philosophical influence toward thinking about the common or unified traits of separation processes, using a variety of fluids, fluid propulsion means, and separation mechanisms, was immense. This is best exemplified by his 1991 book, *Unified Separation Science* (16).

7.2   THE PHASE DIAGRAM VIEW OF
UNIFIED CHROMATOGRAPHY

We must start with fluid behavior to understand the basic concepts of unified chromatography. We must forget most of what we know from common experience about
liquid and gas behavior since this experience is tied with ambient conditions.
Instead, we must embrace the new possibilities afforded by temperatures and pressures that are different from ambient. This new view requires phase diagrams
(17, 18).

We usually think of a transition of a pure material from its liquid state to its
gaseous state as requiring a discontinuous phase change, i.e. boiling. This is depicted
by path *a* in Figure 7.1 in which we transform pure liquid water at ambient conditions, *A*, to its gaseous form at 150°C, *B,* by heating at constant, atmospheric pressure. Once again, this discontinuous change, our normal expectation, results from

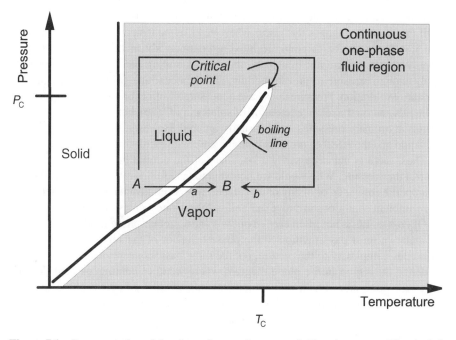

Figure 7.1   Representation of the phase diagram for a pure fluid such as water. The shaded
area is the continuum through which we can continuously vary the properties of the fluid. The
high-pressure and high-temperature limits shown here are arbitrary. They depend only on the
capabilities of the experimental apparatus and the stability of the apparatus and the fluid.

our experiences of living our entire lives at ambient pressure. However, in actuality, there is a continuum of fluid behavior linking the normal liquid and gas states, shown by the shaded area in the figure. In order to avoid discontinuities, we simply need to place the phase transitions off limits.

By traveling within the continuum and avoiding the boiling line, we can convert the ordinary liquid to its gaseous form without undergoing a phase change. One path accomplishing this is path *b* in the figure. Of course, an unlimited number of such paths is possible. The only requirement for continuous change is that we never cross the boiling line. Instead, we will go around the critical point, the point at which liquid and vapor states merge into a single fluid with shared properties.

Now, we should ask ourselves about the properties of water in this continuum of behavior mapped with temperature and pressure coordinates. First, let us look at temperature influence. The viscosity of the liquid water and its dielectric constant both drop when the temperature is raised (19). The balance between hydrogen bonding and other interactions changes. The diffusion rates increase with temperature. These dependencies on temperature provide us with an opportunity to tune the solvation properties of the liquid and change the relative solubilities of dissolved solutes without invoking a chemical composition change on the water.

If the property changes that occur when we raise the temperature are useful to us, and if we would like to continue moving continuously in the direction of these improvements, we are not limited by the normal boiling point. We simply need to apply enough pressure to prevent boiling, and then continue going on with our temperature exploration. Pressure does have a measurable effect on the solvent properties of liquids and on the relative retention of solutes in LC (20), but this effect is small in liquids at ordinary pressures and is usually ignored. However, the compressibility of the liquid and the effect of pressure both increase with increasing temperature.

Thus, there is not necessarily a boundary at the normal boiling point when we control the pressure. Why would we not want to take full advantage of the full range of properties of water, or of any other solvent, whenever advantages discovered away from ambient conditions improve our ability to separate solutes of interest?

Of course, LC is not often carried out with neat mobile-phase fluids. As we blend solvents we must pay attention to the phase behavior of the mixtures we produce. This adds complexity to the picture, but the same basic concepts still hold: we need to define the region in the phase diagram where we have continuous behavior and only one fluid state. For a two-component mixture, the complete phase diagram requires three dimensions, as shown in Figure 7.2. This figure represents a Type I mixture, meaning the two components are miscible as liquids. There are numerous other mixture types (21), many with miscibility gaps between the components, but for our purposes the Type I mixture is sufficient.

The shaded region is that part of the phase diagram where liquid and vapor phases coexist in equilibrium, somewhat in analogy to the boiling line for a pure fluid. The ordinary liquid state exists on the high-pressure, low-temperature side of the two-phase region, and the ordinary gas state exists on the other side at low pressure and high temperature. As with our earlier example, we can transform any Type I mixture

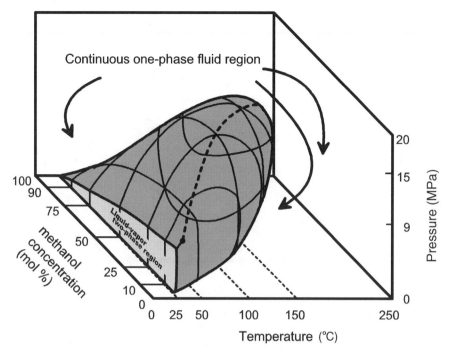

Figure 7.2   A three-dimensional phase diagram for a Type I binary mixture (here, CO_2 and methanol). The shaded volume is the two-phase liquid-vapor region. This is shown truncated at 25 °C for illustration purposes. The volume surrounding the two-phase region is the continuum of fluid behavior.

from its liquid state to its gaseous state in a completely continuous manner by changing the pressure and temperature to simply go around the two-phase region of the figure. We can further continuously change the nature of the fluid by changing its composition. We are completely free to do so in a Type I mixture without realizing a discontinuity of any sort as long as we avoid the two-phase region (and, of course, the solid region, not shown, existing at lower temperatures). Now we have quite a large range of continuous behavior available to us, represented by the volume of the figure outside the forbidden two-phase region and limited only by the temperature and pressure capabilities of our system.

Many chromatographic techniques have been named and are practiced in various regions of the fluid continuum. These regions are identified in Figures 7.3–7.8. We have not specified the mobile-phase components, and not all of these techniques are necessarily practical with the same mobile-phase component choices. However, the general view is valid.

LC is a limiting technique that occurs when the column outlet pressure is near ambient and we choose well-behaved liquids as our mobile phases. Our only means of adjusting solute retention (after selecting the stationary phase and the

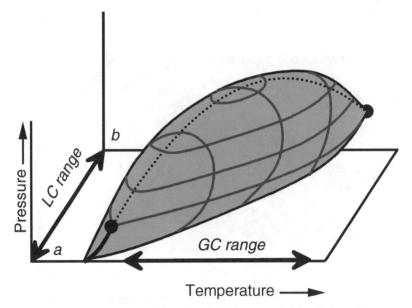

Figure 7.3   The positions occupied by LC and GC in a generic Type I phase diagram representing the mobile phase. Note that the GC mobile phase is shown as being composed of 100% component *a*, but this makes no difference chemically because there are no solute–mobile-phase interactions in GC. Reproduced by permission of the American Chemical Society.

mobile-phase components) is to vary the composition of the mobile phase, as indicated by the LC arrow in Figure 7.3. Although pressure is elevated at the column inlet, its only purpose is to generate mobile-phase flow. Pressure provides virtually no opportunity for adjusting retention or selectivity in conventional high-performance LC (HPLC). High-temperature LC is practiced simply by raising the column temperature. It provides some new benefits, of course, but is limited by the normal boiling point of the mobile phase unless significant pressure is applied at the column outlet.

When the pressure is controlled over a substantial range, we can more fully enter the subcritical fluid chromatography (SubFC) and enhanced-fluidity liquid chromatography (EFLC) (22, 23) regions shown in Figures 7.4 and 7.5, respectively. Both of these techniques are performed below the critical temperature locus of the mobile phase. Both of them frequently take advantage of very volatile mobile-phase components in order to maximize diffusion rates and speed of analysis. The only difference between these techniques is which fluid is denoted as being the main fluid. In SubFC, the more-volatile mobile-phase component is considered to be the main component, while the less-volatile component is considered to be a modifier used to change solvation characteristics. In EFLC, the less-volatile component is considered to be the main component, and the more-volatile component is considered to be a viscosity (and diffusion) enhancer. Clearly, there is no chemically meaningful boundary between SubFC and EFLC.

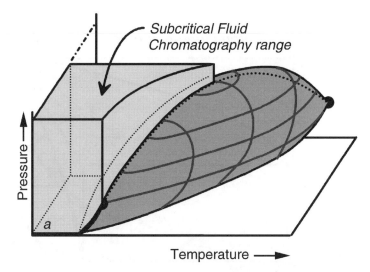

Figure 7.4 The Subcritical Fluid Chromatography range. This occupies the volume in the phase diagram below the locus of critical temperatures, above and below the locus of critical pressures, and is composed mostly of the more volatile mobile-phase component. Reproduced by permission of the American Chemical Society.

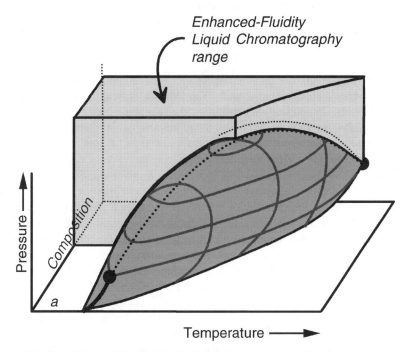

Figure 7.5 The Enhanced Fluidity Liquid Chromatography range. This occupies the volume in the phase diagram below the locus of critical temperatures, above and below the locus of critical pressures, and is composed mostly of the less volatile mobile-phase component. Reproduced by permission of the American Chemical Society.

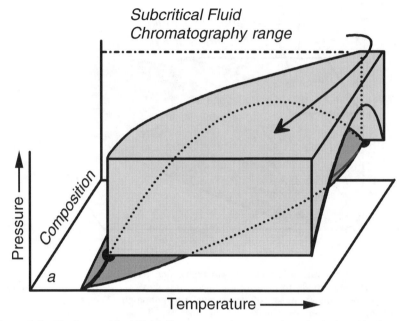

Figure 7.6   The Supercritical Fluid Chromatography range, above both the critical tempera-
ture and pressure at all compositions. (Reprinted with permission from reference 17.
Copyright 1997 American Chemical Society.) Reproduced by permission of the American
Chemical Society.

SFC (see Figure 7.6) occurs when both the critical temperature and critical pressure of the mobile phase are exceeded. (The locus of critical points is indicated in
Figure 7.2 by the dashed line over the top of the two-phase region. It is also visible
or partly visible in Figures 7.3 – 7.8). Compressibility, pressure tunability, and
diffusion rates are higher in SFC than in SubFC and EFLC, and are much higher
than in LC.

We have denoted the next region as *hyperbaric chromatography* (see Figure 7.7)
where temperatures are above the two-phase boundary and pressure is below critical.
In this region, the compressibility of the mobile phase is higher than in the previously described regions. However, the solvent is not as strong, in general, because
the decrease in pressure and the expansion of the compressible mobile phase
increase intermolecular distances and reduce solute – mobile-phase interactions.
Solvating gas chromatography (SGC) is a limiting case of hyperbaric chromatography in which the column outlet is allowed to default to one atmosphere (24).

Mobile phases with some solvating potential, such as CO_2 or ammonia, are necessary in SGC. Even though this technique is performed with ambient outlet pressure,
solutes can be separated at lower temperatures than in GC because the average pressure on the column is high enough that solvation occurs. Obviously, solute retention
is not constant in the column, and the local values of retention factors increase for all
solutes as they near the column outlet.

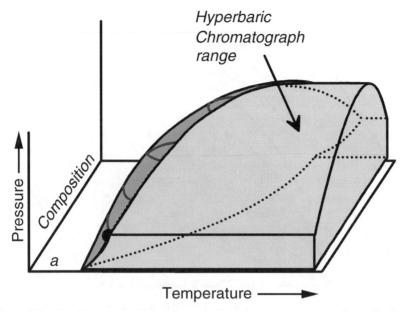

Figure 7.7 The Hyperbaric Chromatography region, at temperatures above the 2-phase region and pressures below the locus of critical pressures. Reproduced by permission of the American Chemical Society.

As we continue lowering the pressure, GC is the final limiting case when the mobile phase has zero solvent strength over the entire column length and where temperature is the only effective control parameter. Gas chromatography is shown in Figure 7.3.

Our view of unified chromatography simply involves realizing that the technique names and the boundaries defining them are arbitrary definitions lacking any real meaning with respect to fluid behavior, except perhaps when ambient conditions are invoked by default. In order to perceive unified chromatography, we simply eliminate the arbitrary boundaries separating the adjacent techniques and merge or unify all these techniques into one, as shown in Figure 7.8. We are also now completely free to select the conditions that produce the best separation for a particular problem from anywhere within this continuum of behavior without regard to naming conventions or arbitrary restrictions. We will probably want to avoid the two-phase region under most circumstances, and are limited in temperature and pressure only by the capabilities and stability of the instrument, stationary phase and solutes.

7.3   INSTRUMENTATION

The basic instrument required for packed-column unified chromatography is shown schematically in Figure 7.9. This is essentially a two-pump HPLC instrument utilizing high-pressure mixing with just a few new components. At least one pump must

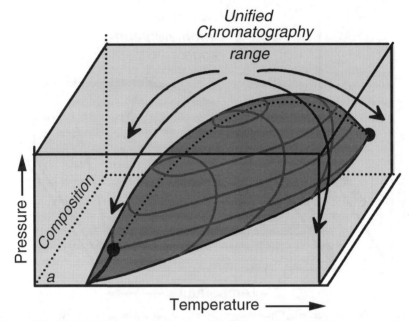

Figure 7.8 Unified Chromatography is achieved by realizing that the apparent barriers between the previously mentioned techniques are only arbitrary naming conventions, and that no real discontinuities exist between the techniques as long as the two-phase region is avoided. Reproduced by permission of the American Chemical Society.

be modified for handling (in addition to ordinary liquids) very volatile fluids such as CO_2, propane, and 1,1,1,2-tetrafluoroethane. Column temperature control must be added if not already provided. Finally, a device for controlling the mobile-phase pressure at the column or detector outlet is necessary.

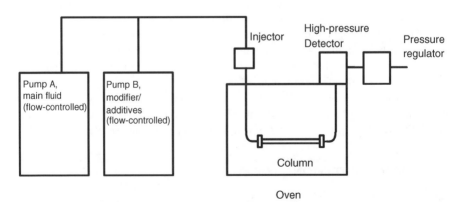

Figure 7.9 Schematic diagram of the basic instrumentation used for Unified Chromatography.

Very volatile fluids (such as CO_2 and others mentioned earlier) are often gases at ambient conditions. However, to be useful as mobile phases in unified chromatography they have to be pumped as liquids and later heated, if necessary. This requires delivering the fluids to the pump inlet under elevated pressure. The vapor pressure of the fluid is usually sufficient if the pump head is cooled $10-20\,°C$ below the temperature of the fluid reservoir. It is also believed that lowering the pump-head temperature reduces the possibility of cavitation occurring in the pump, but there is some disagreement on the overall role of pump temperature (25). Regardless, pumping is accomplished in an HPLC-like fashion very successfully with these fluids by cooling the pump head and compensating for the compressibility of the fluid.

Many HPLC instruments are already furnished with temperature controls for the column. Unified chromatography requires a much wider temperature range than is currently practiced in HPLC. Until better defined by experience, a temperature range from about -60 to about $350\,°C$ seems reasonable as a specification. Since this is well in the range of a GC oven with subambient temperature capability, no new technology is required.

Pressure is controlled by the use of a back-pressure regulator or programmable valve used in conjunction with a pressure transducer and placed under control of a pressure-regulating electronic circuit or a computer. Positioning the pressure-control point downstream (following the column outlet or the detector) allows the mobile-phase fluid to be independently flow-controlled at the column inlet. This provides the ability to carry out volumetric mixing of two fluid components and to program mobile-phase-composition gradients in a fashion exactly like that already done in HPLC (26).

When a flow-through detector that can be operated at outlet pressure is used, such as a UV or fluorescence detector, it is most effective to control pressure at the detector outlet. In these cases, the pressure drop through the detector is negligible, so the column outlet pressure is still controlled. Since this arrangement does not require lowering the pressure before detection, the possibility of solute precipitation or an unwanted phase change occurring during transport to the detector is minimized. In addition, transport through a pressure regulator or valve always introduces some degree of mixing which, when significant, contributes to peak broadening and loss of resolution. This worry is eliminated when pressure control follows detection.

When low-pressure detectors such as an evaporative-light-scattering detector or a mass spectrometer are necessary, other means are preferred for controlling pressure and delivering solutes to the detector to eliminate the possibility for solute precipitation or peak broadening. One of these means involves replacing the pressure regulator with a tee-joint in which the less-volatile mobile-phase component is mixed under pressure control from a third pump (27). A flow restrictor must be added at the inlet of the low-pressure detector to prevent the pressure from dropping until the last instant and to limit the amount of make-up fluid needed to maintain the pressure at the set point. Care must be taken with the order of mixing and temperature change to make sure the mobile phase never enters the two-phase region anywhere on the path between the column outlet and the detector inlet. Other arrangements also work well

(28), especially when the solutes have some degree of volatility. However, attention to mass-transfer is necessary when non-volatile or even low-volatility solutes must be transported to and detected by low-pressure detectors.

All of these changes to the underlying HPLC instrument are already practiced very successfully in the technique we call packed-column SFC. It is possible to perform conventional HPLC on an SFC instrument just by choosing HPLC mobile-phase components, setting the temperature to a value appropriate for conventional HPLC, and setting the outlet pressure to ambient. Of course, this is not a surprise to us if we are convinced that the SFC instrument actually has all of the essential capabilities of a unified chromatograph and has simply been misnamed, and if we agree that HPLC is just one specific application of unified chromatography.

GC is the most abbreviated form of unified chromatography. GC requires the least and cheapest equipment, and provides the fastest analyses for small, volatile solutes. Therefore, it is likely to continue flourishing as a technique practiced separately from the others.

7.4 ADVANTAGES OF AND CHALLENGES FOR UNIFIED CHROMATOGRAPHY TECHNIQUES IN MULTIDIMENSIONAL SYSTEMS

Here we will limit the scope of discussion to non-ionic solutes. Interactions of solutes with stationary and mobile phases are determined by fundamental inter-molecular forces. Orientation forces, when present, are important contributors and often influence selectivity between solutes. These forces include hydrogen bonding (subdivided into hydrogen-bond donating and accepting) and dipole–dipole interactions not involving hydrogen bonding. On many bonded stationary phases, the orientation influence is provided by specific functional groups occupying a small fraction of the entire stationary phase. An example is the hydrogen-bonding interaction of a cyano group at the end of a cyanopropyl chain substituted onto a silicone polymer. Induction forces, such as between a permanent dipole and a polarizable group, are weaker but are often encountered, for example when a phenyl-containing stationary phase interacts with an alcohol. The dispersion force, arising from the temporary dipoles occurring randomly in all molecules, is the weakest intermolecular force, but often dominates retention since all parts of the stationary phase contribute.

In thinking about performing multidimensional separations within the framework of unified chromatography, we must think about using all available tuning opportunities to maximize the differences in the separation mechanisms in the successive parts of the process. The following is just one example.

Normal-phase LC tends to separate according to solute polarity since the stationary phase is polar and retention is often dominated by hydrogen bonding. Thus, normal-phase LC is useful in sorting out classes of materials according to the polarity of the solutes. Fatty acids are easily separated from monoglycerides, but the separation of individual saturated fatty acids from each other on the basis of their carbon content is more difficult. In theory, GC performed with a highly deactivated non-polar column, would provide a high degree of orthogonalilty to normal-phase LC. Retention on a non-polar GC column would correlate strongly with the molecular weight (and thus the carbon number) of the fatty acids.

Now, let us expand this thinking into other realms of unified chromatography. Suppose we have a problem similar to the fatty acids, but with solutes which for some reason are not amenable to GC. What other techniques provide a high degree of orthogonality to normal-phase LC but are more widely applicable than GC? We would need a separation dominated by dispersion interactions. Thus, a highly deactivated, non-polar stationary phase is still appropriate. The mobile phase should also be either non-polar or low in polarity so that solutes are distributed on the basis of dispersion interactions. If we needed to separate quickly in the second dimension, as the case would be in a comprehensive, two-dimensional separation, we would need to select a non-polar mobile phase with the highest possible diffusion rate. If GC were not applicable, our needs would lead us toward a low-polarity mobile phase such as CO_2 or perhaps a small alkane used at the highest temperature and lowest pressure that would achieve our needs. Of course, there may be an overload problem in a case like this where we have purposely mismatched very polar solutes such as fatty acids with non-polar stationary and mobile phases in the second dimension. Some trade-off may be necessary between orthogonality and practicality, here perhaps by selecting a somewhat more polar stationary phase (for example, with a little phenyl substitution) to reduce overloading. Small concentrations of a polar modifier or of an acid or base may also be necessary in the mobile phase to produce symmetric peaks, depending on the characteristics of the stationary phase and the solute retention mechanism.

There is much that needs to be explored here. In general, it seems that to make the most of two-dimensional unified chromatography we will have to choose the two dimensions to have the biggest differences in selectivity that we can manage while trying to maximize the analysis speed.

Rather than looking just for new opportunities for orthogonality, it may also be fruitful to look for ways to make existing approaches work faster. For example, if reversed-phase LC is already used with some degree of success in one dimension of a multidimensional separation, it can often be made faster by increasing the temperature and lowering the viscosity of the mobile phase, such as with the addition of CO_2 or fluoroform (29). Normal-phase LC can benefit from the addition of CO_2 to the mobile phase, up to total replacement, as long as solute retention does not go too high. Additional selectivity may be possible by switching to a mobile phase that will not freeze or become so viscous at low temperatures that pumping becomes impossible, then lowering the column temperature significantly. There are over 40 current references regarding temperature effects on chiral separations using unified chromatography techniques, with several examples at temperatures as low as -50 °C (30, 31).

Therefore, with unified chromatography we are not really reinventing separations so much as taking advantage of the widening possibilities of conditions. It is perfectly suitable to start development of a separation by using HPLC, particularly if there is history of the problem at hand and an earlier solution on which to build. Then we can simply investigate the possibilities of changing temperature, pressure, and mobile-phase composition in our search for achieving our goals for separation in the least possible time.

7.5 COLUMN EFFICIENCY AND PLATE HEIGHTS IN UNIFIED CHROMATOGRAPHY

The plate height in chromatography, H, is a useful indicator of the rate of peak broadening. The local plate height at any point on a column is given by the following:

$$H = d(\sigma^2)/dz \qquad (7.1)$$

where σ^2 is the spatial variance of the peak along the column axis and z is the distance of the peak center from the column inlet (32). If H is constant and there are no gradients, then:

$$H = \sigma^2/L \qquad (7.2)$$

where σ^2 is determined at the column outlet and L is the column length.

There have been a few reports of column efficiency and reduced plate height measurements in several unified chromatography techniques. These have been based on the *apparent* plate height observed at the column outlet. In the notation used by Giddings (32) the apparent plate height, $\hat{H}$, is given by the following:

$$\hat{H} = L\left(\frac{\tau}{t_R}\right)^2 \qquad (7.3)$$

where L is the column length, t_R is the retention time of the peak in question, and τ is the standard deviation of the retention time. (This standard deviation is a measure of the width of one peak in time units, and not our ability to reproduce a t_R value with repeated trials of the same experiment.)

Chromatographers know well that plate height measurements carried out in the manner of equation (7.3) are indicative of the actual column plate height only when the column is completely uniform, that is, when there are no gradients. In practice, this no-gradient rule is often erroneously interpreted as a *no-program* rule: if all of the adjustable experimental parameters are kept constant, then we, too frequently, assume we can accurately measure the column plate height or efficiency by observing the peaks and their widths at the column outlet. Giddings and co-workers showed that this is not true in GC because of the velocity gradient that exists in the column (33).

This velocity gradient arises from the pressure gradient necessary to create mobile-phase flow. The pressure drops as we travel from the column inlet to the outlet. The gaseous mobile phase expands according to Boyle's law as it decompresses. Thus, the mobile-phase velocity increases monotonically from the column inlet to the outlet. Giddings and co-workers derived two new factors to correct the velocity-independent and velocity-dependent terms of the plate-height equations for predicting the observed plate height on columns with pressure drops (33). This correction is small in GC, usually amounting to only a few percent (33).

Pressure-drop correction is totally unnecessary in HPLC because, although there is a huge pressure drop in comparison to GC, the mobile phase is nearly incompressible. Thus, no significant velocity gradient results. However, in unified chromatography the departure of the apparent plate height from any meaningful measure of column performance is more likely than in LC or GC. We must not use equation (7.1) recklessly to estimate the true plate height or any average. Let's look into this a little more closely.

First, we need to distinguish between two kinds of gradients, i.e. temporal and spatial. Temperature programming in GC is an example of a pure temporal gradient. The column temperature is changed as a function of time, and at any particular time the column temperature is the same, spatially, over the entire column. Since local retention factors for a solute vary only as a function of temperature on any particular GC column if it is uniformly coated or packed, the retention factor for a given solute will be the same at all locations on such a column held at a constant temperature. This means that the ratio of the average velocity of the solute molecules relative to the local mobile-phase velocity will also be constant everywhere on the column. Likewise, if we look at a single peak while it is on the column, and examine the local retention factor spatially across the width of the peak, we would find the retention factor is constant over the entire peak. *There is no spatial component in a temperature program in (conventional) GC.*

This is not the case in gradient-elution LC. Here the gradients are obviously temporal since the mobile-phase composition is programmed to change as a function of time. However, there is also a spatial component: at any given time, the mobile phase at a reference point on the column is older than the mobile phase upstream from that point. If a (temporal) gradient of the modifier solvent is programmed at 5%/min, then the mobile phase that is one minute upstream from our reference point will be 5% stronger. If we examine a peak on an LC column at a particular location and point in time while it is in the midst of a gradient, we would find, as we look downstream toward the leading edge of the peak, that the mobile phase would weaken. Similarly, looking upstream toward the trailing edge of the peak, the mobile phase would be increasingly stronger.

The consequence of this *spatial* component of the gradient is that the local retention factor of the solute will be lower on the peak's leading edge than on its trailing edge. If we assume the LC mobile phase is not compressible so that the mobile-phase velocity is uniform, and if we temporarily disregard peak broadening phenomena, then the local value of the solute velocity would be faster on the trailing edge of the peak than on the leading edge. This means that the application of a mobile-phase gradient in HPLC contributes to a narrowing of the peaks spatially and temporarily. Of course, the peak-broadening processes are still at work in the column, opposing this focusing phenomenon. However, the focusing effect of the spatial component of the gradient tends to produce peaks that are narrower than they would otherwise be in a pure temporal gradient. (If the direction of the spatial gradient is reversed from what was described, then the peaks are defocused and broadened.)

Giddings wrote in 1963, '*The plate height theory is particularly limited when gradients exist in a column . . . These limitations do not, of course, make it necessary to abandon the plate height theory as a measure of column efficiency, but they do suggest that severe caution be taken in applying the plate height theory to new theoretical areas*' (34). Giddings' work at the time focused on GC. He explained the effects of several non-uniformities in GC columns on apparent plate height. The non-uniformities included pressure drop and the resulting velocity gradient on a GC column, inhomogeneous packings, and coupled columns. This work included the development of the pressure correction factors for GC mentioned earlier, and an explanation of the apparent plate height when dissimilar columns are coupled in series. However, in all of this work, Giddings treated cases in which retention factors were constant except when affected by the column.

There is a further complication in unified chromatography: even when conditions are not programmed, the pressure drop on a column induces a spatial pressure gradient as in GC, but because the mobile phase is compressible and solvating, its strength also changes spatially. This is a departure from the conditions that Giddings described. Local retention factors are not constant in unified chromatography, in general, except in the extremes (LC and GC). *Local retention factors may vary in unified chromatography even if the stationary phase is completely uniform, the mobile-phase composition is constant, and no form of temporal programming is applied.*

Poe and Martire expanded Giddings' work and derived equations for the observed plate height in general (35). Although these authors only claimed applicability to GC, LC, and SFC, it appears that their equations are applicable to all of the forms of unified chromatography mentioned earlier. Poe and Martire arrived at these equations by expressing Giddings' equations in terms of density rather than pressure (Boyle's law is not followed by the non-ideal fluids often used as unified chromatography mobile phases, although the density-volume product is always constant), and by including the density influence on local retention factors and local plate height. They reported the following:

$$\hat{H} = \langle (H(1 + k)^2 \rho^2)_z / ((1 + k)\rho) \rangle_z^2 \tag{7.4}$$

where H is the local plate height on the column, k is the solute retention factor, ρ is the mobile-phase density, and the brackets, $\langle \ \rangle_z$ denote that the spatial average has been taken along the length of the column. Note that k may change as a function of ρ, H may change as a function of ρ and of k, and ρ will change with z on a column with a significant pressure drop between the inlet and outlet if the fluid is compressible. We should also note that equation (7.4) properly reduces to an agreement with Giddings' treatment in the GC limit.

Expressing the averages in equation (7.4) with integrals, we obtain:

$$H = L \left[\int_0^L (1 + k)^2 r^2 dz \middle/ \int_0^L (1 + k) r dz\right]^2 \qquad (7.5)$$

What conditions would be necessary for the apparent plate height to match the true plate height? The first immediately obvious instance is when H, k and ρ are all constant. Then, the right-hand side of equation (7.5) reduces to H. If k and ρ were constant, or if the product $(1 + k)\rho$ were constant, then $\hat{H}$ would equal the spatial average of H.

The only unified chromatography technique in which k and ρ are both constant is LC. For all of the other unified chromatography techniques, k and ρ are not constant until reaching the other limiting case, i.e. GC, where k becomes constant again. Considerable effort would be required to express H, k, and ρ in terms of z (or to change the variables, if necessary to make integration easier, or even possible) before being able to calculate a value for $\hat{H}$. Thus, it is not obvious if $\hat{H}$ would be smaller or larger than the spatial average H or some particular local H value. However, it is unlikely that $\hat{H}$ would provide any useful measure of column performance without sorting out all of these complications.

The point of all this is simply that we must not use the apparent plate height or the apparent plate number as performance criteria in the unified chromatography techniques on the justification that they already work well for LC and that they work well for GC when a pressure correction is applied. A considerable expansion of theory and an effective means for evaluating equations (7.4) or (7.5) are required first. Likewise, as we consider multidimensional chromatography involving techniques existing between the extremes of LC and GC, we must not build judgments of the multidimensional system on unsound measures of the individual techniques involved.

REFERENCES

1. P. J. Slonecker, X. D. Li, T. H. Ridgway and J. G. Dorsey, 'Informational orthogonality of two-dimensional chromatographic separations', *Anal. Chem.* **68**: 682–689 (1996).
2. R. E. Boehm and D. E. Martire, 'A unified theory of retention and selectivity in liquid chromatography. 1. Liquid-solid (adsorption) chromatography', *J. Phys. Chem.* **84**: 3620–3630 (1980).
3. D. E. Martire and R. E. Boehm, 'A unified theory of retention and selectivity in liquid chromatography. 2. Reversed-phase liquid chromatography with chemically bonded phases', *J. Phys. Chem.* **87**: 1045–1062 (1983).
4. D. E. Martire and R. E. Boehm, 'Unified molecular theory of chromatography and its application to supercritical fluid mobile phases. 1. Fluid-liquid (absorption) chromatography', *J. Phys. Chem.* **91**: 2433–2446 (1987).
5. D. E. Martire, 'Unified theory of absorption chromatography: gas, liquid and supercritical fluid mobile phases', *J. Liq. Chromatogr.* **10**: 1569–1588 (1987).
6. D. E. Martire, 'Unified theory of adsorption chromatography: gas, liquid and supercritical fluid mobile phases', *J. Chromatogr.* **452**: 17–30 (1988).

7. D. E. Martire, 'Unified theory of adsorption chromatography with heterogeneous surfaces: gas, liquid and super-critical fluid mobile phases', *J. Liq. Chromatogr.* **11**: 1779–1807 (1988).

8. D. Ishii and T. Takeuchi, 'Unified fluid chromatography', *J. Chromatogr. Sci.* **27**: 71–74 (1989).

9. D. Ishii, T. Niwa, K. Ohta and T. Takeuchi, 'Unified capillary chromatography. *J. High. Resolut. Chromatogr. Chromatogr. Commun.* **11**: 800–801 (1988).

10. R. E. Robinson, D. Tong, R. Moulder, K. D. Bartle and A. A. Clifford, 'Unified open tubular column chromatography: sequential gas chromatography, at normal pressures and supercritical fluid chromatography on the same column', *J. Microcolumn. Sep.* **3**: 403–409 (1991).

11. D. Tong, K. D. Bartle, A. A. Clifford and R. E. Robinson, 'Unified gas and supercritical fluid chromatography on 50 μm i.d. columns, *J. High Resolut. Chromatogr.* **15**: 505–509 (1992).

12. D. Tong, K. D. Bartle, R. E. Robinson and P. Altham, 'Unified chromatography in petroleum analysis, *J. Chromatogr. Sci.* **31**: 77–81 (1993).

13. D. Tong and K. D. Bartle, 'Band broadening during mobile phase change in unified chromtography (GC–SFC)', *J. Microcolumn Sep.* **5**: 237–243 (1993).

14. D. Tong, K. D. Bartle and A. A. Clifford, 'Principles and application of unified chromatography', *J. Chromatogr.* **703**: 17–35 (1995).

15. D. Tong, K. D. Bartle, A. A. Clifford and R. E. Robinson, 'Unified chromatograph for gas chromatography, supercritical fluid chromatography and micro-liquid chromatography, *Analyst* **120**: 2461–2467 (1995).

16. J. C. Giddings, *Unified Separation Science,* John Wiley & Sons, New York (1991).

17. T. L. Chester, 'Chromatography from the mobile-phase perspective', *Anal. Chem.* **69**: 165A–169A (1997).

18. T. L. Chester, 'Unified chromatography from the mobile phase perspective', in *Unified Chromatography*, J. F. Parcher and T. L. Chester (Eds), ACS. Symposium Series 748, American Chemical Society, Washington, DC, pp. 6–36 (2000).

19. R. C. Reid, J. M. Prausnitz and B. E. Poling, *The Properties of Gases and Liquids*, McGraw-Hill, New York (1987).

20. V. L. McGuffin, C. E. Evans and S. H. Chen, 'Direct examination of separation processes in liquid-chromatography: effect of temperature and pressure on solute retention', *J. Microcolumn Sep.* **5**: 3–10 (1993).

21. P. H. van Konynenburg and R. L. Scott, 'Critical lines and phase equilibriums in binary Van der Waals mixtures', *Philos. Trans. R. Soc. London*, **298**: 495–540 (1980).

22. Y. Cui and S. V. Olesik, 'High-performance liquid chromatography using mobile phases with enhanced fluidity', *Anal. Chem.* **63**: 1812–1819 (1991).

23. S. V. Olesik, 'Applications of enhanced-fluidity liquid chromatography in separation science: an update', in *Unified Chromatography*, J. F. Parcher and T. L. Chester (Eds), ACS. Symposium Series 748, American Chemical Society, Washington, DC, pp 168–178 (2000).

24. Y. F. Shen and M. L. Lee, 'High-efficiency solvating gas chromatography using packed capillaries', *Anal. Chem.* **69**: 2541–2549 (1997).

25. T. A. Berger, 'Practical advantages of packed column supercritical fluid chromatography in supporting combinations chemistry', in *Unified Chromatography*, J. F. Parcher and T. L. Chester (Eds), ACS Symposium Series 748, American Chemical Society, Washington, DC, pp. 203–233 (2000).

26. K. Anton, M. Bach and A. Geiser, 'Supercritical fluid chromatography in the routine stability control of antipruritic preparations', *J. Chromatogr.* **553**: 71–79 (1991).

27. T. L. Chester and J. D. Pinkston, 'Pressure-regulating fluid interface and phase behavior considerations in the coupling of packed-column supercritical fluid chromatography with low-pressure detectors', *J. Chromatogr.* **807**: 265–273 (1998).

28. B. Herbreteau, A. Salvador, M. Lafosse and M. Dreux, 'SFC with evaporative light-scattering detection and atmospheric-pressure chemical-ionisation mass spectrometry for methylated glucoses and cyclodextrins analysis, *Analusis* **27**: 706–712 (1999).

29. J. Zhao and S. V. Olesik, 'Phase diagram studies of methanol–CHF_3 and methanol–H_2O–CHF_3 mixtures, *Fluid Phase Equilib.* **154**: 261–284 (1999).

30. F. Gasparrini, F. Maggio, D. Misiti, C. Villani, F. Andreolini and G. P. Mapelli, 'High performance liquid chromatography on the chiral stationary phase, (*R,R*)-DACH-DNB using carbon dioxide-based eluents', *J. High Resolut. Chromatogr.* **17**: 43–45 (1994).

31. C. Wolf and W. H. Pirkle, 'Enantioseparations by subcritical fluid chromatography at cryogenic temperatures', *J. Chromatogr.* **785**: 173–178 (1997).

32. J. C. Giddings, *Dynamics of Chromatography. Part I: Principles and Theory,* Marcel Dekker, New York (1965).

33. G. H. Stewart, S. L. Seager and J. C. Giddings, 'Influence of pressure gradients on resolution in gas chromatography', *Anal. Chem.* **31**: 1738 (1959).

34. J. C. Giddings, 'Plate height of nonuniform chromatographic columns. Gas compression effects, coupled columns and analogous systems', *Anal. Chem.* **35**: 353–356 (1963).

35. D. P. Poe and D. E. Martire, 'Plate height theory for compressible mobile phase fluids and its application to gas, liquid and supercritical fluid chromatography', *J. Chromatogr.* **517**: 3–29 (1990).

8 Multidimensional Planar Chromatography

Sz NYIREDY

Research Institute for Medicinal Plants, Budakalász, Hungary

8.1 INTRODUCTION

Unquestionably, most practical planar chromatographic (PC) analytical problems can be solved by the use of a single thin-layer chromatographic (TLC) plate and for most analytical applications it would be impractical to apply two-dimensional (2-D) TLC. One-dimensional chromatographic systems, however, often have an inadequate capability for the clean resolution of the compounds present in complex biological samples, and because this failure becomes increasingly pronounced as the number of compounds increases (1), multidimensional (MD) separation procedures become especially important for such samples.

Multidimensional planar separation exploits combinations of different separation mechanisms or systems (2, 3); such methods can generally be developed by combining almost any of the different chromatographic mechanisms or phases (stationary and/or mobile), electrophoretic techniques, and field-flow fractionation sub-techniques (4). According to Giddings (5), the correct definition of multidimensional chromatography includes two conditions. "First, it is one in which the components of a mixture are subjected to two or more separation steps in which their displacements depend on different factors. The second criterion is that when two components are substantially separated in any single step, they always remain separated until the completion of the separative operation." This latter condition, therefore, precludes simple tandem arrangements in which compounds separated in the first separation system can re-merge in the second (3). The following modes have most frequently been used for multidimensional separations (6) involving planar chromatography:

- 2-D development on the same monolayer stationary phase with mobile phases characterized by different total solvent strength (S_T) and selectivity values (S_V);
- 2-D development on the same bilayer stationary phase either with the same mobile phase or with mobile phases of different composition;
- multiple development (^nD) in one, two, or three dimensions on the same monolayer stationary phase with mobile phases characterized by different solvent strengths and selectivity values;

Multidimensional Chromatography, edited by L. Mondello, A. C. Lewis and K. D. Bartle
©2002 John Wiley & Sons Ltd.

- coupled layers with stationary phases of decreasing polarity developed with a mobile phase of constant composition;
- a combination of at least two of the above-mentioned modes;
- automated coupling of two chromatographic techniques in which PC is used as the second dimension and another separation method, e.g. gas chromatography (GC), high-performance liquid chromatography (HPLC), etc., as the first.

In the following discussion, two-dimensional and multidimensional planar chromatography are defined, the above-mentioned possibilities are discussed in detail, and attention is drawn to the different possibilities, as well as the advantages and limitations of the various modes of multidimensional planar chromatography.

8.2   TWO-DIMENSIONAL OR MULTIDIMENSIONAL PLANAR CHROMATOGRAPHY?

Multidimensional (or coupled) column chromatography is a technique in which fractions from a separation system are selectively transferred to one or more secondary separating systems to increase resolution and sensitivity, and/or to reduce analysis time. The application of secondary columns is illustrated schematically in Figure 8.1. The smaller the Δt_r value applied, then the greater is the resolution and number of runs needed to check a certain portion of the sample (5).

In planar chromatography, the fractions are not always transferred to another separation system, but rather a secondary separation is developed, orthogonally on the same chromatographic plate. Therefore, for all substances not completely separated it is possible that baseline separation can be achieved by means of a second separation process with an appropriate mobile (stationary) phase. Figure 8.2 shows that in the second dimension a theoretically unlimited number of secondary 'columns' can be applied. Because of this, the terminology 'two-dimensional PC' is not sufficiently

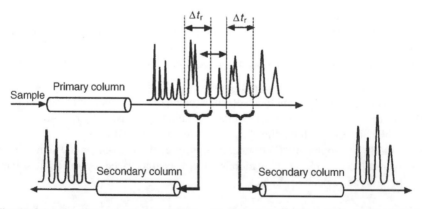

Figure 8.1   Coupled column liquid chromatographic system in which compounds are passed from the primary column to secondary columns in a time Δt_r.

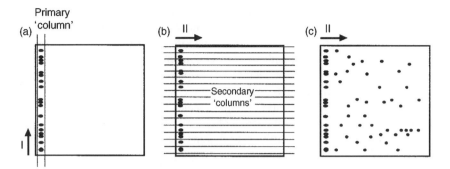

Figure 8.2 Schematic illustration of a bi-directional, multidimensional planar chromato-graphic system in which the number of 'secondary columns' is unlimited.

expressive; instead, the term 'multidimensional planar chromatography' (MD-PC) should be used.

MD-PC is highly important in its own right, because this is the only real multi-dimensional separation method in which all compounds can be passed to a next dimension. It therefore serves as the reference system (7) against which all other multidimensional systems can be compared.

8.3 TWO-DIMENSIONAL DEVELOPMENT ON SINGLE LAYERS

CHARACTERISTICS OF 2-D SEPARATIONS

The term bi-directional or two-dimensional chromatography has its origin in TLC; it was defined as chromatographic development in one direction followed by a second development in a direction perpendicular to the first (8). The method consists of spotting a sample at the corner of a chromatographic plate (Figure 8.3(a)) and enabling migration of the mobile phase (characterized by S_{T1}; S_{V1}) in the first direc-tion (Figure 8.3(b)). After drying, sequential development of the plate, in a direction at right angles to the first development (Figure 8.3(c)), can be started with a second mobile phase, (characterized by S_{T2}; S_{V2}). The overall separation obtained is superior

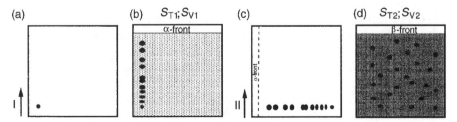

Figure 8.3 Schematic illustration of a classical, bi-directional or two-dimensional planar chromatographic procedure.

if, as a result of the use of mobile phases of different composition, the interactive forces which bring about retention are different for the two consecutive developments. A good separation can be obtained when the surface area of the plate over which the spots are spread is relatively large (Figure 8.3(d)).

Thus, a 2-D separation can be seen as 1-D displacement operating in two dimensions. The 2-D TLC separation is of no interest if selection of the two mobile phases is not appropriate. With this in mind, displacement in either direction can be either selective or non-selective. A combination of two selective displacements in 2-D TLC will lead to the application of different separating mechanisms in each direction. As an extreme, if the solvent combinations are the same ($S_{T1} = S_{T2}$; $S_{V1} = S_{V2}$) or very similar ($S_{T1} \sim S_{T2}$; $S_{V1} \sim S_{V2}$), the compounds to be separated will be poorly resolved or even unresolved, and as a result a diagonal pattern will be obtained. In such circumstances, a slight increase in resolution might occur, because of an increase by a factor of $\sqrt{2}$ in the distance of migration of the zone (4).

The point at which the sample is spotted can be regarded as the origin of a coordinate system (9). The process of development is performed in two steps; the first in the direction of the x-axis to a distance L_x. After evaporation of the solvents used, the second development will be performed in the direction of the y-axis to a distance L_y. The positions of the compounds after development in the x-direction depend on the S_T and S_V values of the first mobile phase being applied. Similarly, the migration distances of the individual compounds also depend on the total solvent strength and total selectivity of the second mobile phase. After development in the x-direction, the ordinates of all compounds are zero. After development in the y-direction, their abscissa values follow from their positions on the x-axis after the first development. The final positions of the spots are thus determined by the coordinates $x(i)$ and $y(i)$, which can be expressed as follows:

$$R_{fxy(i)} = R_{fx(i)}, R_{fy(i)} \tag{8.1}$$

The principle of 2-D TLC separation is illustrated schematically in Figure 8.4. The multiplicative law for 2-D peak capacity emphasizes the tremendous increase in resolving power which can be achieved; in theory, this method has a separating capacity of n^2, where n is the one-dimensional peak capacity (9). If this peak capacity is to be achieved, the selectivity of the mobile phases used in the two different directions must be complementary.

For two reasons, the peak capacity in 2-D TLC is less than the product of those of two one-dimensional developments (10). First, the sizes of the spots of the compounds being separated are always larger in the second development than was the initial sample spot. Secondly, during the second development the spots spread laterally and must therefore be separated with a resolution greater than unity at the beginning of the second development if they are to have a resolution of unity at the end (1). The separation efficiency can be increased by performing multiple development in one or both directions. Therefore, 2-D TLC combined with ^nD is a promising route to real improvements in planar chromatography in the future (see Section 8.13 below).

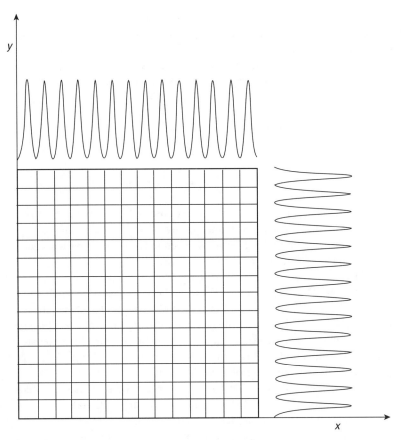

Figure 8.4 Schematic diagram of the peak capacity (n^2) of a 2-D planar chromatographic system (i) the number of squares represents the number of compounds which can theoretically be separated.

8.4 METHODS FOR THE SELECTION OF APPROPRIATE MOBILE PHASES

Nurok and his research group (11) proposed the use of a binary mixture of a strong and a weak solvent, because R_f can be predicted as a function of the solvent composition by the use of the following equation:

$$\log k = a \log X_s + b \tag{8.2}$$

where a and b are empirical constants, X_s is the mole fraction of the strong solvent and k is the capacity factor, which is related to R_f by the relationship:

$$R_f = \frac{1}{1 + k} \tag{8.3}$$

These workers used binary solvent systems over a range of mole fractions to determine, for each solute, the constants a and b of equation (8.2). For methyl and phenacyl esters, TLC was used, while overpressured layer chromatography (OPLC) was used for dansyl amino acids. Nurok and co-workers (11) also evaluated how the quality of a simulated separation varies with changing solvent strength by using the inverse distance function (*IDF*) or planar response function (*PRF*), as follows:

$$IDF = \sum_{i=1}^{k-1} \sum_{j=i+1}^{k} \frac{1}{SD^{ii}} \tag{8.4}$$

$$PRF = \sum_{i=1}^{k-1} \sum_{j=i+1}^{k} \ln \frac{SD^{ii}}{SD^{\text{SPEC}}} \tag{8.5}$$

where SD is the center-to-center spot separation and SD^{SPEC} is the spot separation desired for a mixture of k solutes. Larger values of SD^{SPEC} reflect the easier separation of a mixture containing fewer solutes, when many solute pairs will make a zero contribution to the *PRF* if too small a value of SD^{SPEC} is selected (11).

Härmälä *et al.* (12) reported a very easy method of searching for appropriate mobile phases for 2-D separations. These authors suggested starting the solvent selection according to the first part of the 'PRISMA' system for planar chromatography (13), with a selection of ten solvents chosen from the eight Snyder (14) selectivity groups, which can be evaluated in parallel in unsaturated chromatographic chambers. The compounds to be separated should be applied to the TLC plates in groups of three or four, for ease of identification. The solvent strength (s_i) of each neat solvent is adjusted individually, by the use of n-hexane, so that the zones of the compounds to be separated are distributed in the R_f range 0.2–0.8. The exact positions of the spots must be located densitometrically. If solvents afford good separation, their homologues or other solvents from the same group can also be tested. The R_f values of the compounds to be separated are obtained from the TLC runs by using the most promising binary solvents and must be compared with each other in a correlation matrix. The next step is validation of the solvent systems by regression analysis, where the solvents with the poorest correlation values are selected. After choice of an appropriate pair of solvents, the development technique (e.g. OPLC or rotation planar chromatography (RPC)) and other separation conditions must be selected, according to the third part of the 'PRISMA' system. In this way, an excellent separation can be achieved for closely related compounds in the minimum time, generally within a few hours.

8.5 TWO-DIMENSIONAL DEVELOPMENT ON BILAYERS

Use of a bilayer plate (15) affords the special chromatographic possibility of being able to perform two different multidimensional separations on the same chromatographic plate (Figure 8.5), either with the same mobile phase or with mobile phases of different composition.

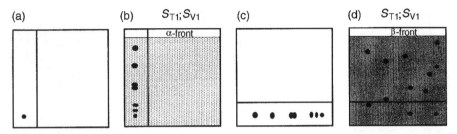

Figure 8.5 Schematic illustration of 2-D multiple development on a bilayer with the same mobile phase (S_{T1}; S_{V1}).

In the first version with a mobile phase of constant composition and with single developments of the bilayer in both dimensions, a 2-D TLC separation might be achieved which is the opposite of classical 2-D TLC on the same monolayer stationary phase with two mobile phases of different composition. Unfortunately, the use of RP-18 and silica as the bilayer is rather complicated, because the solvent used in the first development modifies the stationary phase, and unless it can be easily and quantitatively removed during the intermediate drying step or, alternatively, the modification can be performed reproducibly, this can result in inadequate reproducibility of the separation system from sample to sample. It is therefore suggested instead that two single plates be used. After the reversed-phase (RP) separation and drying of the plate, the second, normal-phase, plate can be coupled to the first (see Section 8.10 below).

8.6 MULTIPLE DEVELOPMENT IN ONE, TWO OR THREE DIMENSIONS

When multiple development is performed on the same monolayer stationary phase, the development distance and the total solvent strength and selectivity values (16) of the mobile phase (17) can easily be changed at any stage of the development sequence to optimize the separation. These techniques are typically fully off-line modes, because the plates must be dried between consecutive development steps; only after this can the next development, with the same or different development distances and/or mobile phases, be started. This method involves the following stages:

- Repeated chromatography of the sample in the same direction with the chromatographic plate being dried before all re-developments.
- Repeated chromatography of the sample in the first direction, followed by repeated development in the second direction, at right angles to the first. Again, the chromatographic plate is dried between re-developments.
- Repeated chromatography in a third dimension after completion of two-dimensional development. Here, development in the first, second, and third dimensions can be envisaged as occurring on three 'plates' arranged in the form of a cube; the plate is again dried between developments.

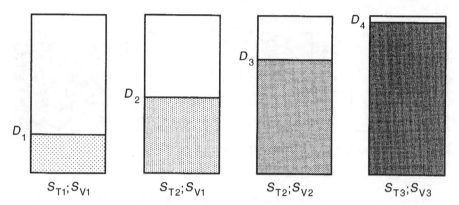

Figure 8.6   The basic possibilities of ^nD in one-dimensional TLC; the shading illustrates variation in the mobile phase composition.

8.7   MULTIPLE DEVELOPMENT IN ONE DIRECTION

Figure 8.6 shows a schematic diagram of linear ^nD on a monolayer stationary phase, in which the composition of the mobile phase, characterized by the total solvent strength (S_T) and total selectivity value (S_V) (16) is changed for each of the four steps and in which the development distance increases linearly. As shown in the figure, between the first and second developments not only is the migration distance changed, but S_T is also changed, at constant mobile phase total selectivity (17). For the third chromatographic step at constant total solvent strength, the S_V was changed (18). For the fourth step, both values characterizing the mobile phase were changed.

Clearly, the number of re-chromatography steps, the development distance, and the total solvent strength and/or selectivity value of the mobile phase can be freely varied, depending on the separation problems (19), as summarized in Table 8.1.

The efficiency of the ^nD is partly a consequence of the zone refocusing mechanism, as depicted in Figure 8.7. Each time the solvent front traverses the stationary sample in multiple development it compresses the zone in the direction of development. The compression occurs because the mobile phase first contacts the bottom edge of the zone, where the sample molecules start to move forward before those

Table 8.1   Characterization of the methods of multiple development

MD method	Abbreviation	Development distance	Mobile phase composition
Unidimensional	UMD	D, constant	S_{T1} and S_{V1}, constant
Incremental	IMD	Increasing, $(D_1 \rightarrow D_n)$	S_{T1} and S_{V1}, constant
Gradient	GMD	D, constant	$S_{T1} \rightarrow S_{Tn}$, $S_{V1} \rightarrow S_{Vn}$
Bivariate	BMD	Increasing, $(D_1 \rightarrow D_n)$	$S_{T1} \rightarrow S_{Tn}$, $S_{V1} \rightarrow S_{Vn}$

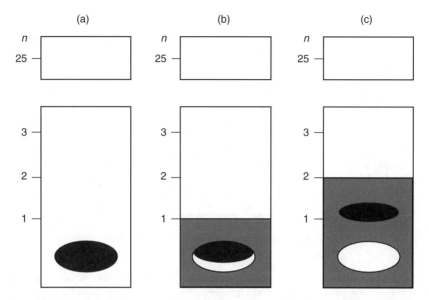

Figure 8.7   Schematic diagram of the zone refocusing mechanism: (a) the applied sample before starting the separation; (b) the solvent front reaches the first stage ($n = 1$), where the sample starts to refocus; (c) the solvent front reaches the second stage ($n = 2$), where all compounds together start to migrate.

molecules still ahead of the solvent front (10). When the solvent front moves beyond the front edge of the zone, the refocused zone starts to migrate and is broadened by diffusion in the normal way. By use of optimum conditions a balance between zone refocusing and zone broadening can be achieved.

A detailed description of the versatility of multiple development techniques in one dimension has been given by Szabady and Nyiredy (18). These authors compared conventional TLC with unidimensional (UMD) and incremental (IMD) multiple development methods by chromatographing furocoumarin isomers on silica using chloroform as the monocomponent mobile phase. The development distance for all three methods was 70 mm, while the number of development steps for both of the ^{n}D techniques was five. Comparison of the effects of UMD and IMD on zone-centre separation and on chromatographic zone width reveals that UMD increases zone-centre separation more effectively in the lower R_f range, while IMD results in narrower spots (Figure 8.8).

For the analysis of multicomponent mixtures spanning a wide polarity range, the simplest form of stepwise gradient development (GMD) is required (18). Separation of components into fractions of increasing polarity can be achieved by ^{n}D with the same chromatographic development distance and a mobile phase gradient of increasing solvent strength. In the first chromatographic step the full length of the layer is developed with the weakest solvent system, optimized for the separation of non-polar compounds. Semi-polar and polar compounds do not migrate with this mobile phase.

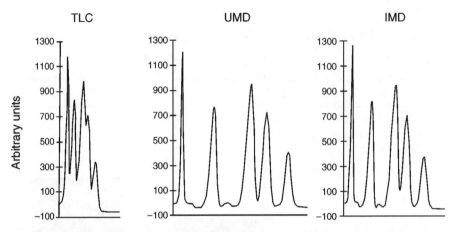

Figure 8.8   Comparison of the effects of UMD and IMD on separation.

After detection of the separated non-polar components, the semi-polar compounds can be separated in the next chromatographic step by the use of an optimized solvent system of medium polarity. With this mobile phase of medium strength, the non-polar components separated in the previous step migrate close to the solvent front, whereas the polar components remain at the origin. The separation of the most polar compounds is executed – after detection of the compounds separated by the preceding chromatographic step – in the next development step with the mobile phase of highest S_T. By use of GMD with a mobile phase gradient of increasing solvent strength, the authors (18) separated furocoumarin isomers, flavonoid aglycones, and flavonoid glycosides from the same sample (Figure 8.9).

Fractionation of components into polarity groups, and their optimized separation (followed by detection) by subsequent development steps increases the separating capacity of the chromatographic system.

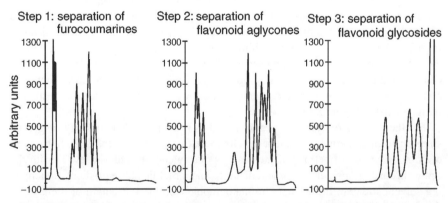

Figure 8.9   GMD separation of various naturally occurring compounds of wide polarity range from the same sample.

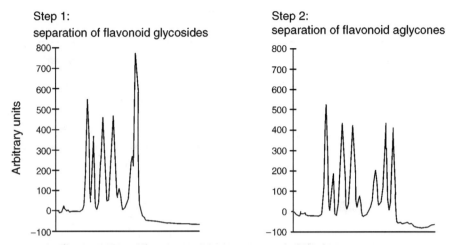

Figure 8.10   Separation of polyphenolic compounds by BMD separation.

Bivariate multiple development (BMD) with a mobile phase gradient of decreasing S_T is preferable for the analysis of less complex mixtures of wide polarity, because the complete chromatographic separation finally obtained can now be detected as a single chromatogram. In BMD, the first development is performed over the shortest distance by using the mobile phase system with the greatest S_T; the development distance is increased and the S_T is then reduced with each subsequent development until the final development is performed over the longest distance with the weakest mobile phase. By use of BMD with a mobile phase gradient of decreasing polarity we have separated (18) polyphenolic compounds spanning a wide polarity range, in two chromatographic steps (Figure 8.10).

In the first development step, the more polar components were separated by a 60 mm chromatographic development with a mobile phase of high S_T; the polar compounds were distributed between the origin and the solvent front of the first chromatographic step (18). The moderately polar compounds were transported close to the solvent front by this mobile phase. After complete drying of the plate, in the second chromatographic step the layer was developed for a greater distance (80 mm) with a mobile phase optimized for the separation of compounds of lower polarity. This second mobile phase with a lower S_T does not significantly affect the position of polar components eluted in the first development step, but resolves the compounds of lower polarity between the fronts of the first and second development steps.

The basis of automated multiple development (AMD) is the use of different modes of multiple development in which the mobile phase composition (S_T and S_V values) is changed after each, or several, of the development steps. Figure 8.11 illustrates the principle of AMD employing a negative solvent-strength gradient (decreasing S_T values).

In the first step, the compounds to be separated were moved and reconcentrated. In subsequent steps, both values characterizing the mobile phase were changed

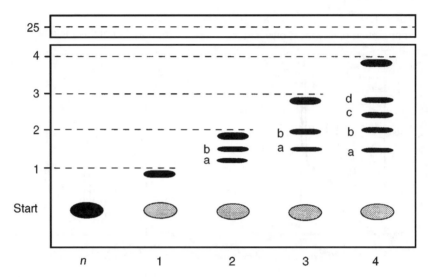

Figure 8.11   Schematic diagram of the course of AMD separation with gradient elution.

and, as a consequence, the more non-polar compounds always migrate near the α-front. By correct selection of the total solvent strength and total selectivity values, all of the compounds can finally be separated. AMD under controlled conditions results in high reproducibility while the convenient facility for vacuum drying of the chromatographic plate and the use of a nitrogen atmosphere reduce the chance of degradation during multiple development. Unfortunately, separations will be slow because of the large number of development and intermediate drying steps; the increase in zone capacity is, however, significant.

In the absence of a suitable method of optimization of the mobile phase for AMD (19), the procedure generally used is to start with 100% methanol ($S_T = 5.1$, $S_V = 2.18$) and to reduce its concentration in 15 stages, during which procedure the amount of solvent B (diethyl ether or dichloromethane) will increase from 0 to 100%. From the 15th to the 25th step, the concentration of solvent B is reduced and the concentration of n-hexane increased, until it reaches 100% (Figure 8.12(a)).

Figure 8.12(b) shows that if dichloromethane ($S_T = 3.1$, $S_V = 1.61$) or diethyl ether ($S_T = 2.8$, $S_V = 4.08$) are used, then the change in solvent strength is not very large. If, however, selectivity values are considered (Figure 8.12(c)) it is apparent that the selectivity value decreases when dichloromethane is used and increases when diethyl ether is used. It can be concluded that selectivity values must also be considered in the search for suitable solvents for ^nD.

8.8   MULTIPLE DEVELOPMENT IN TWO DIRECTIONS

Unfortunately, the fact that, in addition to the 2-D separation, a further increase in separation performance might be obtained by the use of multiple development

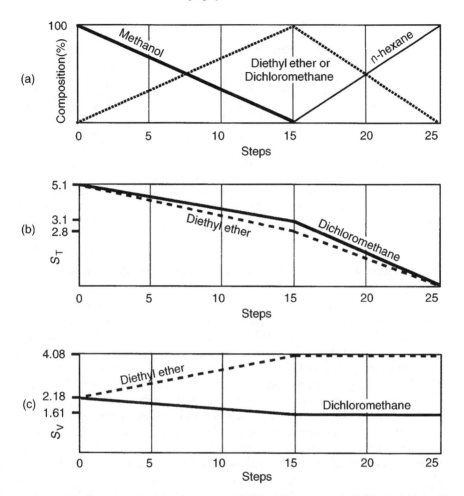

Figure 8.12 The 'universal' elution gradient during the stages of AMD separation: (a) change in mobile phase composition; (b) change of solvent strength; (c) change of selectivity value.

(20–22) for either or both orthogonal developments seems to have been rarely recognized. Figure 8.13(a) shows a sample applied to the chromatographic plate. After 'n' developments, during which the total solvent strength was reduced step-wise at constant selectivity (Figure 8.13(b)), the compounds were separated according to differences in polarity. In the second dimension, perpendicular to the first (Figure 8.13(c)), the chromatographic plate was redeveloped 'm' times such that the selectivity was changed at constant S_T to achieve maximum spot capacity.

2-D TLC combined with multiple development is therefore a promising route which should lead to real improvements in planar chromatography in the near future.

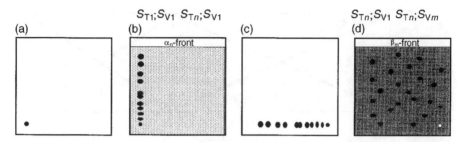

Figure 8.13 Schematic diagram of 2-D planar chromatographic separation using ^nD. In the first dimension (a and b) the total solvent strength was reduced stepwise at constant selectivity (S_{T1}; $S_{V1} \rightarrow S_{Tn}$; S_{V1}) to achieve differences in polarity. In the second dimension (c and d), the selectivity was changed at constant solvent strength (S_{Tn}; $S_{V1} \rightarrow S_{Tn}$; S_{Vm}) to achieve maximum spot capacity.

8.9   MULTIPLE DEVELOPMENT IN THREE DIRECTIONS

A theoretical model whereby maximum peak capacity could be achieved by the use of 3-D planar chromatographic separation was proposed by Guiochon and co-workers (23–27). Unfortunately, until now, because of technical problems, this idea could not be realized in practice. Very recently, however, a special stationary phase, namely Empore™ silica TLC sheets, has now become available for realization of 3-D PC. This stationary phase, developed as a new separation medium for planar chromatography, contains silica entrapped in an inert matrix of polytetrafluoroethylene (PTFE) microfibrils. It has been established that the separating power is only ca. 60% of that of conventional TLC (28); this has been attributed to the very slow solvent migration velocity resulting from capillary action.

The influence of external forces (overpressure or centrifugal force) on the structure of Empore™ silica thin-layer sheets was demonstrated by Botz *et al.* (29) with the aid of scanning electron micrographs. These authors reported that the use of Empore™ sheets for linear overpressured layer chromatography (OPLC) separations demands special preparation. The stretched sheets, placed on a sheet of glass or Teflon™, must be impregnated with polymer suspension along all four of their edges. Afterwards, the inlet channel of the Teflon™ cover plate must be placed on the sheet for development in the first direction. After the first separation step, the chromatographic plate is dried and then developed in a perpendicular direction (the inlet channel of the Teflon™ cover plate must also be rotated through 90°).

On completion of the 2-D OPLC separation, several layers of Empore™ sheets, all with impregnated edges, can be carefully stacked on top of each other, with the developed 2-D sheet resting on the top. After the sheets have been pressed together (by application of hydraulic pressure, as normal) the 3-D OPLC separation can be started by introducing the mobile phase through an appropriate porous cover plate (which serves as the mobile phase inlet) (16). If the number of Empore™ sheets is sufficient, a separation cube can theoretically be constructed (see Figure 8.14), thus enabling a spot capacity of n^3, according to the number of boxes. After separation in

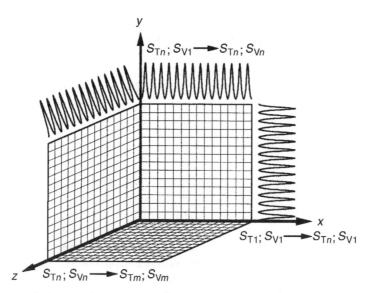

Figure 8.14 Schematic diagram of the peak capacity of a 3-D (n^3) planar chromatographic system employing mobile phases of different composition; the number of cubes represents the number of compounds which can theoretically be separated.

the third dimension, the separated compounds would be found on different sheets. It might, however, also happen that the same spot could be found on more than one sheet. Needless to say, the limitations of such a system would clearly include all the disadvantages of 2-D TLC.

Botz *et al.* (29) also demonstrated, by scanning electron microscopy, that application of overpressure increases the density of the layer, which could be one reason for the higher separation efficiency. These results showed that Empore™ silica TLC sheets enable extremely rapid separations (5–20 min) in one-dimensional OPLC, and gave good resolution. Theoretically, for a 3-D OPLC separations development times of 15–60 min would be required. The separation cube of sheets could be especially useful for micropreparative separations (30).

Combination of 3-D OPLC with multiple development encompasses all of the advantages of three-dimensional and forced-flow planar chromatography, and the separating capacity of multiple development. Favourable conditions could be to start the separation in the first dimension, and then reducing the total solvent strength stepwise at constant mobile phase selectivity to achieve a crude separation on the basis of the polarity of the compounds to be separated. In the second (perpendicular) direction, multiple development could be performed at constant S_T but with variation of the mobile phase selectivity. The third dimension would enable a combination of total solvent strength and mobile phase selectivity for improving the resolution of complex matrices (see Figure 8.14).

Theoretically, 3-D OPLC in combination with multiple development is the most powerful technique of instrumental planar chromatography. Unfortunately, suitable instrumentation is at an early stage of development.

8.10 COUPLED LAYERS WITH STATIONARY PHASES OF DECREASING POLARITY

Chromatographic plates can be connected for both capillary-controlled and forced-flow planar chromatography (FFPC), i.e. irrespective of whether capillary action or forced-flow is the driving force for the separation. The first technique is denoted as 'grafted' planar chromatography (31), while the second is known as 'long distance' (LD) OPLC, which uses heterolayers (32, 33).

8.11 GRAFTED PLANAR CHROMATOGRAPHY

The idea of coupled TLC plates, denoted as graft TLC, was reported in 1979. Graft TLC (31) is a multiple system with layers of similar or different stationary phases for isolation of compounds from natural and/or synthetic mixtures on a preparative scale. Two plates are grafted together and clamped in the fashion of a lap-joint with the edges of their adsorbent layers in intimate contact, so that a compound from a chromatogram developed on the first chromatographic plate can be transferred to the second plate without the usual scraping of bands, extraction, and re-spotting. At that time, the method could be used for preparative purposes only – because home-made plates were employed, compounds could not be transferred quantitatively, as was necessary for analytical separations.

Nowadays, almost all commercially available HPLC stationary phases are also applicable to planar chromatography. In addition to the polar hydroxyl groups present on the surface of native silica, other polar functional groups attached to the silica skeleton can also enter into adsorptive interactions with suitable sample molecules (34). Silica with hydrophilic polar ligands, such as amino, cyano, and diol functions, attached to the silica skeleton by alkyl chains, all of which have been well proven in HPLC, have also been developed for TLC (34).

A simple method for serially connected TLC on coupled chromatographic plates coated with different stationary phases (6) is illustrated schematically in Figure 8.15. An appropriate amount of stationary phase A (light shading in Figure 8.15(a)) must be removed, as must a corresponding part of stationary phase B (dark shading in Figure 8.15(b)). The amounts removed are such that when the two chromatographic plates are turned face to face (Figure 15(c)) and are grafted (pressed) together (Figure 15(d)), the remaining regions of the layers overlap.

The configuration illustrated in Figure 8.15 is an ultra-micro chamber in which the vapour phase is practically unsaturated (35). The distance between the supporting glass plate and the stationary phase of the other chromatographic plate is the layer thickness, which depends on the type of analytical (20–25 μm) or preparative (0.5–2.0 μm) plate applied. The sample to be separated (black zone in Figure 8.15(a)) is applied to stationary phase A after removal of the region of the layer not necessary for the graft TLC separation. The two plates must be clamped in lap-joint fashion with the edges of their stationary phases in close contact (Figure 8.15(d)) so that compounds from the chromatogram developed on the first chromatographic

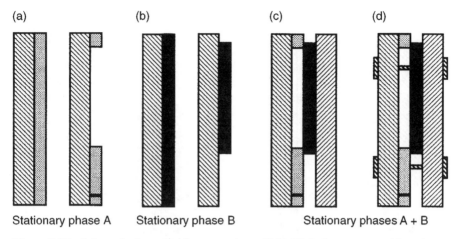

Stationary phase A   Stationary phase B   Stationary phases A + B

Figure 8.15 Schematic diagrams of cross-sections of MD-TLC plates connected in series to ensure multidimensional separation on stationary phases of increasing polarity; hatched lines, glass plate; light shading, stationary phase A; dark shading, stationary phase B.

plate can be transferred to the second plate. This arrangement of the stationary phases can be regarded as multidimensional planar chromatography only if the second stationary phase (B) is much less polar than the first (A); otherwise, the second criterion of multidimensionality is not fulfilled.

On the basis of the principle of grafted TLC, reversed-phase (RP) and normal-phase (NP) stationary phases can also be coupled. The sample to be separated must be applied to the first (2.5 cm × 20 cm) reversed-phase plate (Figure 8.16(a)). After development with the appropriate (S_{T1}; S_{V1}) mobile phase (Figure 8.16(b)), the first plate must be dried. The second (20 cm × 20 cm) (silica gel) plate (Figure 8.16(c)) must be clamped to the first (reversed-phase) plate in such a way that by use of a strong solvent system (S_{Tx}; S_{Vx}) the separated compounds can be transferred to the second plate (Figure 8.16(d)). Figure 8.16(e) illustrates the applied, re-concentrated

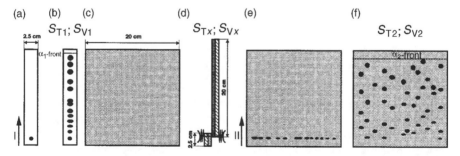

Figure 8.16 Views from above (a–c, e, and f) of coupled RP–NP chromatographic plates (without lines, RP stationary phase; light shading, NP stationary phase), and schematic diagram of a cross-section (d) through the clamped plates using a strong mobile phase (S_{Tx}; S_{Vx}) for the transfer of the sample to be further separated.

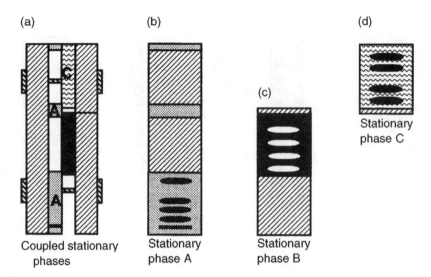

Figure 8.17 Schematic diagram of a cross-section (a) through the clamped plates, and views from above (b–d) of coupled plates serially connected to achieve multidimensional separation with stationary phases with different characteristics (hatched lines, glass plate; light shading, stationary phase A; dark shading, stationary phase B; wavy lines, stationary phase C).

samples, ready for development in the second direction. By use of a mobile phase of appropriate selectivity (S_{T2}; S_{V2}) an effective multidimensional planar chromatographic separation can be observed (Figure 8.16(f)).

Figure 8.17 depicts MD-PC performed on three different types of stationary phase (6). The three grafted chromatographic plates (Figure 8.17(a)) are clamped in lap-joint fashion with the edges of their stationary phases in close contact. The manner in which the three plates are prepared and the separation which can theoretically be achieved are also apparent from the schematic diagrams in Figures 8.17(b–d), in which the most polar stationary phase is phase 'A' and the least polar is stationary phase 'C'.

8.12 SERIALLY CONNECTED MULTILAYER FORCED-FLOW PLANAR CHROMATOGRAPHY

Mincsovics and co-workers (36) found that OPLC is suitable for the development of several chromatographic plates simultaneously if the plates were specially prepared. With this multilayer technique many samples can be separated during a single chromatographic run. On the basis of this concept, Botz *et al.* (32, 33) proposed a novel OPLC technique with a significantly increased separation efficiency, in which the separation distance can be increased as a result of special arrangement of the chromatographic plates. This category of multilayer FFPC, linear development

(LD)-OPLC, involves the serial connection of the chromatographic plates. By application of this technique, it is also possible to use different stationary phases of decreasing polarity during a single development, as shown in Figure 8.18(a).

Specially prepared plates are necessary for this LD-OPLC technique in linear development mode (35) – all four edges of the chromatographic plates must be impregnated with a special polymer suspension. Movement of the mobile phase with a linear front can be ensured by placing a narrow plastic sheet on the layer. Three plates can be placed on top of each other to create the long development distance. The end of the first (top) chromatographic plate has a slit-like perforation to enable transfer of the mobile phase to the second layer, where migration continues until the opposite end of that layer. Development can then either continue to the next chromatographic plate or, if chromatography is complete, the mobile phase can be drained away. On this basis, 60 cm separation distances can be achieved with P-OPLC 50 equipment. As a consequence of the manner in which the layers are prepared, glass-backed plates can be used at the bottom of the stack only.

For this technique, the upper plate has a mobile phase inlet channel on one edge and a slit at the opposite edge for directing the mobile phase toward the next plate (33). The slit (width approx. 0.1 mm) can be produced by cutting the layers with a sharp blade; this enables easy passage of mobile phase and separated compounds without any mixing. The cushion of the OPLC instrument is applied to the uppermost layer only, and each plate presses on to the sorbent layer below.

The potential of serial connection of layers can be used for multidimensional planar chromatographic separations on different stationary phases of decreasing

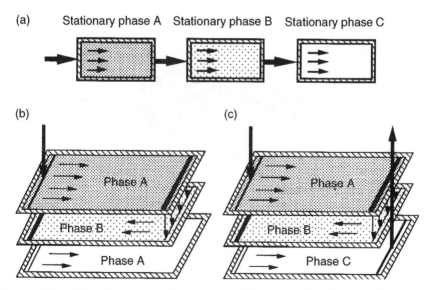

Figure 8.18 Schematic diagrams showing: (a) serially connected multilayer OPLC employing different types of stationary phase of decreasing polarity (A > B > C) as a type of MD-FFPC; (b) fully off-line operational mode; (c) fully on-line operational mode.

polarity. Fully off-line separation (Figure 8.18(b)), in which several samples are applied, is complete when the α-front of the mobile phase reaches the end of the bottom plate. Figure 8.18(c) illustrates the fully on-line operating mode, in which all the compounds of the (single) sample to be separated have to be moved at the same separation distance. The mobile phase can therefore be directed from the lower plate in a manner similar to that in which it is introduced. This gives the possibility of on-line detection. For this fully on-line mode of operation, the length of all layers placed between the uppermost and lowest plates must be reduced by 1 cm, in order to ensure room for the mobile phase outlet.

Figure 8.19 illustrates another example of the versatility of multidimensional OPLC, namely the use of different stationary phases and multiple development ($''$D) modes in combination with circular and 'anticircular' development and both off-line and on-line detection (37). Two different stationary phases are used in this configuration. The lower plate is square (e.g. 20 cm $\times$ 20 cm), while the upper plate (grey in Figure 8.19) is circular with a diameter of, e.g. 10 cm. The sample must be applied on-line to the middle of the upper plate. In the OPLC chamber the plates are covered with a Teflon^{TM} sheet and pressed together under an overpressure of 5 MPa. As the mobile phase transporting a particular compound reaches the edge of the first plate it must–because of the forced-flow technique–flow over to the second (lower) stationary phase, which is of lower polarity.

The place at which this transfer occurs is illustrated in Figure 8.19 as a thin circle on the lower chromatographic plate. Because the overpressure is uniform throughout the whole system, the compounds will be divided into two parts and migrate in both circular (outwards) and 'anticircular' (inwords) directions. A hole at the centre of the

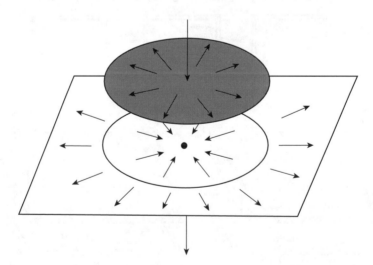

Figure 8.19 Schematic diagram of the combination of multilayers (decreasing polarity) for OPLC with different types of development (circular and 'anticircular') and modes of detection (off-line and on-line).

lower plate enables on-line detection of the compounds after 'anticircular' development. Off-line detection can be used when circular development is complete. Although this combination might seem complicated, MD-PC, partly in combination with multiple development, can easily be realized with little modification of commercially available equipment (37).

8.13 COMBINATION OF MULTIDIMENSIONAL PLANAR CHROMATOGRAPHIC METHODS

In the introduction to this chapter, MD-PC was defined as a procedure in which substances to be separated were subjected to at least two separation steps with different mechanisms of retention (5). Discussion of the basic potential modes of operation showed that because of the versatility which resulted from being able to combine mobile phases of different composition, more than two development steps can easily be realized by the use of ^nD techniques.

Working with the same mobile phase composition, the serial connection of stationary phases under capillary-controlled flow conditions is much more complicated, for two reasons. First, for HPTLC plates the mobile phase velocity might no longer be adequate to maintain optimum separation conditions, and secondly, the handling of more than three chromatographic plates within a separation distance of 8 cm is practically impossible. Unfortunately, only one type of bilayer (silica and RP-18) is commercially available. However, 10 cm $\times$ 20 cm glass-backed plates coated with chemically modified layers are available and can be cut to $(3-5)$ cm $\times$ 20 cm, thus enabling many stationary-phase combinations to be realized by the graft technique.

The best technique is, therefore, a parallel combination of stationary and mobile phases. In order to take advantage of the double effect of MD-PC it is recommended that bilayer plates are combined with multiple development techniques (38) in which total solvent strength and mobile phase selectivity are changed simultaneously. By the use of this technique, in the first direction S_T and S_V are varied in 'n' re-chromatographic steps (Figure 8.20(a)) and in the perpendicular, second, direction (Figure 8.20(b)) S_T and S_V are again varied in 'm' re-chromatographic steps (Figure 8.20(c)). This is an extended version of MD-PC and can be regarded as being 'double' MD-PC, because changes not only of the stationary phases (bilayer) but also of the mobile phase composition ensure the criteria of multidimensional planar chromatography. As is apparent from Figure 8.20(c), the number of compounds separated is clearly more than for any other version.

The versatility of combining stationary phases, ^nD, and forced-flow techniques is illustrated in Figure 8.21. Separation in the first dimension is performed by application of the first stationary and mobile phase combination (S_{T1}; $S_{V1} \rightarrow S_{Tn}$; S_{Vn}) by use of multiple development. After drying of the chromatographic plate, the separation can be continued in a perpendicular direction by using three prepared chromatographic plates, with different characteristics (decreasing polarity), in combination with a second mobile phase (S_{Tx}; S_{Vx}). It must be pointed out that the separating

Figure 8.20 Combination of bilayer plates and multiple development techniques in which total solvent strength and mobile phase selectivity are changed simultaneously: in the first direction (a), S_T and S_V are varied in 'n' re-chromatographic steps, while in the perpendicular, (second) direction (b), S_T and S_V are again varied in 'm' re-chromatographic steps, to give (c).

power can also easily be increased by the use of ^nD in the second dimension. Increasing the dimensions of the separation procedure always increases the number of theoretical plates, which leads to enhancement of the separation number and, therefore, the efficiency of the separation.

Satisfactory combination of more than three chromatographic plates for long-distance OPLC is always difficult, because only the bottom plate can be glass-backed; the others must be aluminium-backed (32, 33).

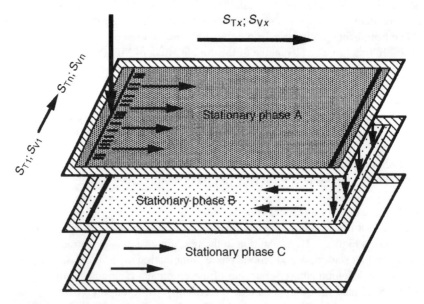

Figure 8.21 Schematic diagram of the combination of bidirectional, multiple development and coupled layers in decreasing polarity (A > B > C).

Naturally, several other possibilities can be used to increase the number of dimensions. Between the first and second developments, or sample, the characteristics of the chromatographic plate or the properties of the sample can also be modified. Although interfacing of on-line OPLC with one- or two-dimensional TLC is not particularly difficult, it is not yet widely practiced. It must be concluded that full exploitation of the versatility of MD-PC is at an early state of development; as a consequence several significant changes in practice might be expected in the next few years (10).

8.14 COUPLING OF PC WITH OTHER CHROMATOGRAPHIC TECHNIQUES

Another means of realizing multidimensional separation is combination of two complementary separation techniques which use different methods of separation. In such multi-modal separation, different techniques can be coupled in which PC is used as the second dimension and another separation method, as the first. Some possible variations are as follows:

- combination of different extraction methods (e.g. thermo and solid phase) with TLC;
- combination of GC and TLC;
- combination of supercritical fluid chromatography (SFC) and TLC;
- combination of HPLC and TLC;
- combination of OPLC and TLC;
- combination of capillary electrophoresis (CE) and TLC;
- combination of different counter-current chromatographic (CCC) methods with TLC.

Current interest is, however, mainly in the coupling of HPLC and TLC, to which considerable attention has been devoted for the solution of difficult separation problems. Since Boshoff et al. (39) first described the direct coupling of HPLC and TLC, several papers (40–43) have been published describing the on-line coupling of liquid chromatographic methods and PC, usually with different interfaces, depending on the first technique applied. If PC is used as the second method, all the MD methods discussed above can be applied to increase the separating power.

8.15 FURTHER CONSIDERATIONS

Multidimensional planar chromatographic separations, as we have seen, require not only a multiplicity of separation stages, but also that the integrity of separation achieved in one stage be transferred to the others. The process of separation on a two-dimensional plane is the clearest example of multidimensional separations. The greatest strength of MD-PC, when properly applied, is that compounds are distributed widely over two-dimensional space of high zone (peak) capacity. Another unique feature of PC is the possibility of using the layer for storage. For quantitation in MD-PC, it is our opinion that we are close to the time when slit-scanning densitometry will be replaced by quantitative image analysers.

On the basis of theory and experimental observations it can be predicted that a zone capacity of ca. 1500 could be achieved by 2-D multiple development. Because the same result can be achieved by application of 2-D forced-flow development on HPTLC plates, it can be stated that the combination of stationary phases, FFPC and ^{n}D offers a fruitful future in modern, instrumental planar chromatography.

REFERENCES

1. C. F. Poole, S. K. Poole, W. P. N. Fernado, T. A. Dean, H. D. Ahmed and J. A. Berndt, 'Multidimensional and multimodal thin-layer chromatography: pathway to the future', *J. Planar Chromatogr.* **2**: 336–345 (1989).

2. J. C. Giddings, 'Two-dimensional separations: concept and promise', *Anal. Chem.* **56**: 1258A–1270A (1984).

3. J. C. Giddings, 'Concepts and comparisons in multidimensional separation', *J. High Resolut. Chromatogr.* **10**: 319–323 (1987).

4. H. J. Cortes (Ed.), *Multidimensional Cromatography. Techniques and Applications,* Marcel Dekker, New York, pp. 1–27 (1990).

5. J. C. Giddings, 'Use of multiple dimensions in analytical separations in multidimensional chromatography', in *Multidimensional Chromatography. Techniques and Application,* H. J. Cortes (Ed.), Marcel Dekker, New York, pp. 1–27 (1990).

6. Sz. Nyiredy, 'Multidimensional planar chromatography' in *Proceedings of the Dünnschicht-Chromatographi (in memoria Prof. Dr. Hellmut Jork),* R. E. Kaiser, W. Günther, H. Gunz and G. Wulff (Eds), InCom Sonderband, Düsseldorf, pp. 166–185 (1996).

7. C. F. Poole and S. K. Poole, 'Multidimensionality in planar chromatography', *J. Chromatogr.* **703**: 573–612 (1995).

8. R. Consden, A. H. Gordon and A. J. P. Martin, ... *Biochem. J.* **38**: 244–... (1944).

9. W. Markowski and K. L. Czapińska, 'Computer simulation of the separation in one- and two-dimensional thin-layer chromatography by isocratic and stepwise gradient development', *J. Liq. Chromatogr.* **18**: 1405–1427 (1995).

10. C. F. Poole, 'Planar chromatography at the turn of the century', *J. Chromatogr.* **856**: 399–427 (1999).

11. D. S. Risley, R. Kleyle, S. Habibi-Goudarzi and D. Nurok, 'Correlations between the ranking of one- and two-dimensional solvent systems for planar chromatography', *J. Planar Chromatogr.* **3**: 216–221 (1990).

12. P. Härmälä, L. Botz, O. Sticher and R. Hiltunen, 'Two-dimensional planar chromatographic separation of a complex mixture of closely related coumarins from the genus *Angelica*', *J. Planar Chromatogr.* **3**: 515–520 (1990).

13. Sz. Nyiredy, K. Dallenbach-Toelke and O. Sticher, 'The "PRISMA" optimization system in planar chromatography', *J. Planar Chromatogr.* **1**: 336–342 (1988).

14. L. R. Snyder, 'Classification of the solvent properties of common liquids', *J. Chromatogr. Sci.* **16**: 223–234 (1978).

15. Sz. Nyiredy, 'Planar chromatography', *Chromatography,* E. Heftmann (Ed.), Elsevier, Amsterdam, A109–A150 (1992).

16. Sz. Nyiredy, 'Solid-liquid extraction strategy on the basis of solvent characterization', *Chromatographia*, **51**: 5288–5296 (2000).

17. Sz. Nyiredy, Zs. Fatér and B. Szabady, 'Identification in planar chromatography by use of retention data measured using characterized mobile phases', *J. Planar Chromatogr.* **7**: 406–409 (1994).

18. B. Szabady and Sz. Nyiredy, 'The versatility of multiple development in planar chromatography', in *Dünnschicht-Chromatographie (in memoriam Prof. Dr. Hellmut Jork)*, R. E. Kaiser, W. Günther, H. Gunz and G. Wulff (Eds), InCom Sonderband, Düsseldorf, pp. 212–224 (1996).

19. Sz. Nyiredy, 'Essential guides to method development in TLC', in: Encyclopedia of Separation Science (eds. I. D. Wilson, T. Addier, M. Cooke, C. F. Poole), Academic Press, London, in press.

20. S. K. Poole, M. T. Belay and C. F. Poole, 'Effective systems for the separation of pharmaceutically important estrogens by thin layer chromatography', *J. Planar Chromatogr.* **5**: 16–27 (1992).

21. G. Matysik and E. Soczewiński, 'Stepwise gradient development in thin-layer chromatography. II. Two-dimensional gradients for complex mixtures', *J. Chromatogr.* **369**: 19–25 (1986).

22. L. V. Poucke, D. Rousseau, C. V. Peteghem and B. M. J. Spiegeleer, 'Two-dimensional HPTLC of sulphonamides on cyanoplates, using the straight-phase and reversed-phase character of the adsorbent', *J. Planar Chromatogr.* **2**: 395–397 (1989).

23. G. Guiochon and A. M. Siouffi, 'Study of the performance of thin-layer chromatography. Spot capacity in thin-layer chromatography', *J. Chromatogr.* **245**: 1–20 (1982).

24. G. Guiochon, M. F. Gonnord, A. M. Siouffi and M. Zakaria, 'Study of the performances of thin-layer chromatography. VII. Spot capacity in two-dimensional thin-layer chromatography', *J. Chromatogr.* **250**: 1–20 (1982).

25. M. Zakaria, M. F. Gonnord and G. Guiochon, 'Applications of two-dimensional thin-layer chromatography', *J. Chromatogr.* **271**: 127–192 (1983).

26. G. Guiochon, M. F. Gonnord, M. Zakaria, L. A. Beaver and A. M. Siouffi, 'Chromatography with a two-dimensional column', *Chromatographia* **17**: 121–124 (1983).

27. L. A. Beaver and G. A. Guiochon, 'System and apparatus for multi-dimensional real-time chromatography', *US Pat. 4 469 601* (1984).

28. S. K. Poole and C. F. Poole, 'Evaluation of the separation performance of Empore™ thin layer chromatography sheets', *J. Planar Chromatogr.* **2**: 478–481 (1989).

29. L. Botz, Sz. Nyiredy, E. Wehrli and O. Sticher, 'Applicability of Empore™ TLC sheets for forced-flow planar chromatography. I. Characterization of the silica sheets', *J. Liq. Chromatogr.* **13**: 2809–2828 (1990).

30. E. Mincsovics, E. Tyihák, Z. Baranyi and B. Tapa, *Hungar. Pat.* (applied for, 14.8.90), patent pending (1990).

31. R. C. Pandey, R. Misra and K. L. Rinehart, Jr., 'Graft thin-layer chromatography', *J. Chromatogr.* **169**: 129–139(1979).

32. L. Botz, Sz. Nyiredy and O. Sticher, 'The principles of long distance OPLC, a new multi-layer development technique', *J. Planar Chromatogr.* **3**: 352–354 (1990).

33. L. Botz, Sz. Nyiredy and O. Sticher, 'Applicability of long distance overpressured layer chromatography', *J. Planar Chromatogr.* **4**: 115–122 (1991).

34. H. E. Hauck and W. Jost, 'Sorbent materials and precoated layers in thin-layer chromatography', *Chromatogr. Sci.* **47**: 251–330 (1990).

35. Sz. Nyiredy, Zs. Fatér, L. Botz and O. Sticher, 'The role of chamber saturation in the optimization and transfer of the mobile phase', *J. Planar Chromatogr.* **5**: 308–315 (1992).

36. E. Tyihák, E. Mincsovics and T. J. Székely, 'Overpressured multi-layer chromato-graphy', *J. Chromatogr.* **471**: 375–387 (1989).

37. Sz. Nyiredy, 'Applicability of planar chromatography in the analysis and isolation of plant substances', in *Proceedings of Biokemia XV*, pp. 146–151 (1991).

38. C. F. Poole and M. T. Belay, 'The influence of layer thickness on the chromatographic and optical properties of high performance silica gel thin layer chromatographic plates', *J. Planar Chromatogr.* **4**: 345–359 (1991).

39. P. R. Boshoff, B. J. Hopkins and V. Pretorius, 'Thin-layer chromatographic transport detector for high-performance liquid chromatography', *J. Chromatogr.* **126**: 35–41 (1976).

40. J. W. Hofstraat, S. Griffioen, R. J. van de Nesse and U. A. Th Brinkman, 'Coupling of narrow-bore column liquid chromatography and thin-layer chromatography. Interface optimization and characteristics for normal-phase liquid chromatography', *J. Planar Chromatogr.* **1**: 220–226 (1988).

41. E. Mincsovics, M. Garami and E. Tyihák, 'Direct coupling of OPLC with HPLC: clean-up and separation', *J. Planar Chromatogr.* **4**: 299–303 (1991).

42. K. Burger, 'Online coupling HPLC–AMD (automated multiple development)', *Analysis* **18**: 1113–1116 (1990).

43. O. R. Queckenberg and A. W. Fraham, 'Chromatographic and spectroscopic coupling: a powerful tool for the screening of wild *amaryllidaceae*', *J. Planar Chromatogr.* **6**: 55–61 (1993).

9 Multidimensional Electrodriven Separations

MARTHA M. DEGEN and VINCENT T. REMCHO
Oregon State University, Corvallis, OR, USA

9.1 INTRODUCTION

Multidimensional separations allow for the analysis of complex mixtures, such as those from biological matrices with thousands of components that would be difficult or impossible to separate by utilizing only one method. Electrodriven separations have been employed to separate biological molecules for many years, due to the charged nature of amino acids and nucleic acids. The addition of an electrodriven component to a multidimensional separation is therefore desirable, especially for the separation of biological mixtures.

This chapter will first cover the nature of electrophoretic separations, especially those concerning capillary electrophoresis. Comprehensive multidimensional separations will then be defined, specifically in terms of orthogonality and resolution. The history of planar and non-comprehensive electrodriven separations will then be discussed. True comprehensive multidimensional separations involving chromatography and capillary electrophoresis will be described next. Finally, the future directions of these multidimensional techniques will be outlined.

9.2 ELECTROPHORETIC SEPARATIONS

Zone electrophoresis is defined as the differential migration of a molecule having a net charge through a medium under the influence of an electric field (1). This technique was first used in the 1930s, when it was discovered that moving boundary electrophoresis yielded incomplete separations of analytes (2). The separations were incomplete due to Joule heating within the system, which caused convection which was detrimental to the separation.

Charged macromolecules, such as proteins or polymers, are often separated electrophoretically. The rate of migration through an electric field increases with net charge and field strength. Molecular size of analytes and viscosity of separation media both have inverse relationships with rate of migration. These variables must all be taken into account in order to optimize the conditions for an efficient electrophoretic separation.

Multidimensional Chromatography, edited by L. Mondello, A. C. Lewis and K. D. Bartle
©2002 John Wiley & Sons Ltd.

Many of the first electrophoretic separations were conducted on planar media. Convection due to Joule heating was reduced in solid support materials, such as cellulose filter paper and polyacrylamide gels, as compared to free solution separations. An open tubular format was desired, but limitations in available materials and tubing diameters presented Joule heating issues. As a result, capillary columns were first employed for electrodriven separation in 1979 (3). In 1981, Jorgenson and De Arman Lukacs introduced free solution electrophoresis in 75 μm glass capillaries, a technique that was named capillary zone electrophoresis (CZE) (4). The main benefits of CZE were the capillary format that reduced Joule heating effects and the free solution separation, which eliminated eddy diffusion as a contributor to zone spreading.

Capillary electrophoresis (CE) today is available in many diversified forms, such as capillary isotachaphoresis and capillary electrochromatography. It is a technique offering very high resolution and efficiency. CE can separate both ionic and non-ionic compounds over an exceedingly broad range of molecular weights. The on-line detection of CE usually yields good quantitative results, but poor mass sensitivity due to the small volumes of sample used. The capillary format in CE yields other benefits aside from the efficient dissipation of heat, including the requirement of only small amounts of sample and minimal solvent usage.

The mechanism by which analytes are transported in a non-discriminate manner (i.e. via bulk flow) in an electrophoresis capillary is termed electroosmosis. Figure 9.1 depicts the inside of a fused silica capillary and illustrates the source that supports electroosmotic flow. Adjacent to the negatively charged capillary wall are specifically adsorbed counterions, which make up the fairly immobile Stern layer. The excess ions just outside the Stern layer form the diffuse layer, which is mobile under the influence of an electric field. The substantial frictional forces between molecules in solution allow for the movement of the diffuse layer to pull the bulk

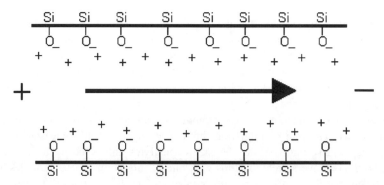

Figure 9.1   The hydrated inner surface of a fused silica capillary is where electroosmotic flow originates.

electrolyte along with it, resulting in a plug-like flow of eluent through the capillary when the capillary inner diameter is in the range (50–150 μm) typically used in CZE.

The flow profiles of electrodriven and pressure driven separations are illustrated in Figure 9.2. Electroosmotic flow, since it originates near the capillary walls, is characterized by a flat flow profile. A laminar profile is observed in pressure-driven systems. In pressure-driven flow systems, the highest velocities are reached in the center of the flow channels, while the lowest velocities are attained near the column walls. Since a zone of analyte-distributing events across the flow conduit has different velocities across a laminar profile, band broadening results as the analyte zone is transferred through the conduit. The flat electroosmotic flow profile created in electrodriven separations is a principal advantage of capillary electrophoretic techniques and results in extremely efficient separations.

9.3   COMPREHENSIVE SEPARATIONS

In order to distinguish the multitude of coupled techniques from truly multidimensional separations, a comprehensive separation must be defined. A genuinely comprehensive multidimensional separation requires a high degree of orthogonality of the coupled techniques and retention of resolution in the interface between the two methods. Orthogonal techniques base their respective separations on sample properties which are as dissimilar as possible. If two techniques are orthogonal and coupling them together causes no loss of resolution, then the peak capacity of the multidimensional system can be defined as the product of the peak capacities for each dimension (5). The high peak capacities that result from multidimensional separations make them applicable to complex sample mixtures containing thousands of analytes.

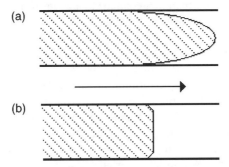

Figure 9.2   Pressure-driven (a) and electrodriven (b) flow profiles. Laminar flow in pressure-driven systems results in a bullet-shaped profile, while the profile of electroosmotic flow is plug-shaped, which reduces band broadening.

Chromatographic and electrophoretic separations are truly orthogonal, which makes them excellent techniques to couple in a multidimensional system. Capillary electrophoresis separates analytes based on differences in the electrophoretic mobilities of analytes, while chromatographic separations discriminate based on differences in partition function, adsorption, or other properties unrelated to charge (with some clear exceptions). Typically in multidimensional techniques, the more orthogonal two methods are, then the more difficult it is to interface them. Microscale liquid chromatography (μLC) has been comparatively easy to couple to capillary electrophoresis due to the fact that both techniques involve narrow-bore columns and liquid-phase eluents.

The sampling of the first dimension by the second is an extremely important aspect in multidimensional separations. In heart-cutting techniques, fractions of eluent from the first column are separated by a second column of different selectivity. Comprehensive sampling is similar to heart-cutting in practice, except with considerably elevated sampling frequency. In a comprehensive sampling arrangement, each peak that elutes from the first column is sampled by the second column numerous times. This results in the maximum retention of resolution and information obtained in the first separation. Another characteristic unique to a comprehensive sampling scheme is that all components are analyzed by the second dimension. Non-comprehensive heart-cutting techniques may not allow for complete gathering of information on all components of the sample, since some analytes may elute between fractions and fail to be sampled by the second dimension. The tremendous resolving power and orthogonality of comprehensive multidimensional techniques are the characteristics that make multidimensionality so attractive in separations.

9.4 PLANAR TWO-DIMENSIONAL SEPARATIONS

Planar two-dimensional separation methods can offer the distinct advantage of truly comprehensive sampling of the first dimension by the second, since transfer of analytes from one medium (or separation conduit) to another is not a requirement. The first comprehensive two-dimensional separations were achieved with paper chromatography in 1944 (6). This discovery led to a number of other multidimensional techniques, with many of these involving electrodriven separations. In 1948, Haugaard and Kroner separated amino acids by coupling paper chromatography with paper electrophoresis, thus performing the original multidimensional electrodriven separation. This study involved the use of a 100 V electric field applied across one dimension of the paper, while a phosphate buffer was used as a chromatographic eluent to move analytes in the orthogonal direction (7).

Two-dimensional planar electrophoresis was first used in 1951 (8), while electrophoresis was coupled with thin-layer chromatography (TLC) in 1964 to separate mixtures of nucleosides and nucleotides (9). These techniques were novel and led to other great discoveries, but did not survive the test of time, and they are no longer commonly used. TLC–electrophoresis in particular was an awkward technique to put into practice. Other planar two-dimensional electrodriven separations, however, are still used today.

In 1975, O'Farrell determined that high-resolution separations of protein mixtures could be achieved with polyacrylamide gel isoelectric focusing (IEF) in one dimension and polyacrylamide slab gel electrophoresis (PAGE) in the other (10). This technique was used to separate 1100 different components from *Escherichia coli*, some of which differed by as little as a single charge, while being similar in molecular weight. Although this technique can be problematic due to the labor-intensive complications of gel preparation and staining, IEF–sodium dodecyl sulfate (SDS)–PAGE continues to be a very popular method for the separation of complex biological samples, especially mixtures of proteins.

In 1988, Burton *et al.* developed a new analytical technique which they dubbed the "chromatophoresis" process. This method coupled reverse-phase (RP) high-performance liquid chromatography (HPLC) with SDS–PAGE in an automated system used to separate proteins. Chromatophoresis involved a separation based on differences in hydrophobicity in the first dimension and molecular charge in the second. The chromatophoresis process is illustrated in Figure 9.3. After eluting from an HPLC column, proteins were passed into a heated reaction chamber, where they were denatured and complexed with SDS. The protein complexes in the eluate stream were then deposited onto the surface of a polyacrylamide gradient gel, and an electrophoretic separation subsequently occurred. The five-hour run-time, inconvenience of gel electrophoresis, and difficult detection were the main disadvantages of this technique (11).

9.5 CHROMATOGRAPHY AND ELECTROPHORESIS COMBINED IN NON-COMPREHENSIVE MANNERS

Many groups have used electrophoresis to enhance a primary chromatographic separation. These techniques can be considered to be two-dimensional, but they are not comprehensive, usually due to the loss of resolution in the interface between the two methods. For instance, capillary electrophoresis was used in 1989 by Grossman and co-workers to analyze fractions from an HPLC separation of peptide fragments. In this study, CE was employed for the separation of protein fragments that were not resolved by HPLC. These two techniques proved to be truly orthogonal, since there was no correlation between the retention time in HPLC and the elution order in CE. The analysis time for CE was found to be four times faster than for HPLC (12), which demonstrated that CE is a good candidate for the second dimension in a two-dimensional separation system, as will be discussed in more detail later.

In 1989, Yamamoto *et al.* developed the first technique that directly coupled chromatography to capillary electrophoresis, although again in a non-comprehensive fashion. Low-pressure gel permeation chromatography, which separates analytes based on differences in molecular size, was combined with capillary isotachophoresis, which separates according to electrophoretic mobility. Capillary isotachophoresis

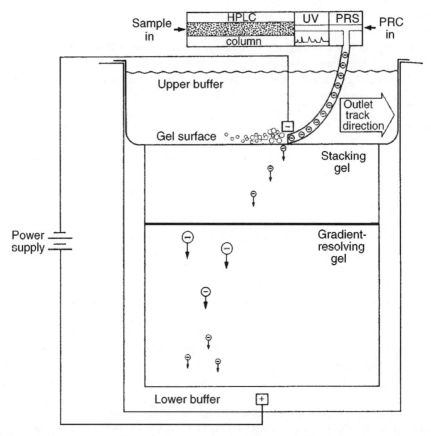

Figure 9.3   Schematic illustration of the electrophoretic transfer of proteins in the chromatophoresis process. After being eluted from the HPLC column, the proteins were reduced with β-mercaptoethanol in the protein reaction system (PRS), and then deposited onto the polyacrylamide gradient gel. (PRC, protein reaction cocktail). Reprinted from *Journal of Chromatography*, **443**, W. G. Burton *et al.*, 'Separation of proteins by reversed-phase high-performance liquid chromatography', pp 363–379, copyright 1988, with permission from Elsevier Science.

is a moving-boundary CE technique that utilizes leading and trailing electrolytes of differing electrophoretic mobility, between which analytes fraction into distinct zones in order of electrophoretic mobility. Each distinct separated zone is equal in concentration (such that analytes in low initial concentration are focused into a sharp, narrow zone of higher concentration), and all separated zones move at the same velocity. Figure 9.4 shows the coupled apparatus, which had the outlet of the microbore chromatographic column connected to the sample injection port of the electrophoresis capillary. The isotachophoretic run time was 18 min, and this analysis was repeated 60 times within the 18 h chromatographic run. This technique was successfully employed to separate proteins in solution with a stated 200 ng detection

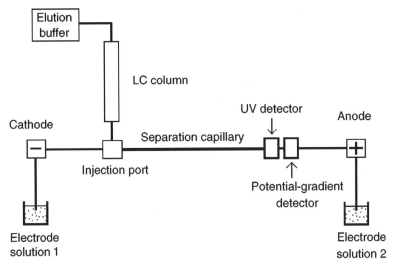

Figure 9.4   General schematic illustration of the apparatus used to combine chromatography with capillary isotachophoresis.

limit, which is comparable to that of polyacrylamide gel electrophoresis. Although this procedure had the possibility of being completely automated, the run time was still quite long (13).

Yamamoto *et al.* also coupled gel permeation HPLC and CE in an on-line fashion in 1990, where capillary isotachophoresis was again used in the second dimension. This technique was also not comprehensive due to the loss of resolution between the techniques. It was also not particularly fast, with a 23 min CE cycle, which was repeated 90 times throughout the HPLC run (14). Volume incompatibility between HPLC and CE was one problem not addressed in this study, in which a large HPLC column was coupled to an electrophoresis capillary.

Other groups have also used LC and CE to perform non-comprehensive multidimensional separations (15, 16). A three-dimensional separation was performed by Stromqvist in 1994, where size exclusion chromatography (SEC), reverse-phase HPLC, and CZE were used in an off-line manner to separate peptides (17). The most useful information gained from all of these non-comprehensive studies was knowledge of the orthogonality and compatibility of LC and CE.

9.6   COMPREHENSIVE TWO-DIMENSIONAL SEPARATIONS WITH AN ELECTRODRIVEN COMPONENT

The most successful multidimensional electrodriven separations to date have been performed by James Jorgenson and his group at the University of North Carolina at Chapel Hill. This group has accomplished several successful comprehensive

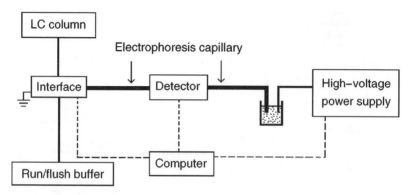

Figure 9.5 The generic setup for two-dimensional liquid chromatography–capillary zone electrophoresis as used by Jorgenson's group. The LC separation was performed in hours, while the CZE runs were on a time scale of seconds.

couplings of different chromatographic methods with capillary zone electrophoresis. Figure 9.5 illustrates the major components of the two-dimensional LC–CZE separation system used by Jorgenson and co-workers (18). The type of LC column (and mode of the separation) and the interface design were both modified many times in order to optimize multidimensional electrodriven separations for various different samples.

9.7 MICROCOLUMN REVERSE PHASE HIGH PERFORMANCE LIQUID CHROMATOGRAPHY–CAPILLARY ZONE ELECTROPHORESIS

In 1990, Bushey and Jorgenson developed the first automated system that coupled HPLC with CZE (19). This orthogonal separation technique used differences in hydrophobicity in the first dimension and molecular charge in the second dimension for the analysis of peptide mixtures. The LC separation employed a gradient at 20 µL/min volumetric flow rate, with a column of 1.0 mm ID. The effluent from the chromatographic column filled a 10 µL loop on a computer-controlled, six-port micro valve. At fixed intervals, the loop material was flushed over the anode end of the CZE capillary, allowing electrokinetic injections to be made into the second dimension from the first.

The HPLC elution time was typically under 260 min, and the CZE analysis took place in 60 s, which led to an overall run time of about 4 h. The 1 min CZE sampling interval was problematic, as the LC column was probably slightly undersampled. A shorter CZE analysis time, which would provide a more frequent sampling rate, would improve this system a great deal. The second-dimension analysis time must be short relative to the first dimension, lest resolution in the first dimension be sacrificed.

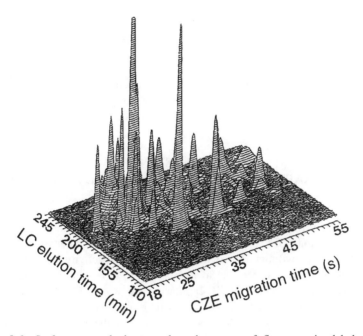

Figure 9.6 Surfer-generated chromatoeletropherogram of fluorescamine-labeled tryptic digest of ovalbumin. Reprinted from *Analytical Chemistry*, **62**, M. M. Bushey and J. W. Jorgenson, 'Automated instrumentation for comprehensive two-dimensional high-performance liquid chromatography/capillary zone electrophoresis, pp 978–984, copyright 1990, with permission from the American Chemical Society.

An example of the results obtained in the form of a "chromatoelectropherogram" can be seen in Figure 9.6. The contour type data display showed the three variables that were studied, namely chromatographic elution time, electrophoretic migration time, and relative absorbance intensity. Peptides were cleanly resolved by using this two-dimensional method. Neither method alone could have separated the analytes under the same conditions. The most notable feature of this early system was that (presumably) all of the sample components from the first dimension were analyzed by the second dimension, which made this a truly comprehensive multidimensional technique.

In 1993, Jorgenson's group improved upon their earlier reverse phase HPLC–CZE system. Instead of the six-port valve, they used an eight-port electrically actuated valve that utilized two 10-μL loops. While the effluent from the HPLC column filled one loop, the contents of the other loop were injected onto the CZE capillary. The entire effluent from the HPLC column was collected and sampled by CZE, making this too a comprehensive technique, this time with enhanced resolving power. Having the two-loop valve made it possible to overlap the CZE runs. The total CZE run time was 15 s, with peaks occurring between 7.5 and 14.8 s. In order to save separation space, an injection was made into the CZE capillary every 7.5s, therefore overlapping the second-dimension runs. The improved injection scheme resulted in enhancement of the apparent rate of the CZE separation (20).

The only other group to have performed comprehensive multidimensional reverse-phase HPLC–CZE separations is at Hewlett-Packard. In 1996, a two-dimensional LC–CE instrument was described at the Frederick Conference on Capillary Electrophoresis by Vonda K. Smith (21). The possibility for a commercial multidimensional instrument may have been explored at that time.

9.8 MICROCOLUMN SIZE EXCLUSION CHROMATOGRAPHY – CAPILLARY ZONE ELECTROPHORESIS

In 1993, Lemmo and Jorgenson used microcolumn size exclusion chromatography coupled to CZE for the two-dimensional analysis of protein mixtures. Under non-denaturing conditions, SEC provides a description of the molecular weight distribution of a mixture, without destroying the biological activity of the analytes. Microcolumn SEC was chosen due to its reasonable separation efficiency and its ability to interface well with the CZE capillary. Here, a μ-SEC column of 250 μm ID was coupled to a CZE capillary of 50 μm ID to generate a comprehensive two-dimensional system. System efficiencies of over 100 000 theoretical plates per meter have been obtained through the coupling of these two techniques.

A six-port valve was first used to interface the SEC microcolumn to the CZE capillary in a valve-loop design. UV–VIS detection was employed in this experiment. The overall run time was 2 h, with the CZE runs requiring 9 min. As in the reverse phase HPLC–CZE technique, runs were overlapped in the second dimension to reduce the apparent run time. The main disadvantage of this μ-SEC–CZE method was the valve that was used for interfacing. The six-port valve contributed a substantial extracolumn volume, and required a fixed volume of 900 nL of effluent from the chromatographic column for each CZE run. The large fixed volume imposed restrictions on the operating conditions of both of the separation methods. Specifically, to fill the 900 nL volume, the SEC flow rate had to be far above the optimum level and therefore the SEC efficiency was decreased (22).

The second interface design that was developed for use with μ-SEC–CZE used the internal rotor of a valve for the collection of effluent from the SEC microcolumn. The volume collected was reduced to 500 nL, which increased the resolution when compared to the valve-loop interface (20). However, a fixed volume again presented the same restrictions on the SEC and CZE operating parameters. An entirely different approach to the interface design was necessary to optimize the conditions in both of the microcolumns.

Lemmo and Jorgenson developed a third interface for μ-SEC–CZE in 1993. This design used a transverse flow of CZE buffer to prevent electromigration injections from occurring into the CZE capillary until the appropriate time. Figure 9.7 shows a block diagram of the "flow-gating interface" (18). The interface consisted of a Teflon

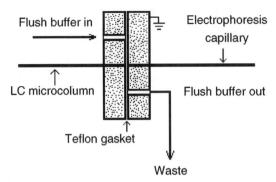

Figure 9.7   Schematic illustration of the flow-gating interface. A channeled Teflon gasket was sandwiched between two stainless steel plates to allow for flow into the electrophoresis capillary, either from the flush buffer reservoir or from the LC microcolumn during an electrokinetic injection.

gasket that separated two stainless steel plates. A channel cut into the Teflon allowed buffer to flow between the two plates, except when an injection was made. In comparison to a valve-loop interface, the flow gating interface delivered a more concentrated sample to the capillary, which resulted in an eight fold improvement in sensitivity. This design also allowed the CZE capillary to sample each of the SEC peaks at least three times and therefore the chromatographic column was not undersampled (23).

9.9   PACKED CAPILLARY REVERSE PHASE HIGH PERFORMANCE LIQUID CHROMATOGRAPHY–CAPILLARY ZONE ELECTROPHORESIS

The increased efficiency observed with μ-SEC–CZE led to the coupling of packed capillary reverse phase HPLC with CZE in order to separate mixtures of peptides. The contents of single cells were separated and detected by using this powerful technique. The reverse phase HPLC microcolumn used was 60 cm long, 50 μm in inner diameter, and was packed with a C8 modified silica packing material. Laser induced fluorescence detection of tetramethylrhodamine isothiocyanate-derived amines was used in this method. UV–VIS could not be used due to the extremely short pathlength and lack of sensitivity. A two-dimensional peak capacity of 20 000 was achieved by using this system (18). The reverse phase HPLC microcolumn used in this setup was more compatible with the CZE capillary in terms of volumetric flow and sampling volume than the larger LC columns used in the previous LC-CZE systems.

9.10   PACKED CAPILLARY REVERSE PHASE HIGH PERFORMANCE LIQUID CHROMATOGRAPHY– FAST CAPILLARY ZONE ELECTROPHORESIS

In order to improve the speed of CZE analysis, Monnig and Jorgenson developed on-column sample gating in 1991. The gating procedure allowed for rapid and automated sample introduction into the CZE capillary. In traditional CZE techniques, a plug of material is mechanically introduced at one end of the capillary, thus yielding a relatively slow sampling rate. In the on-column optical gating method, analytes were first tagged with fluorescein isothiocyanate, a fluorescent label, and then continuously introduced into one end of the capillary. A laser constantly photodegraded the tag near the entrance of the capillary. A sample zone was created by momentarily blocking the laser so that a narrow plug of fluorescent material was created in the column. This method allowed for the rapid injection to be made while the capillary was maintained at operating voltage.

The experimental setup for high-speed CZE can be seen in Figure 9.8. High-speed CZE, or fast CZE (FCZE), yielded 70 000 to 90 000 theoretical plates for the separation of amino acid mixtures. Complete separation was achieved in under 11 s, using a capillary length of 4 cm (24).

In 1995, Moore and Jorgenson used the optically gated CZE system to obtain extremely rapid separations with HPLC coupled to CZE. The rapid CZE analysis made possible more frequent sampling of the HPLC column, thus increasing the comprehensive resolving power. Complete two-dimensional analyses were performed in less than 10 min, with the CZE analyses requiring only 2.5s. A peak

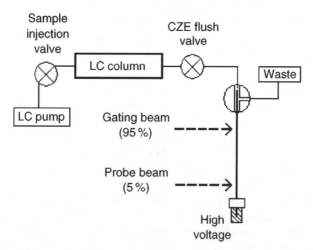

Figure 9.8   Schematic illustration of the two-dimensional HPLC/fast–CZE instrumental setup. The argon laser beam was set at 488 nm, which was used to photodegrade and detect the fluorescein isothiocyanate tag.

capacity of 650 was obtained from this HPLC–CZE system. Rapid fingerprinting of proteins was one of the suggested possible applications for this technique (25). The success of this two-dimensional technique led to the possibility of coupling HPLC-CZE to other techniques to further increase dimensionality and total peak capacity.

9.11   THREE-DIMENSIONAL SIZE EXCLUSION CHROMATOGRAPHY–REVERSE PHASE LIQUID CHROMATOGRAPHY–CAPILLARY ZONE ELECTROPHORESIS

Moore and Jorgenson combined the rapid two-dimensional separation achieved by LC-CZE with SEC to make the first comprehensive three-dimensional separation involving an electrodriven component in 1995. Size exclusion chromatography separated the analytes over a period of several hours while the reverse phase HPLC–CZE combination separated components in only 7 min. A schematic diagram of the three-dimensional SEC–reverse phase HPLC–CZE instrument is shown in Figure 9.9 (18). A dilution tee was placed between the SEC column and the reverse phase HPLC injection loop in order to dilute the eluent from the SEC column, since it contained more methanol than was optimal for the reverse phase HPLC column.

The three-dimensional method was used to separate mixtures of peptides, such as those produced from a tryptic digest of ovalbumin. The three orthogonal characteristics that were used to separate the peptides were size, hydrophobicity, and electrophoretic mobility. A three-dimensional representation of a region of the data gathered can be seen in Figure 9.10. A series of planar slices through the data volume, which form stacks of disks, make it possible to visualize the separation. The peak capacity that resulted from this separation was calculated to be 2800, much greater than the capacities of the individual techniques. The addition of SEC to the original reverse phase HPLC–CZE method increased the peak capacity by a factor of five. One disadvantage of this multidimensional separation system was that extremely concentrated samples had to be used in order to overcome the dilution

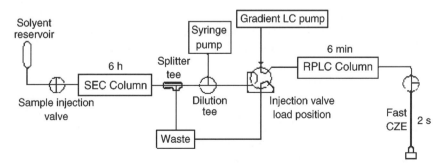

Figure 9.9   Schematic illustration of the instrumental setup used for three-dimensional SEC–RPLC–CZE.

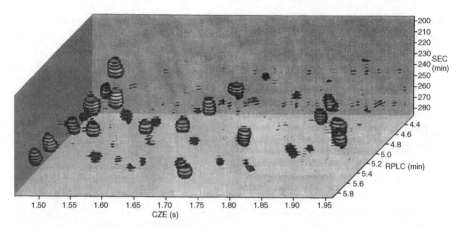

Figure 9.10 Three-dimensional representation of the data 'volume' of a tryptic digest of ovalbumin. Series of planar slices through the data volume produce stacks of disks in order to show peaks. Reprinted from *Analytical Chemistry*, **67**, A. W. Moore Jr and J. W. Jorgenson, 'Comprehensive three-dimensional separation of peptides using size exclusion chromatography/reversed phase liquid chromatography/optically gated capillary zone electrophoresis,' pp. 3456–3463, copyright 1995, with permission from the American Chemical Society.

inherent in the SEC–reverse phase HPLC interface (26). Another problem evident in this separation was the difficulty in analyzing the data, which is a common problem in multidimensional techniques, particularly those with more than two dimensions.

9.12 TRANSPARENT FLOW GATING INTERFACE WITH PACKED CAPILLARY HIGH PERFORMANCE LIQUID CHROMATOGRAPHY–CAPILLARY ZONE ELECTROPHORESIS

In order to observe the junction between micro-HPLC and CZE in two-dimensional systems, a transparent flow gating interface was developed by Hooker and Jorgenson in 1997. This design was similar to the original flow gating interface, except in the fact that it was made from clear plastic. An illustration of the transparent interface can be seen in Figure 9.11. Direct observation and manipulation of the micro-HPLC and CZE capillaries was made possible by the transparent interface. This new interface created a more routine and reproducible way of interfacing the two micro-columns as compared to the one developed in 1993. The split injection/flow system in the new interface delivered a nanoliter per second flow rate to the μ-HPLC column from the gradient LC pump, which was yet another improvement in the system.

The improved design of the gating interface resulted in precise alignment of the two capillaries. A colored dye solution was added to the HPLC eluent to allow for direct observation of the flow gating and injection processes. Through observation of the movement of the dye through the interface, it was possible to ensure that the electrokinetic injections were performed correctly. Troubleshooting had been a

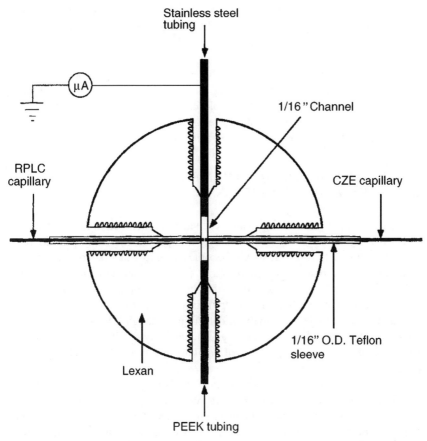

Figure 9.11 Schematic illustration of the transparent interface used to link the HPLC capillary to the CZE capillary. Reprinted from *Analytical Chemistry*, **69**, T. F. Hooker and J. W. Jorgenson, 'A transparent flow gating interface for the coupling of microcolumn LC with CZE in a comprehensive two-dimensional system', pp 4134–4142, copyright 1997, with permission from the American Chemical Society.

problem in the previous μ-HPLC–CZE system, but with the transparent interface this was much easier. This μ-HPLC–CZE separation system yielded high peak capacities, producing over 400 resolved peaks from biological samples (27).

9.13 ONLINE REVERSE PHASE HIGH PERFORMANCE LIQUID CHROMATOGRAPHY–CAPILLARY ZONE ELECTROPHORESIS–MASS SPECTROMETRY

Mass spectrometry (MS) is increasingly being combined with reverse phase HPLC or CZE in order to add an additional dimension to the data that a traditional detection system would not provide. A two-dimensional LC–CZE system with mass spectrometric detection was designed by Jorgenson and co-workers in 1997. In the coupling of the CZE capillary to the MS detector, a new microelectrospray needle was developed. Figure 9.12 shows a diagram of the silica sheath electrospray needle specially designed for this instrument. This needle produced high ionization efficiency, low flow rates, and a sheath flow that enabled the CZE to operate near optimal conditions. A transverse flow gating interface was again used to couple the reverse phase HPLC column to the CZE column. The result of this separation system was the combination of the resolving power of reverse phase HPLC and CZE, with mass spectrometric detection, all within 15 min (28).

9.14 THE FUTURE OF MULTIDIMENSIONAL ELECTROKINETIC SEPARATIONS

Electrodriven separation techniques are destined to be included in many future multidimensional systems, as CE is increasingly accepted in the analytical laboratory. The combination of LC and CE should become easier as vendors work towards providing enhanced microscale pumps, injectors, and detectors (18). Detection is often a problem in capillary techniques due to the short path length that is inherent in the capillary. The work by Jorgenson's group mainly involved fluorescence detection to overcome this limit in the sensitivity of detection, although UV–VIS would be less restrictive in the types of analytes detected. Increasingly sensitive detectors of many types will make the use of all kinds of capillary electrophoretic techniques more popular.

Data analysis is one aspect of multidimensional analyses that must be optimized in the future. The analysis of chromatographic data beyond one dimension is still exceedingly problematic, especially in the analyses of highly complex mixtures. Better software may need to be developed in order to analyze two- and three-dimensional peaks due to their complexity. Three-dimensional data is only useful today in terms of fingerprinting and often that even requires extensive data analysis. A great deal of research must still be carried out to make the interpretation and quantification of multidimensional data easier.

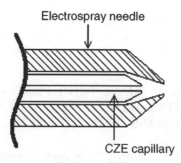

Electrospray needle

CZE capillary

Figure 9.12   Schematic diagram of the silica sheath electrospray needle used to interface capillary zone electrophoresis with a mass spectrometer.

9.15 CONCLUSIONS

Electrodriven techniques are useful as components in multidimensional separation systems due to their unique mechanisms of separation, high efficiency and speed. The work carried out by Jorgenson and co-workers has demonstrated the high efficiencies and peak capacities that are possible with comprehensive multidimensional electrodriven separations. The speed and efficiency of CZE makes it possibly the best technique to use for the final dimension in a liquid phase multidimensional separation. It can be envisaged that multidimensional electrodriven techniques will eventually be applied to the analysis of complex mixtures of all types. The peak capacities that can result from these techniques make them extraordinarily powerful tools. When the limitations of one-dimensional separations are finally realized, and the simplicity of multidimensional methods is enhanced, the use of multidimensional electrodriven separations may become more widespread.

REFERENCES

1. J. W. Jorgenson, 'Overview of electrophoresis', in *New Directions in Electrophoretic Methods*, Jorgenson, J. W. and M. Phillips (Eds), ACS Symposium Series 335, American Chemical Society, Washington, DC, pp. 1–19 (1987).
2. A. Tiselius, 'Electrophoresis of serum globulin', *J. Biochem.* **31**: 313–317 (1937).
3. F. E. P. Mikkers, F. M. Everaerts and P. E. M. Th. Verheggen, 'High-performance zone electrophoresis', *J. Chromatogr.* **169**: 11–20 (1979).
4. J. W. Jorgenson and K. DeArman Lukacs, 'Zone electrophoresis in open-tubular glass capillaries', *Anal. Chem.* **53**: 1298–1302 (1981).
5. J. C. Giddings, *Unified Separation Science,* John Wiley & Sons, New York, pp. 126–128 (1991).
6. R. Consden, A. H. Gordon and A. J. P. Martin, 'Qualitative analysis of proteins: partition chromatographic method using paper', *J. Biochem.* **38**: 224–232 (1944).
7. G. Haugaard and T. D. Kroner, 'Partition chromatography of amino acids with applied voltage', *J. Am. Chem. Soc.* **70**: 2135–2137 (1948).
8. E. L. Durrum, 'Two-dimensional electrophoresis and ionophoresis', *J. Colloid Sci.* **6**: 274–290 (1951).
9. K. Keck and U. Hagen, 'Separation of the DNA [deoxyribonucleic acid] units on cellulose layers', *Biochim. Biophys. Acta.* **87**: 685–687 (1964).
10. P. H. O'Farrell, 'High resolution two-dimensional electrophoresis of proteins', *J. Biol. Chem.* **250**: 4007–4021 (1975).
11. W. G. Burton, K. D. Nugent, T. K. Slattery, B. R. Summers and L. R. Snyder, 'Separation of proteins by reversed-phase high-performance liquid chromatography', *J. Chromatogr.* **443**: 363–379 (1988).
12. P. D. Grossman, J. C. Colburn, H. H. Lauer, R. G. Nielsen, R. M. Riggin, G. S. Sittampalam and E. C. Rickard, 'Application of free-solution capillary electrophoresis to the analytical scale separation of proteins and peptides', *Anal. Chem.* **61**: 1186–1194 (1989).
13. H. Yamamoto, T. Manabe and T. Okuyama, 'Gel permeation chromatography combined with capillary electrophoresis for microanalysis of proteins', *J. Chromatogr.* **480**: 277–283 (1989).

14. H. Yamamoto, T. Manabe and T. Okuyama, 'Apparatus for coupled high-performance liquid chromatography and capillary electrophoresis in the analysis of complex protein mixtures', *J. Chromatogr.* **515**: 659–666 (1990).

15. M. Castagnola, L. Cassiano, R. Rabino, D. V. Rossetti and F. Andreasi Bassi, 'Peptide mapping through the coupling of capillary electrophoresis and high-performance liquid chromatography: map prediction of the tryptic digest of myoglobin', *J. Chromatogr.* **572**: 51–58 (1991).

16. S. Pálmarsdóttir and L. E. Edholm, 'Enhancement of selectivity and concentration sensitivity in capillary zone electrophoresis by on-line coupling with column liquid chromatography and utilizing a double stacking procedure allowing for microliter injections', *J. Chromatogr.* **693**: 131–143 (1995).

17. M. Strömqvist, 'Peptide mapping using combinations of size-exclusion chromatography, reversed-phase chromatography and capillary electrophoresis', *J. Chromatogr.* **667**: 304–310 (1994).

18. T. F. Hooker, D. J. Jeffery and J. W. Jorgenson, 'Two-dimensional separations in high-performance capillary electrophoresis, in High Performance Capillary Electrophoresis, M. G. Khakedi (Ed.), John Wiley & Sons, New York, pp. 581–612 (1998).

19. M. M. Bushey and J. W. Jorgenson, 'Automated instrumentation for comprehensive two-dimensional high-performance liquid chromatography/capillary zone electrophoresis', *Anal. Chem.* **62**: 978–984 (1990).

20. J. P. Larmann-Jr, A. V. Lemmo, A. W. Moore and J. W. Jorgenson, 'Two-dimensional separations of peptides and proteins by comprehensive liquid chromatography–capillary electrophoresis', *Electrophoresis* **14**: 439–447 (1993).

21. R. Weinberger, 'The sixth annual Frederick conference on capillary electrophoresis', *Am. Lab.* **28**: 42–43 (1996).

22. A. V. Lemmo and J. W. Jorgenson, 'Two-dimensional protein separation by microcolumn size-exclusion chromatography–capillary zone electrophoresis', *J. Chromatogr.* **633**: 213–220 (1993).

23. A. V. Lemmo and J. W. Jorgenson, 'Transverse flow gating interface for the coupling of microcolumn LC with CZE in a comprehensive two-dimensional system', *Anal. Chem.* **65**: 1576–1581 (1993).

24. C. A. Monnig and J. W. Jorgenson, 'On-column sample gating for high-speed capillary zone electrophoresis', *Anal. Chem.* **63**: 802–807 (1991).

25. A. W. Moore-Jr and J. W. Jorgenson, 'Rapid comprehensive two-dimensional separations of peptides via RPLC-optically gated capillary zone electrophoresis', *Anal. Chem.* **67**: 3448–3455 (1995).

26. A. W. Moore, Jr and J. W. Jorgenson, 'Comprehensive three-dimensional separation of peptides using size exclusion chromatography/reversed phase liquid chromatography/ optically gated capillary zone electrophoresis', *Anal. Chem.* **67**: 3456–3463 (1995).

27. T. F. Hooker and J. W. Jorgenson, 'A transparent flow gating interface for the coupling of microcolumn LC with CZE in a comprehensive two-dimensional system', *Anal. Chem.* **69**: 4134–4142 (1997).

28. K. C. Lewis, G. J. Opiteck, J. W. Jorgenson and D. M. Sheeley, 'Comprehensive online RPLC–CZE–MS of peptides', *J. Am. Soc. Mass Spectrom.* **8**: 495–500 (1997).

Part 2

Applications

10 Multidimensional Chromatography: Foods, Flavours and Fragrances Applications

G. DUGO, P. DUGO and L. MONDELLO

Università di Messina, Messina, Italy

10.1 INTRODUCTION

Chromatography is the best technique for the separation of complex mixtures. Frequently, samples to be analysed are very complex, so the analyst has to choose more and more sophisticated techniques. Multidimensional separations, off-line and recently on-line, have been used for the analysis of such complex samples.

Food, flavour and fragrance products are a good example of natural complex mixtures. The analysis of these matrices may be carried out to:

- determine the qualitative and/or quantitative composition of a specific class of components;
- control the quality and the authenticity of the product;
- detect the presence of adulteration or contamination.

Sometimes, the monodimensional separation cannot be sufficient to resolve all of the components of interest. Problems of peak overlapping may occur, and a pre-separation of the sample is often necessary. This pre-separation has the aim of reducing the complexity of the original sample matrix, by separating a simpler fraction than the original matrix. The fraction should contain the same amount of the analyte as in the whole sample, ready for analysis and free from substances that can interfere during the chromatographic analysis. Often, the preseparation is carried out off-line because it is easy to operate, although it can present many disadvantages, such as long separation times, the possibility of contamination or formation of artefacts, the difficulty of a quantitative recovery of the components of interest, etc. On the other hand, many on-line pre-separation methods have now been developed that have the advantages of greatly reducing the total analysis time, compared to classical off-line sample preparation techniques, to give a good recovery of the analytes with minimal chance for contamination. The disadvantage is that the equipment is significantly more complex and expensive than for monodimensional chromatography.

Multidimensional Chromatography, edited by L. Mondello, A. C. Lewis and K. D. Bartle
©2002 John Wiley & Sons Ltd.

10.2 MULTIDIMENSIONAL GAS CHROMATOGRAPHY (GC–GC OR MDGC)

A large number of the organic compounds in food and beverages are chiral molecules. In addition, a significant number of the additives, flavours, fragrances, pesticides and preservatives that are used in the food industry are also chiral materials.

The enantiomeric distribution can be very useful for identifying adulterated foods and beverages, for controlling and monitoring fermentation processes and products, and evaluating age and storage effects (1).

The enantiomeric distribution of the components of essential oils can provide information on the authenticity and quality of the oil, on the geographical origin and on their biogenesis (2).

GC using chiral columns coated with derivatized cyclodextrin is the analytical technique most frequently employed for the determination of the enantiomeric ratio of volatile compounds. Food products, as well as flavours and fragrances, are usually very complex matrices, so direct GC analysis of the enantiomeric ratio of certain components is usually difficult. Often, the components of interest are present in trace amounts and problems of peak overlap may occur. The literature reports many examples of the use of multidimensional gas chromatography with a combination of a non-chiral pre-column and a chiral analytical column for this type of analysis.

Mosandl and his co-workers (3–17) have carried out many research studies on the determination of the enantiomeric ratio of various components of food and beverages, as well as plant materials and essential oils. Using a SiChromat 2–8 double-oven system with two independent temperature controls, two flame ionization detectors and a 'live switching' coupling piece, these workers have developed many applications of enantioselective MDGC employing heart-cutting technique from a non-chiral pre-separation column on to a chiral main column. In this way, direct chiral analysis is possible without any further clean-up or derivatization procedure. Table 10.1 summarizes some of these applications. As a typical example, Figure 10.1(a) shows the separation on a Carbowax 20M column of a dichloromethane extract of a 'strawberry' tea (18). As can be seen, the GC profile is very complex. Figure 10.1(b) shows the enantiomeric separation of 2,5-dimethyl-4-hydroxy-3[2H]-furanone, known as 'pineapple ketone', from the tea extract, transferred from the Carbowax 20M pre-column to a modified β-cyclodextrin column. This analysis allowed the detection of the synthetic racemate of 'pineapple ketone' that was added to the tea to give the strawberry flavour.

For the enantioselective flavour analysis of components present in extremely low concentrations, a MDGC–MS method has been developed (19). An example of the application of this technique is the determination of theaspiranes and theaspirones in fruits. These compounds are potent flavour compounds which are widely used in the flavours industry. Figure 10.2 shows the MDGC–MS chromatogram obtained by using multiple ion detection (MID) differentiation between the enantiomers of theaspiranes in an aglycone fraction from purple passion fruit. In fact, using the MID technique, interfering peaks are easily removed and the detection limit is lowered.

Table 10.1 Applications of MDGC reported by Mosandl and his co-workers. These were developed by using a Siemens SiChromat 2 double-oven system with two independent temperature controls, two flame-ionization detectors and a 'live switching' coupling piece, employing the heart-cutting technique

Application	Pre-column	Main column	Reference
Enantiomeric distribution of γ-lactone homologues from different apricot cultivars. Identification of dihydro-actinidiolide (co-eluted with γ-C11 on DB-1701)	Fused silica retention gap (10 m × 0.25 mm i.d.) coupled to a DB-1701 column (15 m × 0.25 mm i.d.; 1 μm 1 μmfilm thickness)	Glass capillary column (38 m × 0.2 mm i.d.) coated with heptakis (3-O-acetyl-2,6-di-O-pentyl)-β-cyclodextrin	3
Determination of chiral-γ-lactones from raw flavour extract of strawberries and other-fruit-containing foods and beverages	Fused silica retention gap (10 m × 0.25 mm i.d.) coupled to a DB-1701 column (15 m × 0.25 mm i.d; 1 μm film thickness)	Glass capillary column (38 m × 0.2 mm i.d.) coated with heptakis (3-O-acetyl-2,6-di-O-pentyl)-β-cyclodextrin	4
Determination of chiral α-pinene, β-pinene and limonene in essential oils and plant extracts	Fused silica Supelcowax™ 10 capillary column (60 m × 0.32 mm i.d; 0.25 μm film thickness)	Glass capillary column (47 m × 0.23 mm i.d.) coated with heptakis (2,3,6-tri-O-methyl)-β-cyclodextrin (10% in OV-1701-vinyl)	5
Stereoanalysis of 2-alkyl-branched acids, esters and alcohols in apple aroma concentrate	Restriction capillary (25 m × 0.23 mm i.d.) coupled to a glass capillary column, (25 m × 0.32 mm i.d.) coated with a 18.8% solution of PS-255 and 1.5% dicumyl peroxide.	Glass capillary column (38 m × 0.23 mm i.d.) coated with heptakis (2,3,6-tri-O-ethyl)-β-cyclodextrin (33% in OV-1701-vinyl)	6
Stereoanalysis of 2-methylbutanoate from apples and pineapples	Duran glass capillary (25 m × 0.23 mm i.d.) coupled to a Superox − 0.6 (0.25 m) glass capillary column (27 m × 0.32 mm i.d.)	Glass capillary column (38 m × 0.23 mm i.d.) coated with heptakis (2,3,6-tri-O-ethyl)-β-cyclodextrin (33% in OV-1701-vinyl)	7
Determination of enantiomeric distribution of some secondary alcohols and their acetates from banana	Fused silica Supelcowax™ 10 capillary column (60 m × 0.32 mm i.d.; 0.25 μm film thickness)	Glass capillary column (47 m × 0.23 mm i.d.) coated with heptakis (2,3,6-tri-O-methyl)-β-cyclodextrin (10% in OV-1701-vinyl)	8

Table 10.1 *(continued)*

Application	Pre-column	Main column	Reference
Determination of enantiomeric distribution of monoterpenoids from geranium oil	Fused silica Supelcowax™ 10 capillary column, (60 m × 0.32 mm i.d.; 0.25 μm film thickness)	Duranglas glass capillary column (26 m × 0.23 mm i.d.) coated with heptakis (2,3-di-O-acetyl-6-O-tert butyldimethyl-1-silyl)-β-cyclodextrin (50% in OV-1701 vinyl)	9
Determination of enantiomeric distribution of the lactone flavour compounds of fruits	OV-1701 fused silica capillary column, (50 m × 0.32 mm i.d., 0.25 μm film thickness)	Glass capillary column (47 m × 0.23 mm i.d.) coated with octakis (3-O-butiryl, 2,6-di-O-pentyl-γ-cyclodextrin	10
Determination of enantiomeric distribution of chiral components of *Cymbopogon* oil	Polyethylene glycol	Glass capillary column (47 m × 0.23 mm i.d.) coated with heptakis (2,3-di-O-acetyl-6-O-tert-butyldimethy-1-1-silyl)-β-cyclodextrin (50% in OV-1701-vinyl)	11
Determination of the enantiomeric distribution of the chiral major compounds of buchu leaf oil	Duranglas glass capillary column, (44 m × 0.23 mm i.d.) coated with a 0.5 m film of OV-215 vinyl	Duranglas capillary column (15 m × 0.23 mm i.d.) coated with heptakis (2,3-di-O-acetyl-6-O-tert-butyldimethy 1-1-silyl)-β-cyclodextrin (25%), octakis (2,3-di-O-acety 1-6-O-tert-butyldimethylsilyl)-γ-cyclodextrin (25%) and OV-1701-vinyl (50%)	12
Determination of the enantiomeric distribution of 2-, 3-, and 4-alkyl-branched acids from Roman Chamomile and Parmesan cheese	Glass capillary column (30 m × 0.32 mm i.d.) coated with SE-52 (0.65 μm film thickness)	Glass capillary column (30 m × 0.23 mm i.d.) coated with 15% heptakis (2,3-di-O-methyl-6-O-tert butyldimethylsilyl)-β-cyclodextrin in PS-268 (0.23 μm film thickness)	13

Application	Column 1	Column 2	Ref.
Determination of the enantiomeric distribution of 4-methyl-5-decanolide in white flowering aochids (*Aerangis confusa*)	Duraglas glass capillary column (30 m × 0.23 mm i.d.) coated with SE52 (0.63 μm film thickness)	Fused silica column (30 m × 0.25 mm i.d.) coated with 15% of heptakis (2,3-di-*O*-methyl-6-*O*-tert-butyl-dimethylsilyl)-β-cyclodextrin in PS-268 (0.25 μm film thickness)	14
Determination of the enantiomeric ratio of linalol and linalyl acetate in bergamot oil	Fused silica Supelcowax™ 10 capillary column (30 m × 0.32 mm i.d.; 0.23 mm film thickness)	Duraglas glass capillary column (26 m × 0.23 mm) coated with heptakis (2,3-di-*O*-acetyl-6-*O*-tert-butyldimethylsilyl 1)-β-cyclodextrin (50% in OV-1701 vinyl)	15
Determination of the enantiomeric ratio of α-pinene, β-pinene and limonene in bergamot oil	Duraglas glass capillary column (30 m × 0.23 mm i.d.) coated with OV-215 (0.23 m film thickness)	Duraglas glass capillary column (25 m × 0.23 mm i.d.) coated with heptakis (2,3-di-*O*-methyl-6-*O*-tert-butyldimethylsilyl)-β-cyclodextrin (50% in OV-1701 vinyl)	15
Determination of the enantiomeric distribution of linalol, linalyl acetate and α-terpineol in neroli and petitgrain oils	Fused silica Supelcowax™ 10 capillary column (30 m × 0.32 mm i.d.; 0.23 m film thickness)	Duraglas glass capillary column (26 m × 0.23 mm) coated with heptakis (2,3-di-*O*-acetyl-6-*O*-tertbutyldimethylsilyl)-β-cyclodextrin (50% in OV-1701 vinyl)	16
Determination of the enantiomeric distribution of α-pinene, β-pinene, limonene, terpinen-4-ol and nerolidol in neroli and petitgrain oils	Duraglas glass capillary column (30 m × 0.23 mm i.d.) coated with OV-215 (0.23 m film thickness)	Duraglas glass capillary column (25 m × 0.23 mm i.d.) coated with heptakis (2,3-di-*O*-methyl-6-*O*-tert-butyldimethylsilyl)-β-cyclodextrin (50% in OV-1701 vinyl)	16
Determination of the enantiomeric distribution of some chiral sulfur-containing trace components of yellow passion fruit [a]	DB-210 fused silica capillary column (30 m × 0.32 mm i.d.; film thickness 0.25 μm)	Fused silica column (30 m × 0.32 mm i.d.; film thickness 0.32 μm) coated with octakis (2,3-di-*O*-butyryl-6-*O*-tertbutyldimethylsilyl)-γ-cyclodextrin (50% in OV-1701 vinyl)	17

[a]Application developed by using a Fisons GC 8000 chromatograph where the two columns were installed and coupled via a moving capillary stream switching (MCSS) system. The chromatograph was equiped with a flame-ionization detector on the MCSS system outlet and a Flame-photometric detector on the main column outlet, and a split/splitless injector.

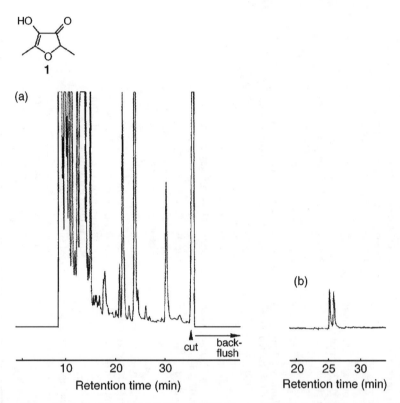

Figure 10.1   Analysis of racemic 2,5-dimethyl-4-hydroxy-3[2H]-furanone (**1**) obtained from a 'strawberry' tea, flavoured with the synthetic racemate of **1** (natural component), using an MDGC procedure: (a) dichloromethane extract of the flavoured 'strawberry' tea, analysed on a Carbowax 20M pre-column (60 m, 0.32 mm i.d., 0.25 μm film thickness; carrier gas H$_2$, 1.95 bar; 170 °C isothermal); (b) chirospecific analysis of (**1**) from the strawberry tea extract, transferred for stereoanalysis by using a permethylated β-cyclodextrin column (47 m × 0.23 mm i.d.; carrier gas H$_2$, 1.70 bar; 110 °C isothermal). Reprinted from *Journal of High Resolution Chromatography*, **13**, A. Mosandl *et al.*, 'Stereoisomeric flavor compounds. XLIV: enantioselective analysis of some important flavor molecules', pp. 660–662, 1990, with permission from Wiley-VCH.

Mondello *et al.* (2, 20–23) have used a multidimensional gas chromatographic system based on the use of mechanical valves which were stable at high temperatures developed in their laboratory for the determination of the enantiomeric distribution of monoterpene hydrocarbons (β-pinene, sabinene and limonene) and monoterpene alcohols (linalol, terpinen-4-ol and α-terpineol) of citrus oils (lemon, mandarin, lime and bergamot). Linalyl acetate was also studied in bergamot oil. The system consisted of two Shimadzu Model 17 gas chromatographs, a six-port two-position valve and a hot transfer line. The system made it possible to carry out fully

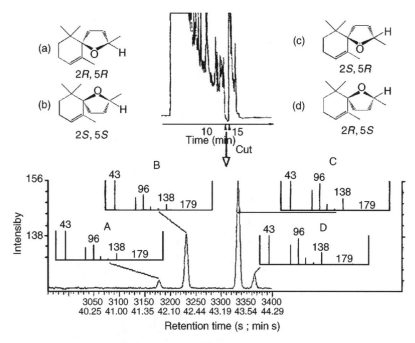

Figure 10.2 MDGC–MS differentiation between the enantiomers of theaspiranes in an aglycone fraction from purple passion fruit: DB5 pre-column (25 m × 0.25 mm i.d., 0.25 μm film thickness; carrier gas He, 0.66 ml/min; oven temperature, 60–300 °C at 10 °C/min with a final hold of 25 min); permethylated β-cyclodextrin column (25 m × 0.25 mm i.d., 0.25 μm film thickness; carrier gas He, 1.96 ml/min; 80 °C isothermal for 20 min and then programmed to 220 °C at 2 °C/min). Reprinted from *Journal of High Resolution Chromatography*, **16**, G. Full *et al.*, 'MDGC–MS: a powerful tool for enantioselective flavor analysis', pp. 642–644, 1993, with permission from Wiley-VCH.

automated multiple transfers, because retention times on the pre-column were reproducible even for those components eluted after numerous transfers. Moreover, when the MDGC system was not used, the two chromatographs can be operated independently without any hardware modification. Figure 10.3 shows chromatograms of a cold pressed lemon oil obtained with an SE-52 pre-column and the system in the stand-by position, the same oil obtained with the SE-52 column and the system in the cut position, and that obtained with the chiral column (modified β-cyclodextrin) for the fractions transferred from the pre-column.

Table 10.2 reports some results obtained for cold-pressed and distilled citrus oils. As can be seen, the values obtained are characteristic of the different oils, and can be used as references for the authenticity and quality of the oil.

In recent years, together with enantioselective analysis, the determination of the natural abundance of stable isotopes by means of stable isotope ratio mass spectrometry (IRMS) can be very useful for the assignment of the origin of foods and food ingredients, and of authenticity evaluation (24).

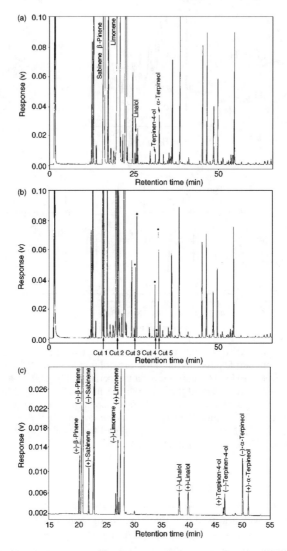

Figure 10.3 Gas chromatograms of a cold-pressed lemon oil obtained (a) with an SE-52 column in the stand-by position and (b) with the same column showing the five heart-cuts; (c) shows the GC–GC chiral chromatogram of the transferred components. The asterisks in (b) indicate electric spikes coming from the valve switching. The conditions were as follows: SE-52 pre-column, 30 m, 0.32 mm i.d., 0.40–0.45 μm film thickness; carrier gas He, 90 KPa (stand-by position) and 170 KPa (cut position); oven temperature, 45 °C (6 min)–240 °C at 2 °C/min: diethyl-*tert*-butyl-*β*-cyclodextrin column, 25 m × 0.25 mm i.d., 0.25 μm film thickness; carrier gas He, 110 KPa (stand-by position) and 5 KPa (cut position); oven temperature, 45 °C (6 min), rising to 90 °C (10 min) at 2 °C/min, and then to 230 °C at 2 °C/min. Reprinted from *Journal of High Resolution Chromatography*, **22**, L. Mondello *et al.*, 'Multidimensional capillary GC–GC for the analysis of real complex samples. Part IV. Enantiomeric distribution of monoterpene hydrocarbons and monoterpene alcohols of lemon oils', pp. 350–356, 1999, with permission from Wiley-VCH.

Table 10.2 Enantiomeric distribution of various components in cold-pressed and distilled citrus oils (2, 20–23)

Component	Cold-pressed key lime (+)	(−)	Cold-pressed Persian lime (+)	(−)	Distilled lime (+)	(−)	Cold-pressed Bergamot (+)	(−)	Distilled Bergamot (+)	(−)
β-Pinene	3.4–3.5	96.6–96.5	9.1–10.3	90.9–89.7	3.2–4.0	96.8–96.0	6.8–9.5	93.2–90.5	8.9–8.2	91.1–91.8
Sabinene	15.1–15.2	84.9–84.8	18.2–23.4	81.8–76.6	–	–	14.1–18.8	85.9–81.2	15.9–15.2	84.1–84.8
Limonene	97.1–98.2	2.9–1.8	96.6–97.3	0.4–2.7	94.5–91.3	5.5–8.7	98.1–97.3	1.9–2.7	98.0–97.7	2.0–2.3
Linalol	29.8–28.5	70.2–71.5	45.6–30.7	54.4–69.3	50.2–50.0	49.8–50.0	0.6–0.3	99.4–99.7	18.4–1.3	81.6–98.7
Terpinen-4-ol	29.2–29.5	70.8–70.5	18.6–24.9	81.4–75.1	42.3–45.0	57.7–55.0	9.7–26.3	90.3–73.7	31.8–27.1	68.1–72.9
α-Terpineol	17.2–14.5	82.8–85.5	25.5–19.2	74.5–80.8	46.7–43.2	53.3–56.8	82.5–30.6	17.5–69.4	73.4–88.8	26.6–11.2
Linalyl acetate	–	–	–	–	–	–	0.3–0.1	99.7–99.9	1.1–0.9	98.9–99.1

Component	Cold-pressed mandarin (+)	(−)	Distilled mandarin (+)	(−)	Cold pressed lemon (+)	(−)	Distilled lemon (+)	(−)
β-Pinene	97.0–98.8	3.0–1.2	96.1–97.7	3.9–2.3	4.2–7.0	95.8–93.0	6.4–6.6	93.6–93.4
Sabinene	76.2–80.5	23.8–19.5	77.8–79.9	22.2–20.1	12.5–15.3	87.5–84.7	14.6–12.7	85.4–87.3
Limonene	98.0–97.7	2.0–2.3	98.3–97.8	1.7–2.2	98.0–98.5	2.0–1.5	98.3–98.3	1.7–1.7
Linalol	86.9–80.2	13.1–19.8	83.7–79.6	16.3–20.4	28.5–49.2	71.5–50.8	40.0–46.9	60.0–53.1
Terpinen-4-ol	10.0–19.2	90.0–81.8	25.3–28.7	74.7–71.3	13.7–26.9	86.3–73.1	28.5–28.4	71.5–71.6
α-Terpineol	30.4–23.2	69.6–76.8	38.8–26.6	61.2–73.4	18.0–35.8	82.0–64.2	23.6–23.0	76.4–77.0
Linalyl acetate	–	–	–	–	–	–	–	–

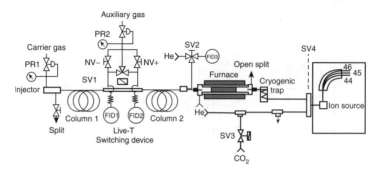

Figure 10.4 Schematic representation of the multidimensional GC–IRMS system developed by Nitz *et al.* (27): PR1 and PR2, pressure regulators; SV1–SV4, solenoid valves; NV− and NV+ , needle valves; FID1–FID3, flame-ionization detectors. Reprinted from *Journal of High Resolution Chromatography*, **15**, S. Nitz *et al.*, 'Multidimensional gas chromatography–isotope ratio mass spectrometry, (MDGC–IRMS). Part A: system description and technical requirements', pp. 387–391, 1992, with permission from Wiley-VCH.

Conventional IRMS requires relatively large sample volumes in a purified gaseous form. Recently, an 'on-line' GC–IRMS system has been developed which combines the high purification effect of GC with the utmost precision of IRMS. Sometimes this system may not be sufficient to determine characteristic minor components from complex matrices, and therefore MDGC–IRMS systems have been developed for the analysis of complex plant extracts and flavour components (25–27).

Figure 10.4 shows a schematic representation of the multidimensional GC–IRMS system developed by Nitz *et al.* (27). The performance of this system is demonstrated with an application from the field of flavour analysis. A Siemens SiChromat 2–8 double-oven gas chromatograph equipped with two FIDs, a live-T switching device and two capillary columns was coupled on-line with a triple-collector (masses 44, 45 and 46) isotope ratio mass spectrometer via a high efficiency combustion furnace. The column eluate could be directed either to FID3 or to the MS by means of a modified 'Deans switching system'.

Figure 10.5 shows the gas chromatograms obtained from a natural *cis*-3-hexen-1-ol fraction after GC–IRMS (a) and MDGC–IRMS (b) analysis. These authors studied the applicability of the multidimensional method by comparing the $\delta\,^{13}C$ values of the components determined by the two methods. The data obtained are reported in Table 10.3. The well-separated compounds (*trans*-2-hexenal and *trans*-2-hexen-1-ol) in both measurements show good congruity. Studies carried out with standard mixtures showed that MDGC–IRMS analysis can give better precision, because standard deviation data are significantly lower than those obtained by conventional GC–IRMS analysis.

In 1998, another application of MDGC–IRMS analysis was developed (25). A Siemens SiChromat 2–8 MDGC system, connected to a Finnigan MAT Delta S

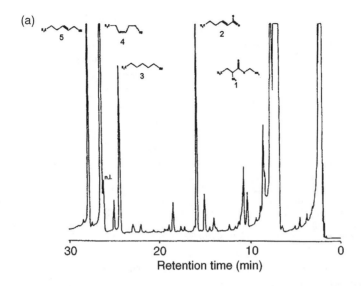

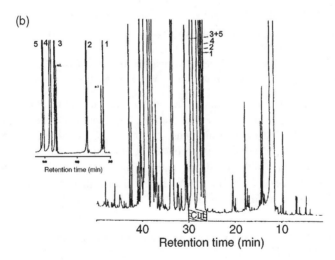

Figure 10.5 (a) Gas chromatogram obtained from a natural *cis*-3-hexen-1-ol fraction after unidimensional separation: CPWax-52-CB column × 25 m, 0.32 mm i.d., 1.2 μm film thickness; carrier gas He, 0.7 bar; oven temperature, 60 °C (5 min) – 180 °C at 2 °C/min). (b) Gas chromatogram obtained from a natural *cis*-3-hexen-1-ol fraction after multidimensional separation: SPB-5 pre-column, 30 m × 0.25 mm i.d., 1 μm film thickness; carrier gas He, 1.8 bar; oven temperature; 60 °C (5 min) – 100 °C at 2 °C/min with a hold of 5 min and then programmed to 220 °C at 5 °C/min; DB-Wax column 30 m × 0.25 mm i.d., 0.25 μm film thickness; carrier gas He, 1.3 bar; oven temperature 60 °C (5 min), then rising to 220 °C at 2 °C/min (cf. Table 10.3). Reprinted from *Journal of High Resolution Chromatography*, **15**, S. Nitz *et al.*, 'Multidimensional gas chromatography – isotope ratio mass spectrometry, (MDGC – IRMS). Part A: system description and technical requirements', pp. 387–391, 1992, with permission from Wiley-VCH.

Table 10.3 $\delta\,^{13}C$ values of the various components of a natural cis-3-hexen-1-ol fraction (cf. Figure 10.5) (27)

		$\delta\,^{13}C$ values ($n = 5$)	
No.[a]	Component	Conventional GC—IRMS[b]	MDGC-IRMS[b]
1	Ethyl 2-methylbutyrate	–	-27.20 ± 0.4
2	trans-2-Hexenal	-29.23 ± 0.3	-29.71 ± 0.3
3	1-Hexanol	-30.91 ± 0.3	-32.49 ± 0.2
4	cis-3-Hexen-1-ol	-32.95 ± 0.2	-34.99 ± 0.4
5	trans-2-Hexen-1-ol	-31.09 ± 0.5	-31.12 ± 0.3

[a] Numbering as shown on Figure 10.5.
[b] Values given relative to the universal standard PDB (*Belemnitella americana* from the *Cretaceous Pedee* formation, South California, USA).
Reprinted from *Journal of High Resolution Chromatography*, **15**, S. Nitz et al., 'Multidimensional gas chromatography–isotype ratio mass spectrometry (MDGC–IRMS). Part A: system description and technical requirements, pp. 387–391, 1992, with permission from Wiley-VCH.

isotope mass spectrometer via a combustion interface was employed. The results confirmed those previously obtained by Nitz et al. (27), as follows:

- the $\delta^{13}C$ values obtained by MDGC–IRMS are in agreement with those obtained by GC–IRMS and also elemental analysis measurements;
- the precision of the measurements expressed as standard deviations is significantly lower than 0.03‰;
- the results demonstrated the absence of isotope discrimination and systematical errors due to the MDGC system with live switching by 'live T-piece' coupling.

The authors demonstrated the importance that correct use of the MDGC–IRMS system is essential for the achievement of precise and accurate measurements. Table 10.4 reports the GC–IRMS measurements of some standard reference materials, obtained with different cut conditions. As can be seen from this table, premature cuts result in $\delta\,^{13}C$ values which are significantly higher than the true values, while delayed cuts give lower $\delta\,^{13}C$ values. This fact indicates that the beginning of the peak is enriched in ^{13}C, while the end is depleted.

Table 10.4 GC–IRMS measurements of standard reference materials obtained under different cut conditions (25)

	$\delta\,^{13}C$ (‰)		
Condition	5-Nonanone	Menthol	(R)-γ-Decalactone
Complete cuts	-27.84	-26.61	-30.05
Premature cuts	-3.58	-14.70	-13.09
Delayed cuts	-57.06	-71.60	-91.28

Reprinted from *Journal of High Resolution Chromatography*, **21**, D. Juchelka et al., 'Multidimensional gas chromatography coupled on-line with isotype ratio mass spectrometry (MDGC–IRMS): progress in the analytical authentication of genuine flavor components', pp. 145–151, 1998, with permission from Wiley-VCH.

Wines and other alcoholic beverages such as distillates represent very complex mixtures of aromatic compounds in an ethanol–water mixture. Once an extract or concentrate of the required compounds is prepared, a suitable chromatographic system must be used to allow separation and resolution of the species of interest. Many applications have been developed that use MDGC.

In 1991, Askari *et al.* (28) carried out the direct determination of the enantiomeric distribution of the monoterpene alcohols (α-terpineol, linalol, and the furanoid linalol oxides) in muscat grapes, musts and wines by using MDGC analysis. This work was carried out on trichlorofluoromethane extracts, using a SiChromat 2 chromatograph equipped with a Carbowax 20M pre-column and differently modified β-cyclodextrin columns. The results obtained for fresh grape juices show characteristic enantiomeric ratios for monoterpene alcohols (stereo-controlled enizmatic biogenesis), while the same alcohols are present as a racemic mixture in the corresponding wines.

Sotolon (4,5-dimethyl-3-hydroxy-2(5H)-furanone) and solerone (4-acetyl-γ-butirrolactone) were claimed to be responsible for some aroma characteristic of flor sherries wines. These compounds are present only as traces, and are chemically unstable. A system of two gas chromatographs coupled with a four-port switching valve was used to quantitate these components without previous fractionation. The first chromatograph was equipped with an on-column injector, in order to avoid thermal degradation of sotolon in the heated injector, a DB-5 column and an FID. The second chromatograph was equipped with an on-column injector, a DB-1701 column and an FID. The method allowed quantification of solerone and sotolon at concentrations as low as a few ppb (29).

The use of specific detectors together with the MDGC system can help in the identification of specific compounds, for example, those containing nitrogen and sulfur, present in very low concentrations in complex matrices. An MDGC system that consists of a temperature-programmed cold injection system, a multicolumn switching system (Gerstel), and two HP 5890 GC ovens connected by a cryotrap interface, has been used for the analysis of nitrogen and sulfur components of a whiskey extract. The second oven was equipped with a mass-selective detector, a chemiluminescence sulfur detector and a nitrogen thermionic detector (30). Figure 10.6 shows a selected cut from the pre-column (Carbowax 20M), cryofocused and passed into the main column (5% polydimethylsiloxane) for simultaneous sulfur and nitrogen MS detection. Some sulfur and nitrogen compounds were easily located in the TIC trace from their respective specific traces, and could thus be readily identified.

Another way to improve the analysis of complex matrices can be the combination of a multidimensional system with information-rich spectral detection (31). The analysis of eucalyptus and cascarilla bark essential oils has been carried out with an MDGC instrument, coupling a fast second chromatograph with a matrix isolation infrared spectrometer. Eluents from the first column were heart-cut and transferred to a cryogenically cooled trap. The trap is then heated to re-inject the components into an analytical column of different selectivity for separation and subsequent detection. The problem of the mismatch between the speed of fast separation and the

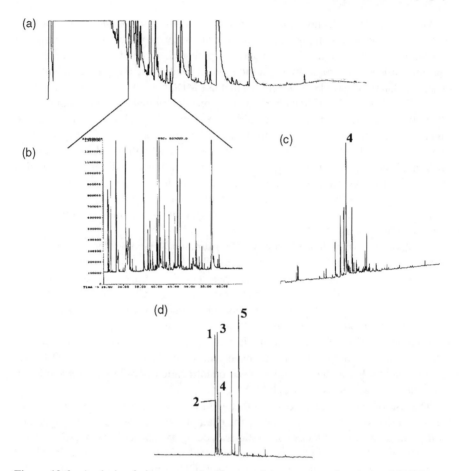

Figure 10.6 Analysis of nitrogen- and sulfur-containing compounds using an MDGC system: (a) pre-column chromatogram; (b) main column chromatogram, TIC; (c) main column chromatogram, nitrogen trace; (d) main column chromatogram, sulfur trace of the selected cut shown in A, of a whiskey extract. The conditions were as follows: precolumn, Carbowax 20M, 15 m × 0.25 mm i.d., 0.25 μm film thickness; main column, 5% diphenylpolysiloxane, 60 m × 0.25 mm i.d., 0.25 μm film thickness; column segments to selective detectors, 5 m; column segment to MS detector, 15 m. Reprinted from *Proceedings of the 15th International Symposium on Capillary Chromatography*, K. MacNamara and A. Hoffmann, 'Simultaneous nitrogen, sulphur and mass spectrometric analysis after multi-column switching of complex whiskey flavour extracts', 1993, with permission from Prof. P. Sandra.

slow response of the IR spectrometer was solved by the use of an efficient off-line interface. The matrix isolated eluents can be analysed by infrared spectrometry in a near real-time analysis. Figure 10.7 shows the separation of a fraction of eucalyptus oil. By using this technique, even minor components can be detected with sufficient IR spectral quality to permit their identification by a spectral library search.

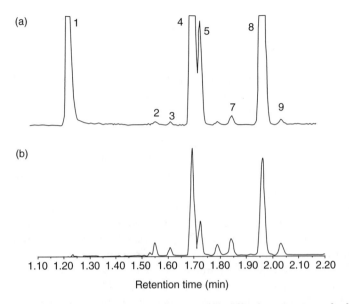

Figure 10.7 (a) IR reconstructed chromatogram and (b) FID chromatogram of a fraction of the eucalyptus oil transferred from a DB-Wax column. The secondary separation was carried out by using an RTX-1701 column, isothermally at 95 °C with a linear velocity of ~90 cm/s. Peak identification is as follows: (1) water; (2) α-pinene; (3) camphene; (4) sabinene; (5) β-pinene; (6) α-phellandrene; (7) α-terpinene; (8) unknown; (9) γ-terpinene. Reprinted from *Analytical Chemistry*, **66**, N. Ragunathan *et al.*, 'Multidimensional fast gas chromatography with matrix isolation infrared detection', pp. 3751–3756, copyright 1994, with permission from the American Chemical Society.

10.3 MULTIDIMENSIONAL HIGH PERFORMANCE LIQUID CHROMATOGRAPHY

Multidimensional HPLC offers very high separation power when compared to monodimensional LC analysis. Thus, it can be applied to the analysis of very complex mixtures. Applications of on-line MD-HPLC have been developed, using various techniques such as heart-cut, on-column concentration or trace enrichment; applications in which liquid phases on both columns are miscible and compatible are frequently reported, but the on-line coupling of columns with incompatible mobile phases have also been studied.

The on-line combination of high performance gel filtration chromatography (GFC), using aqueous compatible, rigid microparticulate exclusion columns, and reversed-phase chromatography (RPC), using C18 columns, have been employed by Enri and Frei (32) to separate senna glicoside extracts. This combination is ideal from the standpoint that the solvents in both techniques are compatible. Another example of this technique is the analysis of vitamin B in a protein food (33). The size-exclusion chromatography (SEC) system consisted of a Micropak TSK-2000

SW column using methanol:water (1 : 9) containing 0.1 M KH_2PO_4 and 0.01 M 1-heptanesulfonic acid for elution, and a MicroPak MCH-10 RP column, eluted with a methanol/water gradient solvent system.

On-column concentration of vitamins in food matrices offers several advantages, i.e. reducing sample preparation times, and increasing detection sensitivity through the use of higher sample loading on the exclusion chromatography (EC) column. Using an analogous method, Apffel et al. (33) carried out the analysis of sugars in a candy formulation. These same authors also analysed sugars in molasses by using a heart-cutting technique and normal-phase HPLC in the second dimension, as shown in Figure 10.8. In this case, the solvents for the two columns are non-compatible, and the volume of solvent that can be passed from the EC column to the normal-phase column is thus greatly restricted.

Johnson et al. (34) coupled SEC in the non-aqueous mode (Micropak TSK gel eluted with tetrahydrofuran) to a gradient RP LC system using acetonitrile/water for the determination of malathion in tomato plants and lemonin in grapefruit peel.

Gel permeation chromatography (GPC)/normal-phase HPLC was used by Brown-Thomas et al. (35) to determine fat-soluble vitamins in standard reference material (SRM) samples of a fortified coconut oil (SRM 1563) and a cod liver oil (SRM 1588). The on-line GPC/normal-phase procedure eliminated the long and laborious extraction procedure of isolating vitamins from the oil matrix. In fact, the GPC step permits the elimination of the lipid materials prior to the HPLC analysis. The HPLC columns used for the vitamin determinations were a 10 μm polystyrene/divinylbenzene gel column and a semipreparative aminocyano column, with hexane, methylene chloride and methyl tert-butyl ether being employed as solvent.

Figure 10.9 shows the chromatograms of fortified coconut oil obtained by using (a) normal-phase HPLC and (b) GPC/normal-phase HPLC. As can be seen from these figures, chemical interferences due to lipid material in the oil were eliminated by using the MD system that was used for quantitative analysis of all of the compounds, except DL-α-tocopheryl acetate, where the latter was co-eluted with a trigliceride compound and needed further separation.

Reversed-phase HPLC employing UV detection has been used for the analysis of polar pesticides in aqueous samples, although the sensitivity is usually insufficient for trace analysis experiments. The selectivity can be improved by using more selective electrochemical or fluorescence detectors, suitable for a restricted number of samples, or an additional sample preparation step can be used. A major advantage can be gained using on-line sample enrichment by large-volume injection in combination with column-switching techniques. Hogendoorn and co-workers (36–38) have used this technique for the analysis of single or groups of polar pesticides in water samples. These authors used two C18 columns with high separation power, increasing the selectivity by applying the cutting technique and the sensitivity by using large-volume injections. Chloroallyl alcohol (CAAL), bentazone, isoproturon, metamitron, pentachlorophenol, and other polar pesticides were determined, with the detection limits being at levels of 0.1 μg/l.

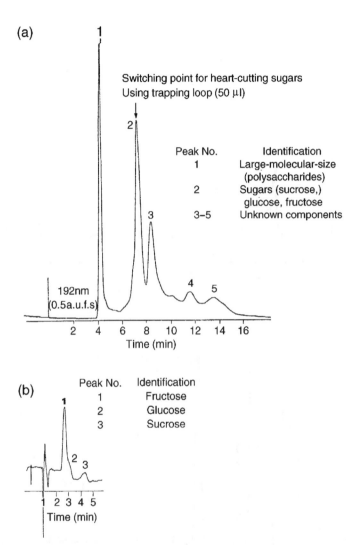

Figure 10.8 Application of the heart-cutting method for the determination of sugars in molasses: (a) exclusion chromatogram of molasses: (b) normal-phase chromatogram of molasses heart-cut. The conditions were as follows: exclusion chromatography, MicroPak TSK 2000 PW column (30 cm × 7.5 mm i.d.); 0.1% acetic acid in water at 1.2 ml/min; detection, UV, 192 nm, 0.5 a.u.f.s.; injection volume, 100 μl: normal-phase chromatography, MicroPak NH2-10 column (30 cm × 4 mm i.d.); acetonitrile–water (65 : 35) at 2.8 ml/min; detection, UV, 192 nm, 0.2 a.u.f.s.; trapping loop volume, 50 μl. Reprinted from *Journal of Chromatography*, **206**, J. A. Apffel *et al.*, 'Automated on-line multi-dimensional high performance liquid chromatographic techniques for the clean-up and analysis of water-soluble samples', pp. 43–57, copyright 1981, with permission from Elsevier Science.

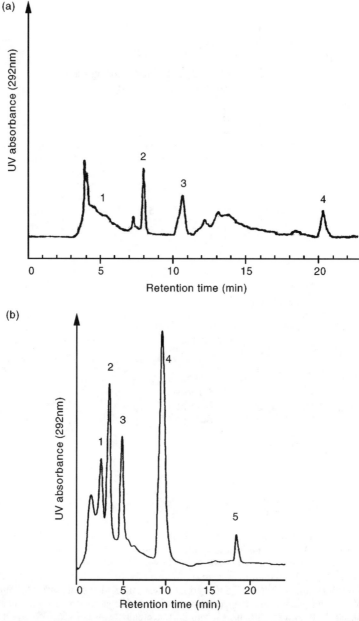

Figure 10.9 Chromatograms of fortified coconut oil obtained by using (a) normal-phase HPLC and (b) GPC/normal-phase HPLC. Peak identification is as follows: 1 (a,b), DL-α-tocopheryl acetate, 2 (b), 2,6-di-*tert*-butyl-4-methylphenol; 2 (a) and 3 (b), retinyl acetate; 3 (a) and 4 (b), tocol; 4 (a) and 5 (b), ergocalciferol. Reprinted from *Analytical Chemistry*, **60**, J. M. Brown-Thomas *et al.*, 'Determination of fat-soluble vitamins in oil matrices by multidimensional high-performance liquid chromatography', pp. 1929–1933, copyright 1988, with permission from the American Chemical Society.

10.4   MULTIDIMENSIONAL CHROMATOGRAPHY USING ON-LINE COUPLED HIGH PERFORMANCE LIQUID CHROMATOGRAPHY AND CAPILLARY (HIGH RESOLUTION) GAS CHROMATOGRAPHY (HPLC–HRGC)

In coupled LC–GC, specific components or classes of components of complex mixtures are pre-fractionated by LC and are then transferred on-line to a GC system for analytical separation. Because of the ease of collecting and handling liquids, off-line LC–GC techniques are very popular, but they do present several disadvantages, e.g. the numerous steps involved, long analysis times, possibility of contamination, etc. The on-line coupled LC–GC techniques avoid all of these disadvantages, thus allowing us to solve difficult analytical problems in a fully automated way.

Different transfer techniques and type of interfaces have been developed. Most of the applications involve normal-phase HPLC conditions, although reversed-phase coupled with capillary GC has also been reported.

On-line coupled LC–GC methods have been developed in food analysis for several reasons, i.e. lower detection limits can be reached, the clean-up is more efficient, and large numbers of samples can be analysed with a minimum of manual sample preparation in shorter times.

Several applications involve the removal of large amounts of triglicerides, including the determination of wax esters in olive oil (39), sterols and other minor components in oils and fats (40, 41), PCBs in fish (42), lactones in food products (43, 44), pesticides (45), and mineral oil products in food (46, 47). Grob *et al.* (47) studied the capacity of silica gel HPLC columns for retaining fats, and concluded that the capacity of such columns is proportional to their size, although the fractions of the volumes that are then transferred to the GC system grow proportionally with the column capacity. For these reasons, 2–3 mm i.d. LC columns are to be preferred for LC–GC applications.

In 1996, Mondello *et al.* (48) published a review article on the applications of HPLC–HRGC developed for food and water analysis over the period from 1986 to 1995. These authors cited 98 references, grouped by following a chronological order and by the subject of the application, as follows:

- composition of edible oils and fats;
- composition of essential oils and flavour components;
- contamination of water and food products;
- mineral oil products in food.

For each application, the LC and GC conditions are listed, together with the type of interface used, and some additional comments on the technique employed and the detection method.

The analysis of sterols, sterols esters, erythrodiol and uvaol, and other minor components of oils and fats, is usually carried out by normal-phase HPLC–HRGC by using a loop-type interface and the concurrent eluent evaporation technique, as reported in the papers cited by Mondello *et al.* (48) (up to 1995) and in more recent papers (49, 50). More recently, reversed-phase LC–GC methods have been

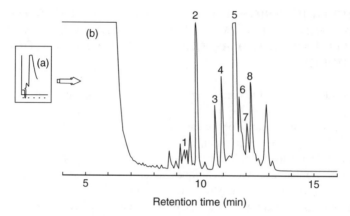

Figure 10.10 Liquid chromatogram (a) of sunflower oil and gas chromatogram (b) obtained after transfer of the indicated fraction (3416 μl of methanol–water eluent (78 : 22). Peak identification is as follows: (1) γ-tocopherol; (2) α-tocopherol; (3) campesterol; (4) stigmasterol; (5) β-sitosterol; (6) Δ⁵-avenasterol; (7) Δ⁷-stigmasterol; (8) Δ⁷-avenasterol. LC conditions: 50 × 4.6 mm i.d. column, slurry packed with 10 μm silica; flow rate, 2000 μl/min; UV detection, 205 nm. GC conditions: 5% phenyldimethylpolysiloxane column (25 m × 0.25 mm i.d.; 0.25 μm film thickness); carrier gas, He; column temperature, from 130 °C to 265 °C at 20 °C/min, and then to 300 °C at 3 °C/min. LC–GC transfer: the discharge of the large volume of vapour resulting from the aqueous eluent during LC–GC transfer is promoted by removing the GC column end from the PTV body; after completion of the transfer step, temperature programming of the GC column was started, and the PTV injector was heated at 14 °C/sec to 350 °C. Reprinted from *Journal of Agricultural and Food Chemistry*, **46**, F. J. Señoráns *et al.*, 'Simplex optimization of the direct analysis of free sterols in sunflower oil by on-line coupled reversed phase liquid chromatography–gas chromatography', pp. 1022–1026, copyright 1998, with permission from the American Chemical Society.

developed for the determination of free sterols in edible oils by using a programmable temperature vaporiser (PTV) (41, 51–53). Figure 10.10 shows the LC–GC separation of the sterol fraction of a sunflower oil obtained by using this system.

Many applications involving the study of the composition of essential oils are based on the use of the on-column interface and retention gap techniques because of the high volatility of the components to be analysed.

In the case of citrus essential oils, LC pre-fractionation can be used to obtain more homogeneous chemical classes of compounds for analysis by GC without any problems of overlapping peaks.

Mondello *et al.* (54) have developed some applications of on-line HPLC–HRGC and HPLC–HRGC/MS in the analysis of citrus essential oils. In particular, they used LC–GC to determine the enantiomeric ratios of monoterpene alcohols in lemon, mandarin, bitter orange and sweet orange oils. LC–GC/MS was used to study the composition of the most common citrus peel, citrus leaf (petitgrain) and flower (neroli) oils. The oils were separated into two fractions, i.e. mono- and sesquiterpene

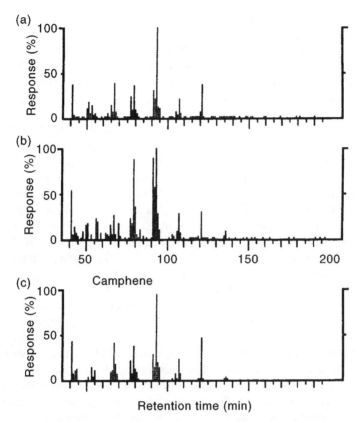

Figure 10.11   Comparison of the mass spectra of a neroli oil peak (camphene) obtained by HPLC–HRGC–MS (a) and GC–MS (b) with a library spectrum of the same compound (c). Reprinted from *Perfumer and Flavorist*, **21**, L. Mondello *et al.*, 'On-line HPLC–HRGC in the analytical chemistry of citrus essential oils', pp. 25–49, 1996, with permission from Allured Publishing Corp.

hydrocarbons, and oxygenated compounds, or into four fractions containing, respectively, mono- and sesquiterpene hydrocarbons, aliphatic aldehydes and esters, and two fractions of alcohols. Since the mass spectra of components of the same class are very similar, and it is necessary to have the spectrum of an extremely pure compound in order to obtain an unambiguous identification when using library matching, the combination of the HPLC–HRGC system with a mass spectrometer allows the components to be more reliably identified. Figure 10.11 shows a comparison of the mass spectra of a peak (camphene) of neroli oil obtained by both HPLC–HRGC–MS and by HRGC–MS with a library spectrum of the same compound.

The use of detection methods such as mass spectrometry (MS) and Fourier-transform infrared (FTIR) spectroscopy can be very useful with respect to the quality and the amount of information yielded. Sometimes, GC/FTIR has a limited diffusion, compared to GC–MS coupling, due to an overall lower sensitivity. These problems may be overcome if the HRGC/FTIR system is coupled on-line to an LC instrument. An on-line LC–GC/FTIR system has been used for the analysis of cit-ropten and bergapten in bergamot oil, using an on-column interface and partially concurrent eluent evaporation with early vapour exit (55). Figure 10.12 shows the LC chromatogram of bergamot oil (a), plus the corresponding GC/FID (b) and GC/FTIR (c) chromatograms after separation of the transferred fraction. The FTIR chromatogram has been constructed by using a selected wavelength chromatogram (SWC) from 1776 to 1779 cm^{-1}, which falls into the range 1760–1780 cm^{-1}, cor-responding to the characteristic C–O stretching band of the lactone ring of citropten and bergapten. In this way, any interaction with interfering peaks is avoided.

On-line LC–GC has frequently been used as a clean-up technique for the analysis of trace levels of contaminants (pesticides, plasticizers, dyestuffs and toxic organic chemicals) in water and food products. Several different approaches have been pro-posed for the analysis of contaminants by on-line LC–GC. Since pesticide residues occur at low concentration in water, soil or food, extraction and concentration is needed before GC analysis is carried out.

Organophosphorus pesticides (OPPs) are extensively used for the protection of olives against several insects. The class of OPPs contains numerous compounds which vary widely in polarity and molecular weight. Capillary GC analysis is the preferred method for the analysis of OPPs, using selective detectors such as the nitrogen–phosphorus (NPD) and flame photometric (FPD) detectors, as well as mass spectrometry (MS). GC analysis has to be performed after removal of the high-molecular-mass fats from the matrix. This pre-separation can be carried out by off-line methods, such as LC or gel permeation chromatography (GPC), although on-line GPC–GC methods have also been developed (56, 57). In the original version of this technique (56), a solvent system of n-decane and the azeotropic mixture of ethyl acetate and cyclohexane was found to give an adequate separation between the fat and the organophosphorus pesticides. The pesticide-containing fraction was transferred to the GC unit by using a loop-type interface with an early vapour exit and co-solvent trapping. The system presented some dif-ficulties during the transfer of the large eluent fraction (3 ml), and optimization of the transfer temperature was critical. An improved version of the on-line GPC–GC (FPD) method for the determination of OPPs in olive oil has been devel-oped (57) which uses an on-column interface with flow regulation. The miniatur-ization of the GPC column (the fraction transferred was now reduced to 1.3 ml) and the selection of another combination of main and co-solvent (methyl acetate/cyclopentane, with n-nonane as co-solvent), both with lower boiling points, thus produced a very robust method for the analysis of 28 OPPs in olive oil with an overall detection limit of 0.002 mg/kg.

The application of automated GPC for the clean-up of various matrices has been demonstrated by other authors (58, 59). As well as organophosphorus pesticides, conventional methods for the analysis of organochlorine pesticides (OPCs) in fatty samples may involve various clean-up methods, such as LC or GPC. The main

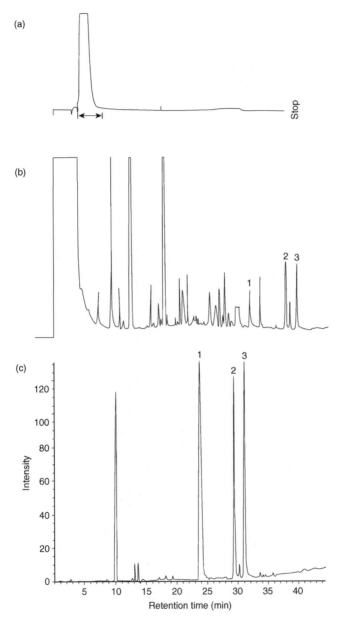

Figure 10.12 (a) LC separation of bergamot oil (305 nm), where the fraction of interest is completely transferred into the GC unit; (b) GC separation (column HP-5, 25 m × 0.32 mm i.d., 0.52 μm film thickness) of the fraction using an FID, and (c) using FTIR detection (SWC 1776–1779 cm^{-1}). Peak identification is as follows: (1) 7-methoxycoumarin (internal standard); (2) citropten; (3) bergapten. Reprinted from *Journal of High Resolution Chromatography*, **14**, G. Full *et al.*, 'On-line coupled HPLC–HRGC: A powerful tool for vapor phase FTIR analysis (LC–GC–FTIR)', pp. 160–163, 1991, with permission from Wiley-VCH.

problem to be solved is the separation of the tryglicerides from the most polar analytes while keeping the transfer volume small. In order to achieve this, van der Hoff *et al.* (60) used a 1 mm i.d. LC silica column (3 μm) to obtain transfer volumes of 100 μl, containing all the analytes of interest, while maintaining the separation between the most polar analytes and the fat matrix. They analysed a number of different organochlorine compounds (PCBs, the DDT group, HCB and the HCHs) at concentration levels of sub-μg/kg in milk fat, by using an on-line LC–GC (ECD) system using partially concurrent eluent evaporation. Figure 10.13 shows the GC chromatogram obtained after on-line LC–GC (ECD) of a human milk analysed for PCBs.

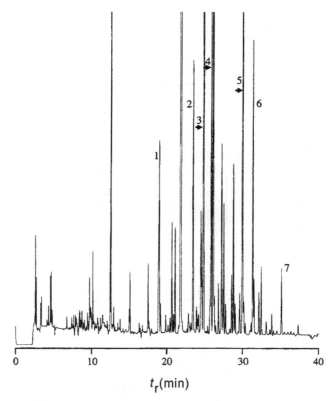

t_r(min)

Figure 10.13   GC chromatogram obtained after on-line LC–GC(ECD) of a human milk sample analysed for PCBs (attenuation × 64). Peak identification is as follows: (1) PCB 28; (2) PCB 118; (3) PCB 153; (4) PCB 138; (5) PCB 180; (6) PCB 170; (7) PCB 207. Reprinted from *Journal of High Resolution Chromatography*, **20**, G. R. van der Hoff *et al.*, 'Determination of organochlorine compounds in fatty matrices: application of normal-phase LC clean-up coupled on-line to GC/ECD', pp. 222–226, 1997, with permission from Wiley-VCH.

10.5 MULTIDIMENSIONAL CHROMATOGRAPHIC METHODS WHICH INVOLVE THE USE OF SUPERCRITICAL FLUIDS

Supercritical fluid extraction (SFE) has been extensively used for the extraction of volatile components such as essential oils, flavours and aromas from plant materials on an industrial as well as an analytical scale (61). The extract thus obtained is usually analysed by GC. Off-line SFE–GC is frequently employed, but on-line SFE–GC has also been used. The direct coupling of SFE with supercritical fluid chromatography (SFC) has also been successfully caried out. Coupling SFE with SFC provides several advantages for the separation and detection of organic substances: low temperatures can be used for both SFE and SFC, so they are well suited for the analysis of natural materials that contain compounds which are temperature-sensitive, such as flavours and fragrances.

Different approaches have been developed for the direct coupling of SFE and GC. Hawthorne *et al.* (62,63) analysed various flavour and fragrance compounds of natural products extracted with supercritical (SC) CO_2 by depositing the extracted analytes coming from the extraction cell outlet restrictor directly inside the GC capillary column through the on-column injector, which was cooled to cryogenically trap the analyte species at an appropriate oven temperature. Another approach was developed by Hartonen *et al.* (64). These authors used a cryotrap/thermal desorption unit external to the GC to collect the extracted compounds and introduce them into the GC column. In this way, the supercritical fluid used for the extraction does not flow through the separation column, thus increasing the lifetime of the latter. The extracted compounds are focused not only by selecting the lowest possible trapping temperature, but also by a low initial GC temperature. In this way, the separation of very volatile compounds is more efficient and more reproducible.

Selected applications of coupled SFE–SFC consider the analysis of tocopherols in plants and oil by-products (65) or the analysis of lipid-soluble vitamins (66) by using a dynamic on-line SFE-SFC coupling, integrated in the SF chromatograph, based on the use of micropacked columns.

A method which uses supercritical fluid/solid phase extraction/supercritical fluid chromatography (SF/SPE/SFC) has been developed for the analysis of trace constituents in complex matrices (67). By using this technique, extraction and clean-up are accomplished in one step using unmodified SC CO_2. This step is monitored by a photodiode-array detector which allows fractionation. Figure 10.14 shows a schematic representation of the SF/SPE/SFC set-up. This system allowed selective retention of the sample matrices while eluting and depositing the analytes of interest in the cryogenic trap. Application to the analysis of pesticides from lipid sample matrices have been reported. In this case, the lipids were completely separated from the pesticides.

Analogous to HPLC–HRGC, the combination of packed column SFC and capillary column GC can be used for the analysis of complex samples. The advantage of SFC/GC with respect to HPLC/GC is the absence of problems associated with the evaporation of the HPLC mobile phase prior to the GC analysis.

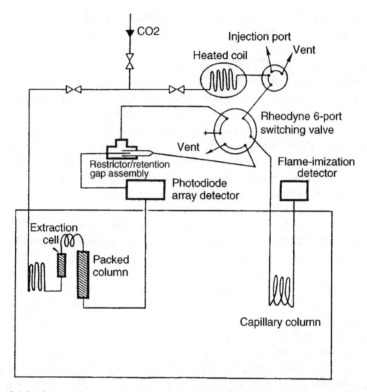

Figure 10.14 Schematic representation of the SFSPE/SFC set-up developed by Murugaverl and Voorhees (67). Reprinted from *Journal of Microcolumn Separation*, **3**, B. Murugaverl and K. J. Voorhees, 'On-line supercritical fluid extraction/chromatography system for trace analysis of pesticides in soybean oil and rendered fats', pp. 11–16, 1991, with permission from John Wiley and Sons, Inc.

In 1994, Nam and King (68) developed a SFE/SFC/GC instrumentation system for the quantitative analysis of organochlorine and organophosphorus pesticide residues in fatty food samples (chicken fat, ground beef and lard). In this way, SFC was used as an on-line clean-up step to remove extracted material. The fraction containing pesticide residues is then diverted and analysed by GC.

10.6 MULTIDIMENSIONAL PLANAR CHROMATOGRAPHY

Thin layer chromatography (TLC) is a very widely used chromatographic technique, and modern high performance (HP)TLC can be advantageously used instead of HPLC or GC, in many analytical situations. Much progress in the field of planar chromatography has been made in recent years, with the literature reporting the use of MD techniques, the coupling with particular detectors or the use of fully computerized image-processing instruments.

Often, planar chromatography is used as a preparative step for the isolation of single components or classes of components for further chromatographic separation or spectroscopic elucidation. Many planar chromatographic methods have been developed for the analysis of food products, bioactive compounds from plant materials, and essential oils.

The separation capacity of a TLC method can be easily improved by use of a two-dimensional high performance TLC technique (2D HPTLC). Various plant essential oils (*menthae*, *thymi*, *anisi*, *lavandulae*, etc.) have been analysed by 2D TLC with florisil (magnesium silicate) as the adsorbent, using dichloromethane/*n*-heptane (4 : 6) in the first direction and ethyl acetate/*n*-heptane (1 : 9) in the second direction (69).

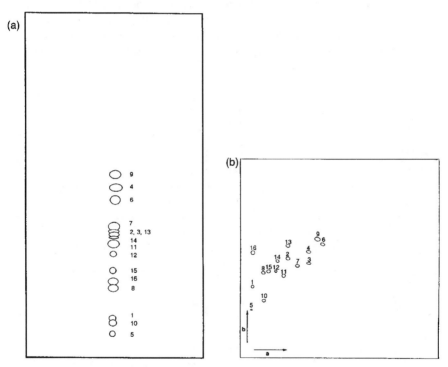

Figure 10.15 (a) One-dimensional OPLC development of the sixteen closely related coumarins: optimised mobile phase, S_t, 1.55, P_s, 271 (7% ethyl acetate, 52.9% chloroform, 20% dichloromethane and 20.1% *n*-hexane); development time, 30 min. (b) Two-dimensional OPLC development of the same coumarins system: mobile phase, 100% chloroform in direction (a) for 55 min, and 30% ethyl acetate in direction (b) for 80 min. Compound identification is as follows: (1) umbelliferone; (2) herniarin; (3) psoralen; (4) osthol; (5) apterin; (6) angelicin; (7) bergapten; (8) oxypeucedanin; (9) isobergapten; (10) scopoletin; (11) sphondin; (12) xanthotoxin; (13) imperatorin; (14) pimpinellin; (15) isopimpinellin; (16) new archangelicin derivate. Reprinted from *Journal of Planar Chromatography*, **3**, P. Härmälä *et al.*, 'Two-dimensional planar chromatographic separation of a complex mixture of closely related coumarins from the genus *Angelica*', pp. 515–520, 1990, with permission from Prof. Sz. Nyiredy, Research Institute for Medicinal Plants, Budakalász, Hungary.

The 2D chromatograms reveal additional components of the natural mixtures. They also give a 'map' of the essential oil, which is helpful in the identification of the components by the position and the characteristic colours of the derivatives on the plate. A further, considerable improvement in the separation performance can be obtained by using overpressured layer chromatography (OPLC). Härmälä *et al.* (70) used 2D OPLC for the separation of coumarins from the genus *Angelica.* Figure 10.15 shows the one-dimensional (a) and two-dimensional (b) OPLC separations of 16 coumarins.

Another way to improve separation in TLC is the use of gradient elution. Programmed multiple development (PMD) is a technique in which the plate is developed over increasing distances with several eluents of decreasing eluent strength, with the mobile phase being evaporated after each development. Other variations of this technique are automated multiple development (AMD) automated version, and the reversed PMD method, multiple gradient development (MGD), where a gradient of increasing strength is applied over decreasing distances. Figure 10.16 shows the

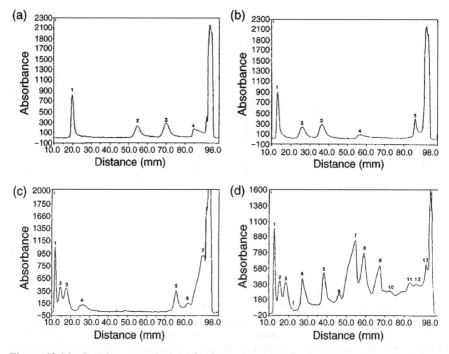

Figure 10.16 Densitograms obtained for four subsequent developments of the extract from *Radix rhei*: (a) first development, 10% (vol/vol) ethyl acetate/chloroform, distance 9 cm; (b) second development, 50% (vol/vol) ethyl acetate/chloroform, distance 9 cm; (c) third development, 100% ethyl acetate, distance 8 cm; (d) fourth development, 15% (vol/vol) methanol/ethyl acetate, distance 5 cm. Reprinted from *Chromatographia*, **43**, G. Matysik, 'Modified programmed multiple gradient development (MGD) in the analysis of complex plant extracts', pp. 39–43, 1996, with permission from Vieweg Publishing.

densitograms obtained for four subsequent developments of the extract from a medicinal plant (*Radix rhei*), obtained according to the reversed PMD (MGD) method (71). The technique has been successfully applied to the separation of compounds from complex plant extracts (71,72), as well as from food and beverages products (73).

REFERENCES

1. D. A. Armstrong, C. Chau-Dung and W. Yong Li, 'Relevance of enantiomeric separations in food and beverage analyses', *J. Agric. Food Chem.* **38**: 1674–1677 (1990).

2. L. Mondello, M. Catalfamo, P. Dugo and G. Dugo, 'Multidimensional capillary GC–GC for the analysis of real complex samples. Part II. Enantiomeric distribution of monoterpene hydrocarbons and monoterpene alcohols of cold-pressed and distilled lime oils', *J. Microcolumn Sep.* **10**: 203–212 (1998).

3. E. Guichard, A. Kustermann and A. Mosandl, 'Chiral flavour compounds from apricots. Distribution of γ-lactones enantiomers and stereodifferentiation of dihydroactinidiolide using multi-dimensional gas chromatography', *J. Chromatogr.* **498**: 396–401 (1990).

4. A. Mosandl, U. Hener, U. Hagenauer-Hener and A. Kustermann, 'Stereoisomeric flavor compounds. 33. Multidimensional gas chromatography direct enantiomer separation of γ-lactones from fruits, foods and beverages', *J. Agric. Food Chem.* **38**: 767–771 (1990).

5. U. Hener, P. Kreis and A. Mosandl, 'Enantiomeric distribution of α-pinene, β-pinene and limonene in essential oils and extract. Part 3. Oils for alcoholic beverages and seasoning', *Flav Fragr. J.* **6**: 109–111 (1991).

6. V. Karl, H.-G. Schmarr and A. Mosandl, 'Simultaneous stereoanalysis of 2-alkyl-branched acids, esters and alcohols using a selectivity-adjusted column system in multi-dimensional gas chromatography', *J. Chromatogr.* **587**: 247–350 (1991).

7. K. Rettinger, V. Karl, H.-G. Schmarr, F. Dettmar, U. Hener and A. Mosandl, 'Chirospecific analysis of 2-alkyl-branched alcohols, acids and esters: chirality evaluation of 2-methylbutanoates from apples and pineapples', *Phytochem. Anal.* **2**: 184–188 (1991).

8. V. Schubert, R. Diener and A. Mosandl, 'Enantioselective multidimensional gas chromatography of some secondary alcohols and their acetates from banana', *Z. C. Naturforsch. C.* **46**: 33–36 (1991).

9. P. Kreis and A. Mosandl, 'Chiral compounds of essential oils. Part XIII. Simultaneous chirality evaluation of geranium oil constituents', *Flav. Fragr. J.* **8**: 161–168 (1992).

10. A. Mosandl, C. Askari, U. Hener, D. Juchelka, D. Lehmann, P. Kreis, C. Motz, U. Palm and H.-G. Schmarr, 'Chirality evaluation in flavour and essential oil analysis', *Chirality* **4**: 50–55 (1992).

11. P. Kreis and A. Mosandl, 'Chirality compounds of essential oils. Part XVII. Simultaneous stereoanalysis of Cymbopogon oil constituents', *Flav. Fragr. J.* **9**: 257–260 (1994).

12. T. Köpke, A. Dietrich and A. Mosandl, 'Chiral compounds of essential oils. XIV: simultaneous stereoanalysis of buchu leaf oil compounds', *Phytochem. Anal.* **5**: 61–67 (1994).

13. V. Karl, J. Gutser, A. Dietrich, B. Maas and A. Mosandl, 'Stereoisomeric flavour compounds. LXVIII. 2-, 3- and 4-alkyl-branched acids. Part **2**: chirospecific analysis and sensory evaluation', *Chirality* **6**: 427–434 (1994).

14. D. Bartschat, D. Lehmann, A. Dietrich, A. Mosandl and R. Kaiser, 'Chiral compounds of essential oils. XIX. 4-methyl-5 decanolide: chirospecific analysis, structure and properties of the stereoisomers', *Phytochem. Anal.* **6**: 130–134 (1995).

15. D. Juchelka and A. Mosandl, 'Authenticity profiles of bergamot oil', *Pharmazie* **51**: 417–422 (1996).

16. D. Jukelka, A. Steil, K. Witt and A. Mosandl, 'Chiral compounds of essential oils. XX. Chirality evaluation and authenticity profiles of neroli and petitgrain oils', *J. Essential Oil Res.* **8**: 487–497 (1996).

17. B. Weber, B. Maas and A. Mosandl, 'Stereoisomeric flavor compounds. 72. Stereoisomeric distribution of some chiral sulfur containing trace components of yellow passion fruits', *J. Agric. Food Chem.* **43**: 2438–2441 (1995).

18. A. Mosandl, G. Bruche, C. Askari and H.-G. Schmarr, 'Stereoisomeric flavor compounds. XLIV: enantioselective analysis of some important flavor molecules', *J. High Resolut. Chromatogr.* **13**: 660–662 (1990).

19. G. Full, P. Winterhalter, G. Schmidt, P. Herion and P. Schreier, 'MDGC–MS: a powerful tool for enantioselective flavor analysis', *J. High Resolut. Chomatogr.* **16**: 642–644 (1993).

20. L. Mondello, M. Catalfamo, P. Dugo and G. Dugo, 'Multidimensional capillary GC–GC for the analysis of real complex samples. Part I. Development of a fully automated GC–GC system', *J. Chromatogr. Sci.* **36**: 201–209 (1998).

21. L. Mondello, M. Catalfamo, A. R. Proteggente, I. Bonaccorsi and G. Dugo, 'Multidimensional capillary GC–GC for the analysis of real complex samples. Part III. Enantiomeric distribution of monoterpene hydrocarbons and monoterpene alcohols of mandarin oils', *J. Agric. Food Chem.* **46**: 54–61 (1998).

22. L. Mondello, A. Verzera, P. Previti, F. Crispo and G. Dugo, 'Multidimensional capillary GC–GC for the analysis of real complex samples. Part V. Enantiomeric distribution of monoterpene hydrocarbons, monoterpene alcohols and linalyl acetate of bergamot (*Citrus bergamia* Risso *et* Poiteau) oils', *J. Agric. Food Chem.* **46**: 4275–4282 (1998).

23. L. Mondello, M. Catalfamo, A. Cotroneo, G. Dugo, Giacomo Dugo and H. McNair, 'Multidimensional capillary GC–GC for the analysis of real complex samples. Part IV. Enantiomeric distribution of monoterpene hydrocarbons and monoterpene alcohols of lemon oils', *J. High Resolut. Chromatogr.* **22**: 350–356 (1999).

24. H. Casabianca, J.-B. Graff, P. Jame, C. Perrucchietti and M. Chastrette, 'Application of hyphenated techniques to the chromatographic authentication of flavours in food products and perfumes', *J. High Resolut. Chromatogr.* **18**: 279–285 (1995).

25. D. Juchelka, T. Beck, U. Hener, F. Dettmar and A. Mosandl, 'Multidimensional gas chromatography coupled on-line with isotope ratio mass spectrometry (MDGC–IRMS): progress in the analytical authentication of genuine flavor components', *J. High Resolut. Chromatogr.* **21**: 145–151 (1998).

26. S. Nitz, H. Kollmannsberger, B. Weinreich and F. Drawert, 'Enantiomeric distribution and $^{13}C/^{12}C$ isotope ratio determination of γ-lactones: appropriate methods for the differentiation between natural and non-natural flavours?', *J. Chromatogr.* **557**: 187–197 (1991).

27. S. Nitz, B. Weinreich. and F. Drawert, 'Multidimensional gas chromatography–isotope ratio mass spectrometry (MDGC–IRMS). Part A: system description and technical requirements', *J. High Resolut. Chromatogr.* **15**: 387–391 (1992).

28. C. Askari, U. Hener, H.-G. Schmarr, A. Rapp and A. Mosandl, 'Stereodifferentiation of some chiral monoterpenes using multidimensional gas chromatography', *Fresenius J. Anal. Chem.* **340**: 768–772 (1991).

29. B. Martin and P. Etiévant, 'Quantitative determination of solerone and sotolon in flor sherries by two dimensional capillary GC', *J. High Resolut. Chromatogr.* **14**: 133–135 (1991).

30. K. MacNamara and A. Hoffmann, 'Simultaneous nitrogen, sulphur and mass spectrometric analysis after multi-column switching of complex whiskey flavour extracts', in

Proceedings of the 15th International Symposium on Capillary Chromatography, Riva del Garda, Italy, May 24–27, P. Sandra and G. Devos (Eds), Hüthig, Heidelberg, Germany, pp. 877–884 (1993).

31. N. Ragunathan, T. A. Sasaki, K. A. Krock and C. A. Wilkins, 'Multidimensional fast gas chromatography with matrix isolation infrared detection', *Anal. Chem.* **66**: 3751–3756 (1994).

32. F. Enri and R. W. Frei, 'Two dimensional column liquid chromatographic technique for resolution of complex mixtures', *J. Chromatogr.* **149**: 561–569 (1978).

33. J. A. Apffel, T. V. Alfredson and R. E. Majors, 'Automated on-line multi-dimensional high performance liquid chromatographic techniques for the clean-up and analysis of water-soluble samples', *J. Chromatogr.* **206**: 43–57 (1981).

34. E. L. Johnson, R. Gloor and R. E. Majors, 'Coupled column chromatography employing exclusion and a reversed phase. A potential general approach to sequential analysis', *J. Chromatogr.* **149**: 571–585 (1978).

35. J. M. Brown-Thomas, A. A. Moustafa, S. A. Wise and W. E. May, 'Determination of fat-soluble vitamins in oil matrices by multidimensional high-performance liquid chromatography', *Anal. Chem.* **60**: 1929–1933 (1988).

36. E. A. Hogendoorn and P. van Zoonen, 'Coupled-column reversed-phase liquid chromatography in environmental analysis (review)', *J. Chromatogr.* **703**: 149–166 (1995).

37. E. A. Hogendoorn, R. Hoogerbrugge, R. A. Baumann, H. D. Meiring, A. P. J. M. de Jong and P. van Zoonen, 'Screening and analysis of polar pesticides in environmental monitoring programmes by coupled-column liquid chromatography and gas chromatography–mass spectrometry', *J. Chromatogr.* **754**: 49–60 (1996).

38. E. A. Hogendoorn, E. Dijkman, S. M. Gort, R. Hoogerbrugge, P van Zoonen and U. A. Th. Brinkman, 'General strategy for multiresidue analysis of polar pesticides in ground water using coupled column reversed phase LC and step gradient elution', *J. Chromatogr.* **31**: 433–439 (1993).

39. K. Grob and Th. Läubli, 'Determination of wax esters in olive oil by coupled HPLC–HRGC', *J. High Resolut. Chromatogr. Chromatogr. Commun.* **9**: 593–594 (1986).

40. K. Grob, M. Lanfranchi and C. Mariani, 'Determination of free and esterified sterols and of wax esters in oils and fats by coupled liquid chromatography–gas chromatography', *J. Chromatogr.* **471**: 397–405 (1989).

41. F. J. Señoráns, J. Tabera and M. Herraiz, 'Rapid separation of free sterols in edible oils by on-line coupled reversed phase liquid chromatography–gas chromatography', *J. Agric. Food. Chem.* **44**: 3189–3192 (1996).

42. K. Grob, E. Müller and W. Meier, 'Coupled HPLC–HRGC for determining PCBs in fish', *J. High Resolut. Chromatogr. Chromatogr. Commun.* **10**: 416–417 (1987).

43. H.-G. Schmarr, A. Mosandl and K. Grob, 'Stereoisomeric flavour compounds. XXXVIII: direct chirospecific analysis of γ-lactones using on-line coupled LC–GC with a chiral separation column', *Chromatographia* **29**: 125–130 (1990).

44. A. Artho and K. Grob, 'Determination of γ-lactones added to foods as flavours. How far must "nature-identical" flavours be identical with the nature?', *Mitt. Gebiete Lebensm. Hyg.* **81**: 544–558 (1990).

45. R. Barcarolo, 'Coupled LC–GC: a new method for the on-line analysis of organochlorine pesticide residue in fat', *J. High Resolut. Chromatogr.* **13**: 465–470 (1990).

46. K. Grob, M. Biedermann and M. Bronz, 'Results of a control of edible oils: frauds by admixtures, contaminations', *Mitt. Gebiete Lebensm. Hyg.* **85**: 351–365 (1994).

47. K. Grob, I. Kaelin and A. Artho, 'Coupled LC–GC: the capacity of silica gel (HP)LC columns for retaining fat', *J. High Resolut. Chromatogr.* **14**: 373–376 (1991).

48. L. Mondello, G. Dugo and K. D. Bartle, 'On-line microbore high performance liquid chromatography–capillary gas chromatography for food and water analysis. A review', *J. Microcolumn Sep.* **8**: 275–310 (1996).

49. M. Lechner, C. Bauer-Plank and E. Lorbeer, 'Determination of acylglycerols in vegetable oil methyl esters by on-line normal phase LC–GC', *J. High Resolut. Chromatogr.* **20**: 581–585 (1997).

50. F. Lanuzza, G. Micali and G. Calabrò, 'On-line HPLC–HRGC coupling and simultaneous transfer of two different LC fractions: determination of aliphatic alcohols and sterols in olive oil', *J. High Resolut. Chromatogr.* **19**: 444–448 (1996).

51. F. J. Señoráns, M. Herraiz and J. Tabera, 'On-line reversed-phase liquid chromatography using a programmed temperature vaporizer as interface', *J. High Resolut. Chromatogr.* **18**: 433–437 (1995).

52. F. J. Señoráns, J. Villén, J. Tabera and M. Herraiz, 'Simplex optimization of the direct analysis of free sterols in sunflower oil by on-line coupled reversed phase liquid chromatography–gas chromatography', *J. Agric. Food Chem.* **46**: 1022–1026 (1998).

53. G. P. Blanch, J. Villén and M. Herraiz, 'Rapid analysis of free erythrodiol and uvaol in olive oils by coupled reversed phase liquid chromatography–gas chromatography', *J. Agric. Food Chem.* **46**: 1027–1030 (1998).

54. L. Mondello, G. Dugo, P. Dugo and K. Bartle, 'On-line HPLC–HRGC in the analytical chemistry of citrus essential oils', *Perfumer Flavorist* **21**: 25–49 (1996).

55. G. Full, G. Krammer and P. Schreier, 'On-line coupled HPLC–HRGC: A powerful tool for vapor phase FTIR analysis (LC–GC–FTIR)', *J. High Resolut. Chromatogr.* **14**: 160–163 (1991).

56. J. J. Vreuls, R. J. J. Swen, V. P. Goudriaan, M. A. T. Kerkhoff, G. A. Jongenotter and U. A. Th Brinkman, 'Automated on-line gel permeation chromatography–gas chromatography for the determination of organophosphorus pesticides in olive oil', *J. Chromatogr.* **750**: 275–286 (1996).

57. G. A. Jongenotter, M. A. T. Kerkhoff, H. C. M. Van der. Knaap and B. G. M. Vandeginste, 'Automated on-line GPC–GC–FPD involving co-solvent trapping and on-column interface for the determination of organophosphorus pesticides in olive oils', *J. High Resolut. Chromatogr.* **22**: 17–23(1999).

58. K. Grob and I. Kälin, 'Attempt for an on-line size exclusion chromatography method for analyzing pesticide residues in foods', *J. Agric. Food Chem.* **39**: 1950–1953(1991).

59. M. De Paoli, M. Barbina Taccheo, R. Mondini, A. Pezzoni and A. Valentino, 'Determination of organophosphorus pesticides in fruits by on-line size-exclusion chromatography–flame photometric detection', *J. Chromatogr.* **626**: 145–151 (1992).

60. G. R. van der Hoff, R. A. Baumann, P. van Zoonen and U. A. Th. Brinkman, 'Determination of organochlorine compounds in fatty matrices: application of normal-phase LC clean-up coupled on-line to GC/ECD', *J. High Resolut. Chromatogr.* **20**: 222–226 (1997).

61. R. M. Smith and M. D. Burford, 'Optimization of supercritical fluid extraction of volatile constituents from a model plant matrix', *J. Chromatogr.* **600**: 175–181 (1992).

62. S. B. Hawthorne, M. S. Krieger and D. J. Miller, 'Analysis of flavor and fragrance compounds using supercritical fluid extraction coupled with gas chromatography', *Anal. Chem.* **60**: 472–477 (1988).

63. S. B. Hawthorne, D. J. Miller and M. S. Krieger, 'Coupled SFE–GC: a rapid and simple technique for extracting, identifying and quantitating organic analytes from solids and sorbent resins', *J. Chromatogr. Sci.* **27**: 347–354 (1989).

64. K. Hartonen, M. Jussila, P. Manninen and M.-L. Riekkola, 'Volatile oil analysis of *Thymus vulgaris* L. by directly coupled SFE/GC', *J. Microcolumn Sep.* **4**: 3–7 (1992).

65. E. Ibañez, J. Palacios and G. Reglero, 'Analysis of tocopherols by on-line coupling super-critical fluid extraction–supercritical fluid chromatography', *J. Microcolumn Sep.* **11**: 605–611 (1999).

66. E. Ibañez, M. Herraiz and G. Reglero, 'On-line SFE–SFC coupling using micropacked columns', *J. High Resolut. Chromatogr.* **18**: 507–509 (1995).

67. B. Murugaverl and K. J. Voorhees, 'On-line supercritical fluid extraction/chromatography system for trace analysis of pesticides in soybean oil and rendered fats', *J. Microcolumn Sep.* **3**: 11–16 (1991).

68. K.-S. Nam and J. W. King, 'Coupled SFE/SFC/GC for the trace analysis of pesticide residues in fatty food samples', *J. High Resolut. Chromatogr.* **17**: 577–582 (1994).

69. M. Waksmundzka-Hanios and M. Markowski, 'Chromatography of some essential plant oils on thin layers of florisil', *Chem. Anal. (Warsaw)* **40**: 163–174 (1995).

70. P. Härmälä, L. Botz, O. Sticher and R. Hiltunen, 'Two-dimensional planar chromato-graphic separation of a complex mixture of closely related coumarins from the genus *Angelica*', *J. Planar Chromatogr.* **3**: 515–520 (1990).

71. G. Matysik, 'Modified programmed multiple gradient development (MGD), in the analy-sis of complex plant extracts', *Chromatographia* **43**: 39–43 (1996).

72. E. Menziani, B. Tosi, A. Bonora, P. Reschiglian and G. Lodi, 'Automated multiple development high-performance thin-layer chromatographic analysis of natural phenolic compounds', *J. Chromatogr.* **511**: 396–401 (1990).

73. M. T. Belay and C. F. Poole , 'Determination of vanillin and related flavor compounds in natural vanilla extracts and vanilla-flavored foods by thin layer chromatography and automated multiple development', *Chromatographia* **37**: 365–373(1993).

11   Multidimensional Chromatography: Biomedical and Pharmaceutical Applications

G.W. SOMSEN and G.J. de JONG

University of Groningen, Groningen, The Netherlands

11.1   INTRODUCTION

Today, separations play a central role in the analysis of pharmaceutical and biological samples. Because many samples are very complex and the concentration levels of interest decrease, systems with high efficiency and selectivity are needed. Another strong requirement is the reduction of analysis time, because the number of samples continues to increase and information has to become available rapidly. This also means that attention is needed for sample pretreatment procedures which preferably have to be integrated with the analysis step, if possible in an automated fashion. The coupling of chromatographic systems in the same mode or even the combination of different chromatographic modes has shown a high potential in order to reach these goals. These so-called multidimensional chromatography systems offer many possibilities to increase the selectivity, and thus the sensitivity of the total analytical system, especially when more or less orthogonal techniques are combined.

This present chapter describes the application of multidimensional chromatography (including capillary electrophoresis) in the biomedical and pharmaceutical field. Because liquid chromatography (LC) is the main technique in these areas, LC is often at least one of the components of such a coupled-column system. The coupling of two LC columns is now routine and widely used, and at present LC coupled to gas chromatography (GC) is applied for bioanalysis as well. The on-line coupling of solid-phase (micro)extraction (SPE and SPME) with LC or GC for the determination of substances of biological and/or pharmaceutical interest will also be discussed. In this respect, special attention will be paid to the use of selective sorbents in on-line SPE. The biomedical application of supercritical fluid extraction (SFE) in coupled chromatographic systems will also be treated briefly. Finally, applications of multidimensional systems involving capillary electrophoresis (CE), which has a high potential for the separation of drugs, peptides and proteins, will be described. The goal of this chapter is not to present an overview of all relevant applications, but rather to stress the potential of the various multidimensional systems in the biomedical and pharmaceutical field by means of typical examples. In this respect it should be noted that multidimensional separation systems are much more important in biomedical analysis than in pharmaceutical analysis, because the usefulness of such systems strongly depends on the complexity of the samples to be analysed. No attention will be paid here to thin-layer chromatography (TLC). Over the past decade, new developments in two-dimensional TLC for biomedical purposes have been scarce and in addition the amount of research in coupled chromatographic systems with TLC as one dimension has been very small (1, 2).

The discussions in essence will be confined to the application of on-line (i.e. directly coupled) chromatographic techniques, since with respect to off-line coupling, these systems offer improved precision and accuracy as a result of the elimination of intermediate steps such as fraction collection, evaporation and manual transfer. Moreover, on-line coupling offers shorter analysis times and good possibilities for automation, with both of these being important aspects in modern bioanalysis. In addition, with off-line systems commonly only an aliquot of the fraction collected from the first chromatographic dimension is introduced into the second dimension. It should be added that efficient on-line coupling requires thorough attention to the instrumental aspects, and that the interfacing is not always easy and/or can give some limitations to the total system. The main advantages of an off-line combination of two separation systems are the higher flexibility in the choice of the operating conditions and the possibility of independent optimization. At present, the availability of efficient robotic analysers allows full automation of off-line methods. However, generally the other disadvantages of the off-line mode are still valid for these so-called 'at-line' systems.

11.2 LIQUID CHROMATOGRAPHY–LIQUID CHROMATOGRAPHY

LC is an important and widely used analytical separation technique in the pharmaceutical, biomedical and clinical field. The technique is well established with a variety of sophisticated instrumentation and high performance columns readily available. The range of biologically interesting substances that can be analysed directly by LC is very large as it includes charged, polar, thermolabile, non-volatile and high-molecular-weight compounds. These type of substances are not amenable to GC without using time-consuming (off-line) chemical derivatization procedures. Furthermore, various modes of LC are carried out using (partly) aqueous eluents and, therefore, are highly suitable for the processing of aqueous samples, a frequently encountered sample type in biomedical analysis. However, biological samples are generally highly complex mixtures in which the compounds of interest may appear as minor constituents. As a consequence, a large excess of interferences frequently hinders the direct determination of target compounds. Moreover, the sensitivity and specificity of UV absorbance detection–this being the standard detection principle in LC–is often too low to allow the analysis of trace-level components of biological samples. Hence, in biomedical analysis usually some kind of off-line or on-line sample preparation procedure is required to achieve sample clean-up and analyte enrichment prior to LC analysis. In LC, a popular way towards enhanced selectivity and sensitivity is the use of small solid-phase extraction (SPE) cartridges filled with a more or less selective sorbent for the preconcentration of samples (3). The application of the on-line combination of SPE and LC for bioanalytical purposes will be discussed to some extent in the next section. Another viable approach towards more chromatographic selectivity is coupled-column LC (LC–LC) in which two or more analytical columns are combined in an on-line fashion to accomplish the isolation of the compound(s) of interest. LC–LC can be used either for the profiling of a complete sample or for the analysis of target compounds (4).

11.2.1   COMPREHENSIVE LC–LC

The objective of the profiling mode of LC–LC is to fractionate all components of the analysed mixture. This may be accomplished by so-called comprehensive two-dimensional LC in which the entire chromatogram eluting from the primary column is submitted to the secondary column. The secondary instrument must operate fast enough to preserve the information contained in the primary signal. That is, it should be able to generate at least one chromatogram during the time required for a peak to elute from the primary column. Until now, a limited number of studies involving comprehensive LC–LC for the analysis of compounds of biological interest have been reported. Most of these studies were carried out within the group of Jorgenson and mainly deal with the design, construction and implementation of comprehensive LC–LC systems for the separation of either a complex protein mixture or an enzymatic digest of a protein (i.e. a mixture of peptides) (5–10). For these purposes, orthogonal on-line combinations of ion-exchange chromatography, size-exclusion chromatography (SEC) or reversed-phase (RP) LC are used. In a recent study, peptide fragments generated in the tryptic digests of ovalbumin and serum albumin are separated by SEC in the first dimension (run time, 160 min) and fast RPLC in the second (run time, 240 s) (7). Following RPLC, the peptides flow to an electrospray mass spectrometer for on-line identification. The complete LC system yields a peak capacity of almost 500, thereby maximizing the chance of completely resolving each peptide of the digest and, thus, permitting highly reliable peptide mapping. In addition, a comprehensive ion-exchange LC–RPLC–mass spectrometry (MS) system for the analysis of proteins was demonstrated in which a 120-min ion-exchange LC run is sampled by 48 RPLC runs of 150 s, leading to a peak capacity of over 2500 (8). The system was succesfully applied to the screening of an *Escherichia coli* lysate without any prior knowledge of the characteristics (e.g. molecular weight, isoelectric point, hydrophobicity, etc.) of its individual components.

Other bioanalytical applications of systems in which the eluate of a first LC column is sampled in continuous and repetitive intervals and subjected to a second LC dimension are, for example, described by Wheatly *et al.* (11) and Matsuoka *et al.* (12). Wheatly coupled gradient affinity LC with RPLC for the determination of the isoenzymatic- and subunit composition of glutathione *S*-transferases in cytosol extracts of human lung and liver tissues. Matsuoaka combined anion-exchange LC with RPLC for the analysis of the enzymatic digests of bovine calmodulin and D59 protein, and for the separation of a crude peptide mixture which was extracted from bovine brain tissue.

With comprehensive two-dimensional LC, indeed truly orthogonal separations with very high efficiency and peak capacity can be obtained, although the interfacing can be quite complicated. Additionally, in bioanalysis large peak capacities are often not needed, especially when the objective is the determination of only a few target compounds (see next section). The usefulness of comprehensive LC–LC lies particularly in the entire profiling of protein and peptide mixtures of unknown composition (protein/peptide mapping). In this respect, comprehensive LC–LC might be an alternative for two-dimensional gel electrophoresis, with LC–LC having the advantage that it can be on-line coupled to MS so that molecular-weight information becomes available fast and without the need for analyte transfer.

11.2.2 HEART-CUT LC–LC

In contrast to comprehensive LC–LC, the goal of the targeted mode of multidimensional LC is to isolate a single analyte or small group of components of a complex mixture. Target analysis by LC–LC is normally carried out by using the principle of 'heart-cutting'. In this approach, an eluting zone from the first LC column (heart-cut) is switched to a second column for subsequent analysis of the transferred compounds. Thus, the first column is used to extract and enrich analytes, e.g. from complex biological fluids, while the second column is used to separate the molecule(s) of interest. This combination leads to an enhanced selectivity and sensitivity, because the target compounds are preconcentrated and/or can be detected in the absence (or reduced presence) of interfering compounds. As a result, lower detection limits are obtained and the reliability of quantitative analysis is strongly improved. When a very high selectivity is required, the dimensionality (number of subsequent columns) may be increased to three or more by using different retention modes. Of course, on-line LC–LC requires the use of mobile phases which, to some extent, are compatible with the columns involved. Therefore, LC–LC is restricted to certain chromatographic combinations, which often include RPLC, ion-exchange, polar bonded-phase, chiral or affinity columns. It is evident that next to the selection of the proper columns and eluents, correct timing of the column switching and optimization of the transfer volume are essential aspects, and, therefore, systematic method development is an important issue in LC–LC (13–15).

In biomedical analysis, LC–LC has been used most extensively and successfully in the heart-cut mode for the analysis of drugs and related compounds in matrices such as plasma, serum or urine. Table 11.1 gives an overview of analytes in biological matrices which have been determined by heart-cut LC–LC systems. A typical example of such an approach is the work of Eklund et al. (16) who determined the free concentration of sameridine, an anaesthetic and analgesic drug, in blood plasma

Table 11.1 Biomedical applications of on-line heart-cut LC–LC

Analyte(s)	Sample matrix	Sample preparation[a]	First LC mode[b]	Subsequent LC mode(s)[b]	Detection[c]	Reference
Drugs and Related Compounds						
Ampicillin	human plasma	deproteination	C18	C18	post-Flu	28
	human urine	buffer dilution	C18	C18	post-Flu	28
β-agonists	human/bovine urine	none	C18	C18	UV	22
	bovine urine	none	C18	C18	MS–MS	23
Basic drugs	human serum	liq–liq extraction	ion-pair C18	C18	UV	37
Bupivacaine	human plasma	ultrafiltration	C8	CIEX	UV	24
Chloramphenicol	pig tissue	liq–liq extraction	PRP	C18	UV	32
Dexamethasone	bovine tissue	liq–liq extraction	phenyl	silica; CN	UV	29
Dopa and its metabolite	human/rat plasma	deproteination	CIEX	C18	EC	66
Efletirizine	human plasma	SPE	C18	C18	UV	69
	human urine	dilution	C18	C18	UV	69
Glycyrrhizin and its metabolite	human plasma	SPE	C18	C18	UV	39
Ibuprofen	human serum	liq–liq extraction	ion-pair C18	C18; C18	UV	34
Manidipine and its metabolite	human serum	liq–liq extraction	ion-pair C18	C18	UV	38
Mefenamic acid	human serum	liq–liq extraction	C18	C18; C18	UV	34
Melengestrol acetate	bovine tissue	liq–liq extraction	phenyl	silica; silica	UV	31

Table 11.1 (*continued*)

Analyte(s)	Sample matrix	Sample preparation [a]	First LC mode[b]	Subsequent LC mode(s) [b]	Detection [c]	Reference
Methandrostenolone and its metabolites	equine plasma/urine	liq–liq extraction	phenyl	C18; C8	UV and MS–MS	30
Metyrapone	human plasma/urine	liq–liq extraction	silica	Chiralcel OJ	UV	42
Phenyl-propanololamine	human plasma/urine	liq–liq extraction	ion-pair C18	C18	UV	36
Probenicid	rat plasma	deproteination	C18	ion-pair C18	UV	63
Propanolol	human plasma	liq–liq extraction	ion-pair C18	C18	UV	35
Methotrexate	human urine	SPE	AIEX	C18	UV	26
Remoxipride metabolite	human plasma	liq–liq extraction	CN	C18	EC	17
Ro 23-7637	dog plasma	SPE	CIEX	C18	UV	27
Ro 24-0238	human plasma	SPE	CIEX	C8	UV	25
Ropivacaine	human plasma	ultrafiltration	C8	CIEX	UV	24
Sameridine	human plasma	ultrafiltration	C18	CIEX; CIEX	UV	16
TCV-116 and its metabolites	human serum/urine	liq–liq extraction	C18	C18	Flu	64
Tipredane metabolites	rat urine	none	CN	C18	UV	18
Vanillylmandelic acid	human urine	none	C18	C1	UV	19
	human urine	acidification, centrifugation	ion-pair C18	AIEX	EC	65
Various drugs	human serum/plasma	none	micel C8 or CN	C18	UV or Flu	33

Compound	Sample	Sample preparation			Detection	Ref.
Zidovudine-β-D-glucuronide	rat plasma	deproteination	C18	ion-pair C18	UV	63
Enantiomers of Drugs and Related Compounds						
Amino acids	protein hydrolysates, bacterial cultures, food, urine	various	CIEX	Crownpak CR(+)	post-Flu	43
Artilide fumarate	human/anim. plasma	SPE	C8	C18; Pirkle	pre-Flu	44
Bupivacaine	human plasma	liq–liq extraction	AGP	C18; C8	UV	54
Chlortalidone	human whole blood	liq–liq extraction	CN	β-CD phenyl	UV	51
Dihydropyridine calcium blocker	dog plasma	liq–liq extraction	AIEX	Ovomucoid	EC	49
Ibutilide fumarate	human/animal plasma	SPE	C8	C18; Pirkle	pre-Flu	44
Ifosamide	human plasma	liq–liq extraction	D,L-naphthyl-alanine	Chiralcel OD	UV	55
Ketoprofen	human plasma	deproteination	C18	C18; Ovomucoid	UV	47
Leucovorine	dog plasma	deproteination	C18	C18	UV	53
	human plasma	deproteination	BSA-silica phenyl	BSA-silica	UV	56
Manidipine	human serum	liq–liq extraction Chiralcel OJ	C18	C18	UV	45
Mefloquine	human plasma/whole blood	liq–liq extraction	CN	silica; (S)-naphthylurea	UV	52
Metoprolol	human plasma	SPE	AGP	C18; C18	Flu	54
Metyrapol	human plasma/urine	liq–liq extraction	silica	Chiralcel OJ	UV	42

Table 11.1 (continued)

Analyte(s)	Sample matrix	Sample preparation [a]	First LC mode [b]	Subsequent LC mode(s) [b]	Detection [c]	Reference
Oxazepam	human plasma	SPE	BSA-silica	C18; C18	UV	54
p-HPPH [d]	rat liver microsomes	liq–liq extraction	C18	C18 with L-prolinamide	UV	46
Pimobendan and its metabolite	human plasma	liq–liq extraction	silica	Chiralcel OD	UV	48
Terbutaline	human plasma	SPE	phenyl	β-CD C18	EC	51
	human plasma	SPE	AGP	C18; C18	EC	54
	human plasma	SPE	phenyl	β-CD	EC or MS	57
Verapamil and its metabolites	human plasma	liq–liq extraction	C18	Ovomucoid	UV	50
Endogenous Compounds						
Creatinine	human serum/urine	buffer dilution	SEC	SEC; CIEX CIEX;	UV	58
Glutathione S-transferases	cytosol of human lung tissue	homogenization, centrifugation	affinity	C8 or C18	UV	11
Hydroxy-deoxyguanosine	human urine	SPE	PRP	C18	EC	59
Hydroxytryptamine and its metabolite	rabbit blood	centrifugation on-line SPE	CIEX	C18	EC	68
Immunoglobulin G and its multimers	bovine serum	none	SEC	affinity	UV	60
Neopterin	human serum	centrifugation	C18	ion-pair C18	Flu	62
Proinsulin fusion protein	E. coli cells	sulfitolysis	SEC	AIEX	UV	61
Riboflavin	human blood/plasma	deproteination	PRP	C18	Flu	67

| Thiamine- and pyridoxal-phoshates | human blood/plasma | deproteination | PRP | ion-pair PRP | post-Flu | 67 |
| Uric acid | human serum/urine | buffer dilution | SEC | AIEX; SEC | UV | 58 |

[a] liq–liq, liquid–liquid; SPE, solid-phase extraction.
[b] affinity, affinity chromatography; AGP, immobilized α1-acid glycoprotein; AIEX, anion-exchange; β-CD, β-cyclodextrines; BSA, bovine serum albumin; C1, methylsilica; C8, octylsilica; C18, octadecylsilica; CIEX, cation-exchange; Crown pak CR(+), octadecyl silica coated with chiral crown ether; CN, cyanopropylsilica; ion pair, ion-pair chromatography; micel, micellar chromatography; phenyl, phenylsilica; Pirkle, 3,5-dinitrobenzoyl-D-phenylglycine-bonded silica; PRP, polymeric reversed-phase; SEC, size-exclusion chromatography.
[c] EC, electrochemical detection; Flu, fluorescence detection; MS, mass spectrometric detection; pre-Flu, fluorescence detection after pre-column derivatization; post-Flu, fluorescence detection after post-column derivatization; UV, UV absorbance detection.
[d] p-HPPH, p-hydroxyphenyl phenylhydantoin.

by coupling RPLC and ion-exchange LC. After ultrafiltration of the blood sample, 400 µl of the ultrafiltrate was injected into an LC–LC system (Figure 11.1(a)) without further pretreatment. A heart-cut (ca. 1 ml) of the eluate of the first column (octadecylsilica (ODS)) which contained the sameridine, was transferred to and enriched on, a cation-exchange extraction column. In the next step, the concentrated sample was desorbed into the second analytical column (cation-exchange) for the final separation and subsequent UV detection at 205 nm. In this way the interference-free and sensitive determination of sameridine could be achieved (Figures 11.1(b) and (c)) with a limit of quantitation (LOQ) of 1 nM and a within-day precision of 2–8 %. In a similar approach, Baker et al. (18) determined the major

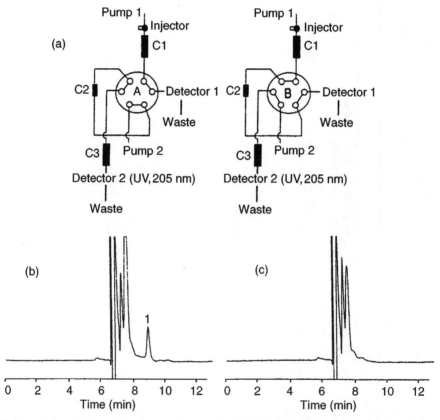

Figure 11.1 (a) Schematic representation of a coupled-column LC system for sameridine analysis consisting of a reversed-phase analytical column (C1), a cation-exchange extraction column (C2) and a cation-exchange analytical column (C3); (b) chromatogram of plasma sample after intrathecal administration of sameridine (plasma concentration, 11.2 nM); (c) chromatogram of a blank plasma sample. Reprinted from *Journal of Chromatography, B* **708**, E. Eklund *et al.* 'Determination of free concentration of sameridine in blood plasma by uttrafiltration and coupled-column liquid chromatography,' pp. 195–200, copyright 1998, with permission from Elsevier Science.

metabolite of the glucocorticoid tipredane in rat urine which was injected directly into the first LC column. In this case, the initial separation was performed on a cyanopropylsilica column and a portion of the eluate was switched to an ODS column where the final separation was carried out. This method allowed quantification of the metabolite down to 25 ng/ml, showed acceptable linearity, precision and accuracy in the 25–5000 ng/ml range, and was successfully used for a long-term toxicology study. Lanbeck-Vallén *et al.* (28) used post-column derivatization with fluorescamine for the analysis of the penicillin ampicillin in plasma by LC with fluorescence detection. However, since fluorescamine reacts with most primary amines, the selectivity of the LC system had to be increased. This was achieved by the use of a coupled column system (Figure 11.2(a)). The total combination yielded a highly selective system which permitted the sensitive determination of ampicillin (Figure 11.2(b)) with a detection limit of 14 nM for 0.5 ml plasma samples.

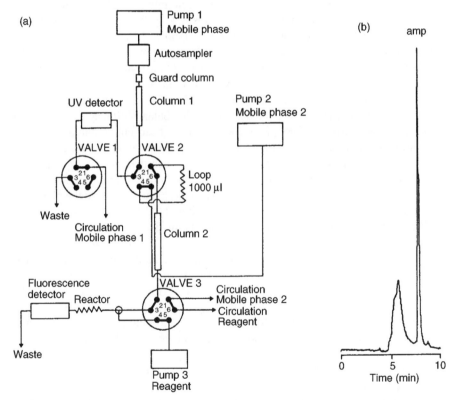

Figure 11.2   (a) LC–LC system with post-column reaction detection for the determination of ampicillin in plasma; (b) Chromatogram of plasma sample (collected 10 min after oral administration of 670 µmol of ampicillin) containing 1.26 µM ampicillin (amp). Reprinted from *Journal of Chromatography*, **567**, K. Lanbeck-Vallén *et al.*, 'Determination of ampicillin in biological fluids by coupled-column liquid chromatography and post-column derivatization,' pp. 121–128, copyright 1991, with permission from Elsevier Science.

In its true sense, the term 'multidimensional' refers to LC systems in which there is a distinct difference in retention mechanisms in the columns used. Still, column switching without change in retention mode can be a relatively easy to implement approach for separating complex mixtures. This is a practical substitute for linear-gradient elution and is particularly useful for analyte enrichment and sample clean-up, as has been extensively demonstrated by Hoogendoorn and co-workers (14, 20, 21) for environmental analysis. The concept is also useful in bioanalysis, as was shown by Polettini *et al.* (22) who applied coupled-column RPLC using two identical ODS columns for the automated analysis of β-agonists in urine samples. With direct injection of large volumes (1.5 ml) of human urine and careful adjustment of transfer volumes and mobile phase conditions, LC–LC with UV detection allowed the determination of clenbuterol at the low-ng/ml level. In a subsequent study (23), the RPLC–RPLC system was coupled to a tandem MS system in order to gain further selectivity and sensitivity. With this set-up, β-agonists could be analysed in bovine urine with a LOQ of 0.1 ng/ml. For the determination of clenbuterol at the 1 ng/ml level, the inter-day reproducibility was 8.4 %. With similar hydrophobic columns, still different retention modes can be established by using one in the ion-pairing mode and the other in the RP mode. This approach has been used by Yamashita and co-workers (34–38) for the analysis of basic and acidic drugs in biological fluids. For example, phenylpropanololamine was determined in human plasma and urine by first pre-separating the basic drug from endogenous compounds on an ODS column using a low-pH eluent containing butanesulfonate as the ion-pair reagent. Subsequently, after column switching of the heart-cut containing the analyte, further separation is carried out on a second, identical column in the absence of butanesulfonate. The selectivity of the method was such that short-wavelength UV detection could be used, thus yielding a detection limit of 0.4 ng/ml in plasma. The method was applied to the determination of phenylpropanololamine in the plasma of human volunteers after oral administration of 25 mg of the drug.

The on-line coupling of achiral and chiral columns is now a proven concept for the bioanalytical separation of enantiomers (40, 41). Table 11.1 summarizes a number of these LC–LC applications. Frequently, first an achiral column is used to separate enantiomeric pairs from matrix components, and then the enantiomers are transferred to a chiral column for selective separation (Figure 11.3). In this way, the chiral separation is often enhanced by excluding interfering substances from the second column. Van de Merbel *et al.* (43) used two-dimensional LC for the determination of D- and L-enantiomers of amino acids in biological samples. The amino acids are first separated by ion-exchange chromatography and subsequent enantioseparation is achieved by injection of 3 μl heart-cuts on to a second column with a chiral crown ether stationary phase. The selectivity and sensitivity of the system was further increased by using fluorescence detection after on-line post-column labelling of the amino acids with *o*-phthalaldehyde. In this way, small quantities of D-enantiomers could be determined in complex biological samples such as protein hydrolysates, urine, bacterial cultures and yoghurt, with hardly any additional pretreatment. Another typical example of achiral–chiral coupled LC is the

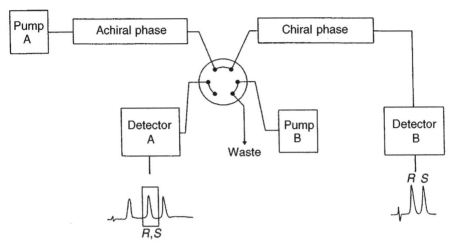

Figure 11.3 Typical configuration for the on-line coupling of an achiral and chiral chromatographic system by means of a switching valve. The non-enantio-resolved solute is isolated on the achiral phase and then stereochemically separated on the chiral phase. Reprinted from G. Subramanian, *A Practical Approach to Chiral Separation by Liquid Chromatography*, 1994, pp. 357–396, with permission from Wiley-VCH.

determination of metyrapone (a steroid biosynthesis inhibitor) and the enantiomers of its chiral metabolite metyrapol in human plasma and urine (42). A short silica column was used to separate metyrapone and metyrapol, followed by a Chiralcel OJ column for the enatioselective separation of (−) and (+)-metyrapol (Figure 11.4). The assay was validated for metabolic studies which indicated that the enzymatic reduction of myterapone is enantiospecific. Some authors have circumvented the poor performance (e.g. broad asymmetric peaks) of particular chiral columns by reversing the order of columns, i.e. carrying out a chiral–achiral coupling and exploiting peak compression effects. After chiral resolution in broad zones, the enantiomers are reconcentrated into sharply defined bands using a separate hydrophobic column for each enantiomer, thereby substantially improving the overall sensitivity and selectivity. This concept has been applied for the enantioseparation of substances such as leucovorine (53), manidipine (45), bupivacaine, metoprolol, oxazepam and terbutaline (54) in plasma.

LC–LC of endogenous compounds in biological fluids has also been reported (see Table 11.1) and does not differ essentially from the bioanalysis of drugs by coupled-column LC. For example, Tagesson *et al.* (59) determined the DNA adduct 8-hydroxydeoxyguanosine in human urine by on-line injecting a sample fraction eluting from a first polymeric RPLC column into a second ODS column which was connected to an electrochemical detector. This system was used for assaying *in vivo* oxidative DNA damage in cancer patients. High levels of the urinary adduct were found in patients subjected to body irradiation and chemotherapy.

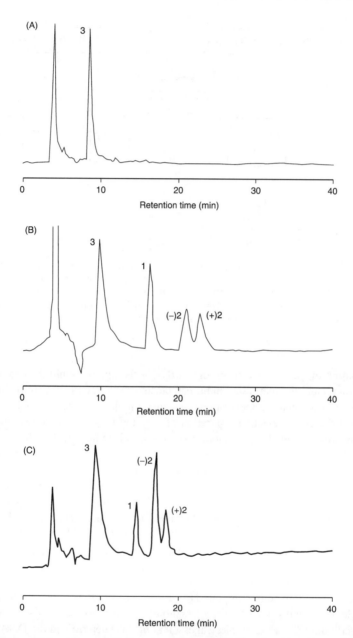

Figure 11.4 Chromatograms of plasma samples on a silica–chiralcel OJ coupled column system: (a) plasma spiked with oxprenolol (internal standard); (b) plasma spiked with 0.40 μg/ml metyrapone and 0.39 μg/ml metyrapol (racemate); (c) plasma sample obtained after oral administration of 750 mg metarypone. Peaks are as follows: 1, metyrapone; 2, metyrapol enantiomers; 3, oxprenolol. Reprinted from *Journal of Chromatography*, **665**, J. A. Chiarotto and I. W. Wainer, 'Determination of metyrapone and the enantiomers of its chiral metabolite metyrapol in human plasma and urine using coupled achiral–chiral liquid chromatography,' pp. 147–154, copyright 1995, with permission from Elsevier Science.

Considering the numerous applications, heart-cut LC–LC has convincingly proven its value. Nevertheless, in LC–LC specific method development is generally needed for each analyte. Moreover, heart-cut procedures require accurate timing and, therefore, the performance of the first analytical column in particular should be highly stable to thus yield reproducible retention times. This often means that in LC–LC some kind of sample preparation remains necessary (see Table 11.1) in order to protect the first column from proteins and particulate matter, and to guarantee its lifetime.

11.3 SOLID-PHASE EXTRACTION–LIQUID CHROMATOGRAPHY

Today, solid-phase extraction (SPE) is probably the most popular sample preparation method used for sample enrichment and/or clean-up (3). During the last decade, many improvements in the SPE field such as new formats (e.g. sophisticated cartridges and discs, pipette tips and 96-well plates), new sorbents and the development of automated systems, have led to an extensive use of SPE. Moreover, the on-line combination of SPE and LC using column switching techniques is now routine. SPE, can in principle, be described as a simple liquid chromatographic process in which the sorbent is the stationary phase. Therefore, the direct coupling of SPE and LC can be considered to be a multidimensional chromatographic technique. However, in contrast to LC–LC, in SPE–LC no continuous elution is carried out during the first dimension. Ideally, the analytes are fully retained on the SPE phase during the sorption and washing steps, and, subsequently, quickly and completely desorbed in a small volume during the elution step. In an on-line SPE–LC set-up the desorption solvent is entirely injected on to the LC column.

Numerous studies involving the on-line coupling of SPE columns to LC analytical columns have been reported, including a significant number of biomedical applications (70). Various drugs and endogenous compounds have been analysed in biological matrices by employing this technique, including antibiotics (71), retinoids (72, 73), methotrexate (74), codenie (75), aspirin (76), psilocin (77), almokalant (78), anabolic steroids (79), morphine (80), ceftazidime (81), terbinafine (82), clozapine (83), mizolastine (84), cebaracetam (85) and piroxicam (86). In many of these applications ODS sorbents are used for SPE. Typically, 0.1–1 ml of plasma or ca. 1–10 ml of urine are loaded on a preconditioned SPE cartridge (2–15 mm long, 1–4.6 mm i.d., packed with 15–40 μm particles) which is mounted on a six- or ten-port valve and replaces the conventional injection loop. After washing (clean-up), the pre-column is desorbed with mobile phase which is led to the analytical column (often packed with an RPLC phase) for separation of the analytes. Quantification and/or identification is frequently carried out using UV or MS detection. During analysis, the SPE cartridge may be cleaned, reconditioned and loaded with the next sample. Figure 11.5 shows a typical example of on-line SPE–LC for the determination of drugs in human plasma (87). Many of the on-line SPE–LC applications for bioanalysis described in the literature are quite similar and we will not further discuss these studies. In the remaining part of this section, some selected at-line and

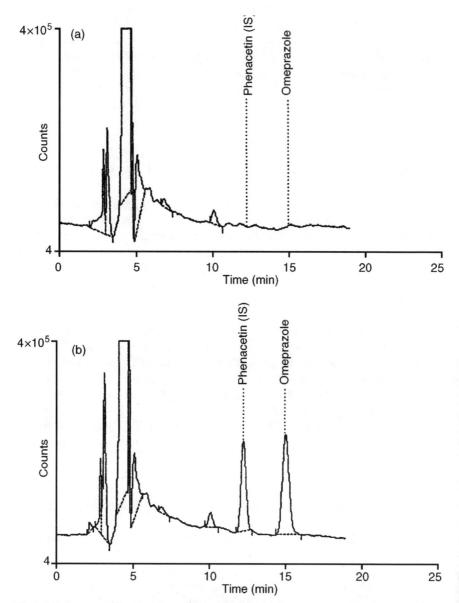

Figure 11.5 Chromatograms of plasma samples obtained with fully automated on-line SPE–LC: (a) drug-free human plasma; (b) human plasma spiked with omeprazole (100 ng/ml) and phenacetin (internal standard; 1000 ng/ml). Reprinted from *Journal of Pharmaceutical and Biomedical Analysis*, **21**, G. Garcia-Encina *et al.*, 'Validation of an automated liquid chromatographic method for omeprazole in human plasma using on-line solid-phase extraction,' pp. 371–382, copyright 1999, with permission from Elsevier Science.

on-line applications will be treated in which special formats and/or selective SPE sorbents are used in combination with LC.

11.3.1 AUTOMATED SPE–LC

In order to achieve more cost-effective analyses, biomedical laboratories aim for a higher sample throughput and, thus, also for faster sample pretreatment procedures. This aspiration has led to an increasing interest in the automation of sample preparation by using robotic devices. The whole off-line SPE sequence can be automated with instruments such as the ASPEC (Gilson), Microlab (Hamilton) and Rapid Trace (Zymark). Some of these devices can be coupled to LC in an at-line fashion, as they provide the possibility for automated injection of the final extract. A good illustration of the high degree of automation that can be accomplished today is the determination of the glucocorticosteroid budesonide in plasma samples, as described by Kronkvist *et al.* (88). Their automated method comprised a Tecan robot for initial sample manipulation (e.g. pipetting and mixing), an ASPEC XL for SPE of the samples, an on-line trace enrichment system for further purification and concentration of the sample extracts, and a gradient LC–MS–MS system for separation and selective detection of the analyte and the internal standard. The total system can analyse up to 800 samples a week with satisfactory accuracy, yielding an LOQ of 15 pM for 1 ml plasma samples.

In 1996, the 96-wells SPE technology was introduced for high-throughput analysis; this allows the simultaneous processing of 96 samples in a standard microtiter plate format (89). Such technology, in combination with a pipetting robot, was used to design a high-throughput SPE system followed by RPLC with UV detection for the determination of cimetidine (a histamine H2-receptor antagonist) in plasma (90). Validation results were found to be adequate and a good correlation between the results obtained with the automated method and with a manual method was demonstrated. The average sample preparation time decreased from 4 min to 0.6 min per sample and a sample throughput of 160 samples a day was achieved. Recently, Jemal *et al.* (91) made a comparison of plasma sample pretreatment by manual liquid–liquid extraction, automated 96-well liquid–liquid extraction and automated 96-well SPE for the analysis by LC–MS–MS. The 96-well methods were three times faster than the manual method. The time required for the 96-well SPE could be further reduced by 50 % when the extracts were injected directly on to the LC (i.e. not incorporating drying and reconstitution steps in the SPE procedure).

In principle, on-line SPE–LC can be automated quite easily as well, for instance, by using such programmable on-line SPE instrumentation as the Prospekt (Spark Holland) or the OSP-2 (Merck) which have the capability to switch to a fresh disposable pre-column for every sample. Several relevant applications in the biomedical field have been described in which these devices have been used. For example, a fully automated system comprising an autosampler, a Prospekt and an LC with a UV detector was applied for the analysis of omeprazole, a drug for the treatment of gastric ulcers, in human plasma (87). The extraction was carried out on-line by using disposable SPE cartridges filled with an ODS sorbent, which, after washing, yielded sufficient sample clean-up and enrichment to permit measurement of drug levels at the low ng/ml level (see Figure 11.5). The automated system was validated and used for a large number of plasma samples from bioequivalence studies. In another study, adapalene (a retinoid) and retinol were simultaneously determined in human plasma and mouse tissue by combining an autosampler and an OSP-2 unit with a gradient LC with UV and fluorescence detection (92). The high degree of automation of the on-line SPE–LC system, its high sensitivity (LOQ of 0.25 ng/ml for adapalene) and its good reproducibility, made this method convenient for the determination of pharmacokinetic drug profiles.

11.3.2 RESTRICTED-ACCESS SORBENTS IN SPE–LC

In on-line SPE–LC, deproteination of plasma and serum is often required before extraction, especially when the same SPE cartridge has to be used for repeated analyses. However, today there is strong interest in on-line sample pretreatment techniques which permit the handling of untreated biological samples. For this purpose, special SPE sorbents (so-called restricted-access materials (RAMs)) have been developed which allow sorption (enrichment) of low-molecular-weight analytes at the inner pore surface of the particles, but at the same time exclude high-molecular-weight matrix compounds (e.g. proteins) from the pores (Figure 11.6(a)). In recent years, RAMs have become quite popular for the direct injection of biological fluids into on-line SPE–LC systems (93–98). For instance, Yu and Westerlund (93) selected a RAM pre-column filled with an alkyl-diol silica to rapidly separate bupivocaine (a local anaesthetic) from proteins and polar endogenous compounds in human plasma. A 500 µl plasma sample was directly introduced on to the SPE column and the fraction containing the analyte was on-line transferred to an RPLC column for final separation. A single alkyl-diol silica pre-column could withstand over 50 ml of plasma injections. The same type of RAM pre-columns appeared to be very useful for on-line clean-up and enrichment during the determination of several drugs and metabolites in biological fluids (e.g. serum, urine, intestinal aspirates and supernatants of cell cultures) which could be directly injected (94).

LC–MS with on-line SPE using a RAM pre-column with an internal ODS phase was described by van der Hoeven *et al.* (95) for the analysis of cortisol and prednisolone in plasma, and arachidonic acid in urine. The samples were injected directly and the only off-line pretreatment required was centrifugation. By using the on-line SPE–LC–MS system, cortisol and related compounds could be totally recovered and quantified in 100 µl plasma within 5 min with a typical detection of 2 ng/ml (Figure 11.6(b)). The RAM-type of sorbents, in which the outer surface of the particles is covered with α_1-acid glycoprotein, also appear to be useful for direct SPE of

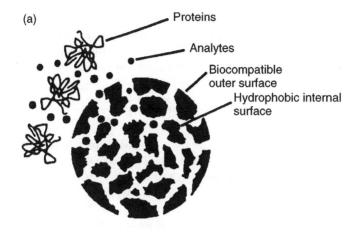

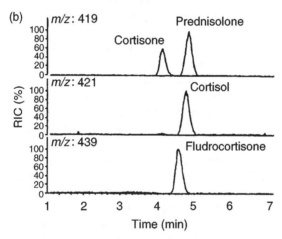

Figure 11.6 (a) Schematic illustration of a restricted-access sorbent particle. (b) On-line SPE–LC–MS using an alkyl-diol RAM pre-column for the determination of cortisol, cortisone, prednisolone and fludrocortisone (100 ng/ml each) in untreated plasma. Part (a) reprinted from *Journal of Chromatography*, *A* **797**, J. Hermansson *et al.*, 'Direct injection of large volumes of plasma/serum of a new biocompatible extraction column for the determination of atenolol, propanolol and ibuprofen. Mechanisms for the improvement of chromatographic performance,' pp. 251–263, copyright 1998, with permission from Elsevier Science. Part (b) reprinted from *Journal of Chromatography*, *A* **762**, R. A. M. van der Hoeven *et al.*, 'Liquid chromatography–mass spectrometry with on-line solid-phase extraction by a restricted-access C_{18} precolumn for direct plasma and urine injection,' pp. 193–200, copyright 1997, with permission from Elsevier Science.

relatively large volumes (200–500 µl) of plasma and serum. By using these pre-columns in an on-line SPE–LC set-up with fluorescence detection, Hermansson *et al.* (96) could quantitatively determine atenolol and propanolol in serum down to the low ng/ml level. The stability and performance of the system was good, even after the injection of 200 serum samples of 500 μl. An on-line sample clean-up using a Pinkerton-GFF2 RAM pre-column was combined with LC on a non-porous ODS phase for the analysis of six cardiovascular drugs in serum (97). Matrix interferents could be removed efficiently and were not detected in the serum chromatograms.

11.3.3 IMMUNOAFFINITY SORBENTS IN SPE–LC

In the large majority of all SPE–LC applications, SPE pre-columns are packed with a hydrophobic packing material such as an ODS phase or a polystyrene–divinyl-benzene copolymer. With these packing materials, in principle, sensitivity can be increased dramatically (high analyte retention), but selectivity may only be slightly improved (non-selective hydrophobic interaction). If enhanced selectivity is a major requirement, packing materials with immobilized antibodies (immunoaffinity sorbents) can be used which provide extraction based on molecular recognition. After some exploration work of others (99, 100), Farjam and co-workers were the first to extensively study the on-line application of immunoaffinity pre-columns in combination with LC for the selective determination of low-molecular-weight compounds (steroids and aflatoxins) in biological samples (101–105). A system consisting of a pre-column packed with Sepharose-immobilized polyclonal antibodies against β-19-nortestorenone (β-19-NT), a second ODS-filled pre-column, an analytical ODS column and a UV detector (Figure 11.7(a)), was used for the determination of β-19-NT and α-19-NT in calf urine which was injected directly (101). The analytes were subsequently trapped on the immunoaffinity pre-column, selectively desorbed by using the cross-reacting steroid norgestrel, reconcentrated on the ODS precolumn and transferred to the analytical column. Recoveries of β-19-NT were over 95% and the detection limit was 50 ng/l for a 25 ml urine injection (Figure 11.7(b)). A similar procedure was also used for the determination of β-19-NT and α-19-NT in calf bile, liver, kidney and meat samples (102), and for the analysis of clenbuterol in calf urine (106). In a subsequent study, an anti-aflatoxin immunoaffinity pre-column was used on-line for the selective preconcentration of aflatoxin M1 from urine and milk samples (103, 104). By using LC with fluorescence detection following the immuno-SPE step, detection limits of 20 ng/ml for aflatoxin M1 were found if 2.4-ml milk samples are used. The lifetime of the immunoaffinity pre-column was strongly reduced when milk samples were analysed, due to proteolic enzymes in milk which degrade the antibodies. This problem was circumvented by adding an on-line dialysis unit to the system to prevent interfering macromolecular compounds from entering the pre-column (104). A single immunoaffinity pre-column could now be used for over 70 milk analyses. Another example of the use of on-line immunoaffinity-SPE–LC is the selective quantification of Δ^9-tetrahydrocannabinol, the major indicator of cannabis intoxication, in human saliva (107).

Immunoaffinity extraction combined on-line with LC in conjunction with MS (108–110) or tandem MS (111, 112) has also been demonstrated for the determination of analytes in biological fluids. Obviously, such systems offer a very high

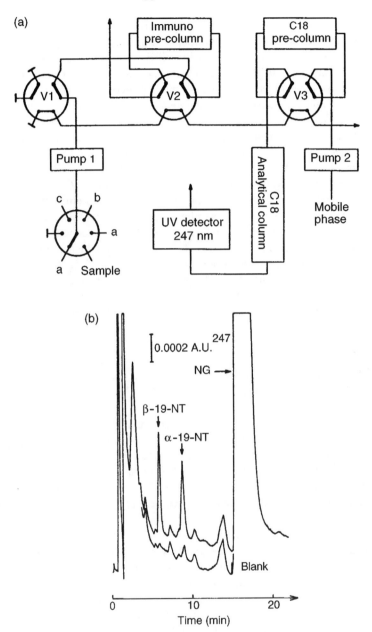

Figure 11.7 (a) Set-up used for immunoaffinity SPE–LC. (b) The analysis of β-19-nortestosterone (β-19-NT) and α-19-nortestosterone (α-19-NT) (300 ng/l each) in 26.5 ml calf urine by using on-line SPE–LC with an immunoaffinity pre-column against β-19-NT. Norgestrel (NG) is used for desorption of the immunoaffinity pre-column. Reprinted from *Journal of Chromatography*, **452**, A. Farjam *et al.*, 'Immunoaffinity pre-column for selective on-line sample pre-treatment in high-performance liquid chromatography determination of 19-nortestosterone,' pp. 419–433, copyright 1988, with permission from Elsevier Science.

selectivity which may be needed when analytes have to be determined at the trace level and or characterization and confirmatory information is required. Cai and Henion (111) used an immunoaffinity–SPE–SPE–LC–MS–MS system for the analysis of LSD and its analogues and metabolites in human urine. The on-line chromatographic system involved an immunoaffinity pre-column, a packed capillary trapping column and a packed capillary analytical column. With the trapping column as intermediate the immuno pre-column could be operated at high flow rates, while the analytical column was maintained at 3.5 µl/min, thereby facilitating the coupling with MS by electrospray. The urine of LSD users was analysed and concentrations of LSD and analogues at the low-ppt level could be measured. A similar system was used for the analysis of five β-agonists in bovine urine yielding LOQs in the 10–50 ppt range, and applied to the determination of the bovine renal elimination of clenbuterol over a period of 15 days after administration (112). For the selective extraction of propanolol from urine, a protein G pre-column was primed with drug-specific antibodies and coupled on-line to an SPE–LC–MS system which comprised a CN analytical column for separation and an ion-spray source for LC–MS interfacing. By using the single-ion-monitoring mode, propanolol could be specifically detected down to 1 ng/ml in 20 ml urine. Creaser et al. (108) used on-line immunoaffinity-SPE–LC–ion-trap MS with a particle beam interface for the determination of corticosteroids in equine urine.

11.3.4 METAL-LOADED SORBENTS AND MOLECULARLY IMPRINTED POLYMERS IN SPE–LC

An interesting–but not often used–option to increase SPE selectivity is the application of metal-loaded sorbents that are able to bind certain organic compounds which form specific complexes with the immobilized metal ions such as Cu(II), Ag(I), Hg(II) and Pd(II). Nielen et al. (113) showed that metal-loaded phases can be used for selective on-line sample handling and trace enrichment in LC. Some bioanalytical applications were reported by Irth and co-workers 114–116). A pre-column with a silver-loaded thiol phase was used in an on-line fashion for the extraction of 5-fluorouracil (a chemotheurapetic agent) from plasma (114). SPE sorption of the analyte by complexation with Ag(I) occured at high pH, while desorption to the ODS analytical column was based on analyte protonation at low pH. For plasma samples, either a second pre-column packed with PLRP-S was placed before the Ag(I)-thiol pre-column in order to remove macromolecular and apolar components, or the samples were deproteinated before analysis. By using UV detection, a detection limit of 3 nM was obtained for 5-fluorouracil for direct injection of 1 ml samples. The same SPE–LC–UV system with a silver-loaded pre-column was used for the trace-level determination of 3′-azido-3′-deoxythymidine (AZT), a potential drug for the treatment of AIDS, in plasma (115), and also for the analysis of four barbiturates in plasma (116).

Recently, molecularly imprinted polymers (MIPs) have gained attention as new, selective sorbents for chromatography and SPE. The cavities in the polymer selectively recognize and bind a specific compound in a similar manner as immunoaffinity sorbents, but can be tailor-made (117). A disadvantage of most current MIPs is that the analyte needs to be dissolved in a non-protic organic solvent and, as a consequence, the use of MIPs in on-line SPE of aqueous biofluids has been very limited so far. In order to circumvent this problem, Boos *et al.* (118) have proposed an on-line SPE–SPE–LC scheme which comprises a RAM and a MIP pre-column. After sample loading, the RAM sorbent is desorbed with a pure organic solvent which is led to the MIP pre-column for molecular recognition of the target analyte. A practical evaluation of this approach, as well as other multidimensional set-ups involving MIPs, can be expected in the near future.

11.4 LIQUID CHROMATOGRAPHY–GAS CHROMATOGRAPHY

LC is not only a powerful analytical method as such, but it also allows effective sample preparation for GC. The fractions of interest (heart-cuts) are collected and introduced into the GC. The GC column can then be used to separate the fractions of different polarity on the basis of volatility differences. The separation efficiency and selectivity of LC is needed to isolate the compounds of interest from a complex matrix.

Traditionally, LC and GC are used as separate steps in the sample analysis sequence, with collection in between, and then followed by transfer. A major limitation of off-line LC–GC is that only a small aliquot of the LC fraction is injected into the GC μ (e.g. $1-2$ μl from 1 ml). Therefore, increasing attention is now given to the on-line combination of LC and GC. This involves the transfer of large volumes of eluent into capillary GC. In order to achieve this, the so-called on-column interface (retention gap) or a programmed temperature vaporizor (PTV) in front of the GC column are used. Nearly all on-line LC–GC applications involve normal-phase (NP) LC, because the introduction of relatively large volumes of apolar, relatively volatile mobile phases into the GC unit is easier than for aqueous solvents. On-line LC–GC does not only increase the sensitivity but also saves time and improves precision.

11.4.1 NPLC–GC

The first bioanalytical application of LC–GC was presented by Grob *et al.* (119). These authors proposed this coupled system for the determination of diethylstilbestrol in urine as a replacement for GC–MS. After hydrolysis, clean-up by solid-phase extraction and derivatization by pentafluorobenzyl bromide, the extract was separated with normal-phase LC by using cyclohexane/1% tetrahydrofuran (THF) at a flow-rate of 260 μl/min as the mobile phase. The result of LC–UV analysis of a urine sample and GC with electron-capture detection (ECD) of the LC fraction are shown in Figures 11.8(a) and (b), respectively. The practical detection limits varied between about 0.1 and 0.3 ppb, depending on the urine being analysed. By use of

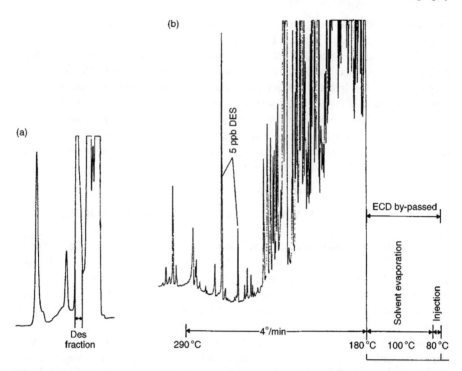

Figure 11.8 (a) LC–UV trace of a typical sample after derivatization of DES to the dipentafluorobenzyl ether. The transferred DES fraction almost corresponds to a peak in the LC trace but, of course, this peak does not represent DES. (b) GC–ECD of the DES fraction from LC separation, involving the pentafluorobenzyl ether derivative of the two isomers of DES. The sample wash spiked with DES (5 ppb) before sample preparation. During passage of the solvent vapour through the column, the column eluent was driven away from the detector cell through the line usually used for feeding the make-up gas into the detector. Reprinted from *Journal of Chromatography*, **357**, K. Grob Jr *et al.*, 'Coupled high-performance liquid chromatography–capillary gas chromatography as a replacement for gas chromatography–mass spectrometry in the determination of diethylstilbestrol in bovine urine,' pp. 416–422, copyright 1986, with permission from Elsevier Science.

derivatization and electron-capture detection, the sensitivity of the LC–GC method exceeded that obtained by GC–MS by at least a factor of ten.

Gianesello *et al.* (120) described the determination of the bronchodilator broxaterol in plasma by on-line LC–GC. After deproteination and extraction, the LC separation was carried out by using a mixture of *n*-pentane and diethyl ether (55 : 45 (vol/vol) as mobile phase. A small cut of the LC chromatogram (shown in Figure 11.9(a)) was introduced at 85 °C into the GC via so-called concurrent solvent evaporation. Figure 11.9(b) demonstrates that a detection limit of about 0.03 ng/ml was obtained. A fully automated LC–GC instrument was described by Munari and Grob (121) and its applicability was demonstrated by the determination of heroin metabo-

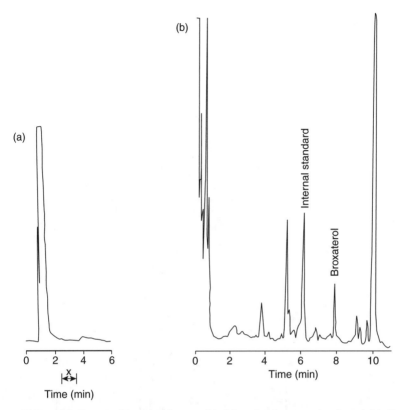

Figure 11.9 (a) LC trace of human plasma with 0.1 μg/ml broxaterol, where 'x' indicates the cut introduced into the GC run; (b) chromatogram of human plasma with 0.1 ng/ml broxaterol and 5 ng/ml internal standard. Reprinted from *Journal of High Resolution Chromatography and Chronatographical Communication*, **11**, V. Gianesello *et al.*, 'Determination of broxaterol in plasma by coupled HPLC–GC,' pp 99–102, 1988, with permission from Wiley-VCH.

lites. The LC eluent consisted of ethyl ether/methanol/diethylamine (91.5 : 8 : 0.5, vol/vol) at a flow-rate of 400 μl/min and a fraction of 500 μl was transferred to the GC unit through a loop-type interface. Ghys *et al.* (122) have used the same set-up for the coupling of micro-SEC and GC. The interesting aspect of a micro LC column of 320 μm i.d. with a flow-rate of about 1 μl/min is that a relatively large fraction can be introduced into the GC unit. For the analysis of steroid esters in a pharmaceutical formulation, a volume of only 4 μl THF was transferred to the GC system. Figure 11.10 shows the potential of this system. The heart-cut chromatogram of the sample is very similar to the chromatogram of a standard solution of the steroid esters.

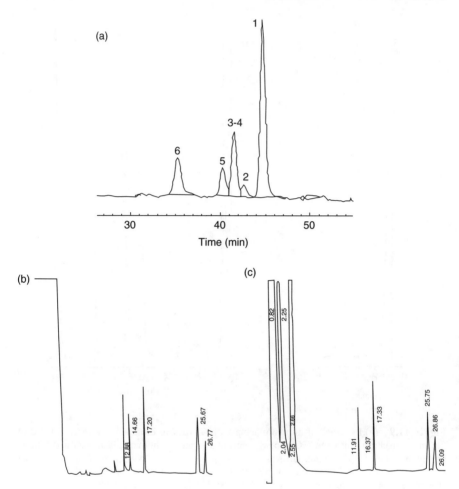

Figure 11.10 (a) Micro-SEC–UV trace of Sustanon, where peak 2-5 were transferred to the GC unit. Peak identification is as follows: 1, benzylalcohol; 2, testosterone propionate; 3, testosterone isocaproate; 4, testosterone phenylpropionate; 5, testosterone decanoate; 6, oil matrix; (b) GC analysis of the transfer (4 μl) from the micro SEC system; (c) Direct GC analysis of a standard solution of the steroid esters. Reprinted from *Proceedings of the 10th Symposium on Capillary Chromatography*, M. Ghys *et al.*, 'On-line micro size-exclusion chromatography–capillary gas chromatography,' 1989, with permission from Wiley-VCH.

Presently, the on-line coupling of NPLC and GC via heart-cutting is an established procedure which has been used successfully for several bioanalytical applications. Obviously, direct analysis of aqueous samples is not possible by NPLC, and therefore, a solvent switch by a sample pretreatment step (e.g. liquid–liquid extraction or SPE) is always required when biological samples are analysed by NPLC–GC.

11.4.2 RPLC–GC

For pharmaceutical and biomedical analysis, RPLC is much more important than NPLC. However, the interfacing techniques used in NPLC–GC do not generally work well when used for RPLC–GC. The main difficulties encountered when transferring water or water-containing eluents to a GC unit are due to the large vapour volume of water, its high surface tension, poor wetting characteristics, high boiling point and aggressive hydrolytic reactivity. Two approaches have been described for interfacing RPLC and GC on-line, i.e. (i) direct introduction of the aqueous LC fraction by miniaturization of the LC step or by use of special retention gaps, and (ii) phase-switching techniques, i.e. the analytes are first transferred to an organic solvent and subsequently introduced into the GC system.

For drug analysis, Goosens *et al.* (123) used a Carbowax-deactivated retention gap to transfer eluents from the LC unit to the GC part of the system. Up to 200 µl of eluent (acetonitrile-water) were introduced into a Carbowax-coated retention gap by using an on-column interface and solvent vapour exit (124). It was found that the water content of the eluent should not exceed that of the azeotropic mixture, or otherwise water, which is left in the gap after evaporation of the azeotropic mixture, will mar the analysis. In order to deal with the presence of buffers or ion-pairing agents, an anion-exchange micromembrane device was inserted between the LC and GC parts of the system to remove the ion-pairing agent methanesulphonic acid from an acetonitrile–water LC eluent (125). The applicability of the on-line LC–micromembrane–GC system was illustrated for the potential drug eltroprazine (125) and for an impurity profile of the drug mebeverine (126). Before the LC fraction was introduced on-line into the GC–MS system, acetonitrile was added to achieve an azeotropic acetonitrile/water ratio, and, therefore only a part of the LC peak could be transferred. Nevertheless, electron impact and chemical ionization spectra of an impurity could be obtained at a level of 0.1 % with respect to the drug.

Ogorka *et al.* (127) have coupled RPLC and GC via on-line liquid–liquid extraction of the aqueous mobile phase and used the system for the impurity profiling of drugs. The instrumental set-up is shown in Figure 11.11, where a main critical part is the phase separator. These authors optimized the extraction for mobile phases consisting of methanol–water and acetonitrile–water by using *n*-pentane, *n*-hexane and dichloromethane as extraction solvents. The extraction yield depended on the water content of the mobile phase and the polarity of the organic phase. Transfer volumes of 500 µl of aqueous mobile phase have been used. The usefulness was extended via on-column derivatization by the introduction of a reagent via the loop-type interface or by derivatization during the extraction (128). By the use of MS as the GC detector, the identification of various unknown impurities in pharmaceutical products was achieved. Contrary to direct LC–MS, the composition of the LC eluent is less limited because non-volatile buffers can also be chosen. The same system was also applied for the analysis of biological samples, i.e. the determination of β-blockers in human serum and urine (129) and the determination of morphine and its analogues in urine (130). In the latter case, the analytes were silylated with bis (trimethylsilyl)

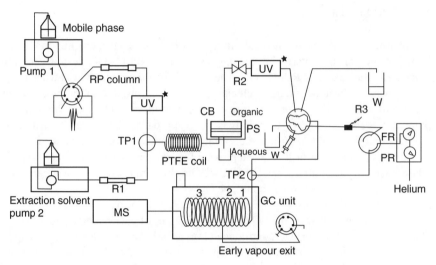

Figure 11.11 Schematic diagram of the instrumental set-up used for on-line coupled LC–GC–MS: 1, retention gap; 2, retaining pre-column; 3, analytical capillary; R1, LC column (restriction); FR, flow regulator; R2, needle-valve restrictor; R3, capillary (75 μm i.d.); PS, phase separator (sandwich type); CB, cooling bath; PR, pressure regulator; W, waste; TP1 and TP2, T-pieces; UV, UV detector. (Note that the UV detector* can be positioned either before or after the liquid–liquid extraction unit.) Reprinted from *Journal of Chromatography*, **626**, J. Ogorka *et al.*, 'On-line coupled reversed-phase high-performance liquid chromatography–gas chromatography–mass spectrometry. A powerful tool for the identification of unknown impurities in pharmaceutical products', pp. 87–96, copyright 1992, with permission from Elsevier Science.

acetamide in the retention gap before the GC separation. After the reaction, the solvent vapour exit was closed and the GC run was started. The yield of the on-line derivatization was comparable with off-line derivatization. It should be noted that the liquid–liquid extraction step in the applied system can offer extra selectivity. The extraction yield was increased by the use of higher temperatures. The high total selectivity is illustrated by Figure 11.12. The total analysis time was less than 60 min, which is much shorter than with more traditional analytical methods. The limits of quantification were 61–92 ng/ml.

11.5 SOLID-PHASE EXTRACTION–GAS CHROMATOGRAPHY

In LC–LC and SPE–LC, the presence of water is commonly no problem at all. Actually, the reverse is true because eluents in RPLC are typically water–methanol or water–acetonitrile mixtures, and a high water content is mandatory during trace enrichment in order to ensure strong retention. However, when such an SPE precolumn or analytical column is coupled to a GC system, the introduction of water should be avoided completely or, at best, be permitted under strictly controlled conditions (see above). It will be clear that on-line trace enrichment (and clean-up) by SPE

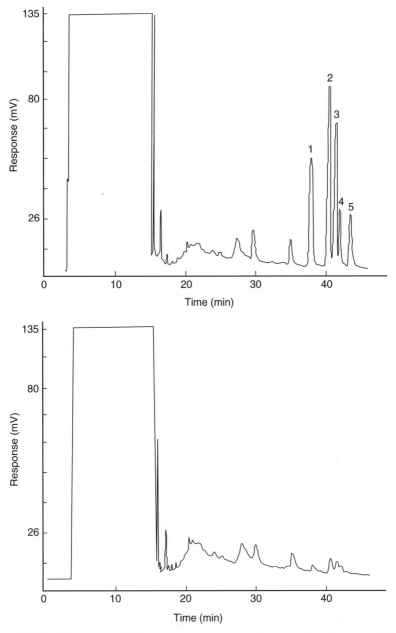

Figure 11.12   GC analysis of (a) urine sample spiked with opiates 3 μg/ml) and (b) blank urine sample. Peak identification is as follows: 1, dihydrocodeine; 2, codeine; 3, ethylmorphine; 4, morphine; 5, heroin. Reprinted from *Journal of Chromatography, A* **771**, T. Hyötyläinen *et al.*, 'Determination of morphine and its analogues in urine by on-line coupled reversed-phase liquied chromatography–gas chromatography with on-line derivatization,' pp. 360–365, copyright 1997, with permission from Elsevier Science.

for GC analysis, i.e. SPE–GC, is a highly interesting approach for the rapid trace-level detection and quantitation of the wide range of GC-amenable compounds that have to be monitored. This convincingly justifies the rather special interest in this particular area of on-line LC–GC, but so far the applications have been mainly in the environmental field and not many bioanalytical applications have been described.

On-line dialysis–SPE–GC was developed for the determination of drugs in plasma, with benzodiazepines as model compounds (131). Clean-up was based on dialysis of 100 μl samples for 7 min by using water as the acceptor, and trapping the diffused analytes on a PLRP-S column. After drying, the analytes were desorbed with 375 μl of ethyl acetate, which were injected on-line into the GC via a loop-type interface. This system provides a very efficient clean-up and offers the possibility of adding chemical agents which can help to reduce drug–protein binding. In order to demonstrate the potential of the approach, benzodiazepines were determined in plasma at their therapeutic levels. Flame-ionization (FID), nitrogen–phosphorus (NPD) and MS detection were used.

The selectivity of the trace-enrichment procedure can be improved by using an immunoaffinity precolumn: 19-β-nortestosterone was used as the test compound (132). Desorption from the antibody-loaded pre-column had to be carried out with about 2 ml of methanol–water (95 : 5, vol/vol), which obviously could not be transferred to the retention gap. The eluate was therefore diluted with HPLC-grade water and the mixture led through a conventional ODS pre-column. As a result of so-called reconcentration by dilution–which means that the gain in breakthrough volume due to increased retention caused by the decrease of the modifier percentage distinctly outweighs the volume increase–the analyte was quantitatively trapped on this second pre-column. Desorption and transfer to the GC system were coried out in a similar way to that described above. The method was applied to the determination of steroid hormones in 5–25 ml human urine. The detection limit for 19-β-nortestosterone was about 0.1 ppb with an RSD of 6 % (see Figure 11.13).

Examples of SPE–GC of biological samples are few, while the usefulness of SPE–GC for the analysis of surface and drinking water has been demonstrated many times (133). This might be due to the fact that biological samples are often considerably more complex than environmental water samples. In addition, various biomedically and pharmaceutically interesting analytes will not be amenable to GC. Nevertheless, because many of the initial SPE–GC interfacing problems have now been solved (133), it seems appropriate and worthwhile to explore its utility in the bioanalytical field more thoroughly.

11.6 SOLID-PHASE MICROEXTRACTION COUPLED WITH GAS OR LIQUID CHROMATOGRAPHY

Although solid-phase microextraction (SPME) has only been introduced comparatively recently (134), it has already generated much interest and popularity. SPME is based on the equilibrium between an aqueous sample and a stationary phase coated on a fibre that is mounted in a syringe-like protective holder. For extraction, the fibre

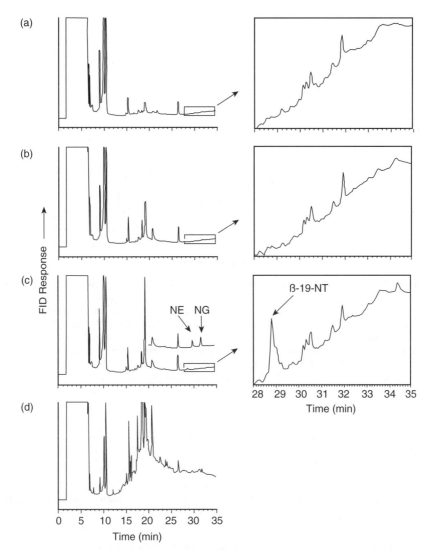

Figure 11.13 (a–c) Immunoaffinity extraction-SPE–GC-FID traces of (a) HPLC-grade water (b) urine (c) urine spiked with β-19-nortestostrone (0.5 µg/l) or norethindrone and norgestrel (both 4 µg/l); (d) SPE–GC-FID trace of urine. Reprinted from *Analytical Chemistry*, **63**, A. Farjam *et al.*, 'Direct introduction of large-volume urine samples into an on-line immunoaffinity sample pretreatment-capillary gas chromatography system,' pp. 2481–2487, 1991, with permission from the American Chemical Society.

is exposed to the sample by suppressing the plunger. Sorption of the analytes on the fibre takes place in either the sample by direct-immersion or the headspace (HS) of the sample. After equilibrium or a well-defined time, the fibre is withdrawn in the septum piercing needle and introduced into the analytical instrument where the

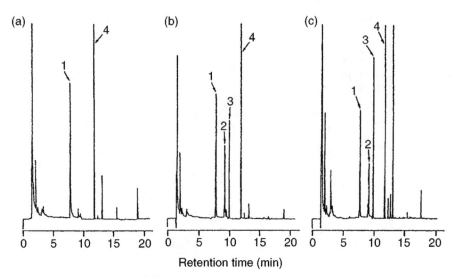

Figure 11.14 Analysis of amphetamines by GC-NPD following HS-SPME extraction from human hair: (a) Normal hair; (b) normal hair after addition of amphetamine (1.5 ng) and methamphetamine (16.1 ng); (c) hair of an amphetamine abuser. Peak identification is as follows: 1, α-phenethylamine (internal standard); 2, amphetamine; 3, methamphetamine; 4, N-propyl-β-phenethyamine (internal standard). Reprinted from *Journal of Chronatography, B* **707**, I. Koide *et al.*, 'Determination of amphetamine and methamphetamine in human hair by headspace solid-phase microextraction and gas chromatography with nitrogen-phosphorus detection,' pp. 99–104, copyright 1998, with permission from Elsevier Science.

analytes are either thermally desorbed into the GC unit or re-dissolved in a proper solvent for LC. Coupling to the LC system requires an appropriate interface and was first reported in 1995. The technique was commercialized in 1993 by Supelco. The initial work was exclusively done with SPME–GC (135–137), due to the direct and convenient sample introduction into the GC system, while the main application area being environmental analysis. Recently, SPME is being increasingly used in bioanalysis. Successful coupling with LC systems enables the analysis of pharmaceuticals, proteins and surfactants that cannot be analysed by GC. Up until now, only a few papers have described the use of direct-immersion SPME for plasma analysis (138–141). For plasma and blood samples, the relevant drug partitions between the fibre, sample and proteins. Models for the relationship between the total amount of the drug present in the plasma and the amount of drug extracted have been developed (138, 139). In this way, a good approximation of the drug–protein binding can also be obtained. A few other typical bioanalytical examples will be discussed below.

In a recent report, HS-SPME was used for the extraction of amphetamines from human hair (142). Human hair analysis is gaining interest in the analysis of drugs of abuse, since it offers attractive features: easy and 'unlimited' sampling, and as the

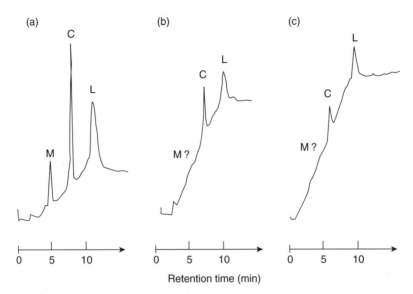

(a)

(b)

(c)

Retention time (min)

Figure 11.15 Cation-exchange micro-LC analysis of a mixture of model proteins: (a) the original sample consisting of myoglobin (M), cytochrome C (C) and lysozyme (L); (b) and (c) proteins adsorbed on to and then released from the polyacrylic acid coated fibre with extraction times of 5 and 240 s, respectively. Reprinted from *Journal of Microcolumn Separations*, **8**, J.-L. Liao *et al.*, 'Solid phase micro extraction of biopolymers, exemplified with adsorption of basic proteins onto a fiber coated with polyacrylic acid,' pp. 1–4, 1996, with permission from John Wiley & Sons, New York.

most important aspect, the possibility to measure the drug after months of use. Drugs are incorporated into hair and remain there for several months. Thus, long term abuse and also the history of the abuse can be ascertained. Hair was alkalinized with NaOH and heated to 55 °C. SPME adsorption from the headspace lasted 20 min and analysis was performed by GC-NPD. Figure 11.14 depicts the potential of the method for the identification of amphetamine abuse.

For the extraction of proteins, SPME was coupled to micro-LC by using columns based on a new continuous polymer bed technology (143). A very short extraction time (a few seconds) was used to ensure that the capacity of the home-made poly(acrylic acid)-coated fibre was sufficient. Because of the low protein binding capacity, the amount of basic proteins adsorbed on to the fibre was found to be proportional to the concentration of the protein. Proportionality was also obtained for longer extraction times provided that the protein content does not exceed the binding capacity; otherwise, the extraction of strongly adsorbed proteins was favoured. Figure 11.15 shows chromatograms for the analysis of proteins obtained by using the micro-LC system with and without SPME. Because myoglobuline was almost in its neutral form under the extraction conditions used, it was not (or only slightly) adsorbed on the cation-exchanger-coated fibre. In addition to the selectivity,

Figure 11.15 also shows that cytochrome C is displaced by lysozyme during extraction, i.e. at longer extraction times (cf. Figure 11.15(b) and (c)) the amount of lysozyme is increased as the amount of cytochrome C is decreased.

Generally speaking, SPME is still relatively slow and/or yields are relatively low, but significant improvements are curently being made. Research effort is directed at the present time towards the development of new SPME fibre coatings in the search for new selectivities. Incorporation of other principles as, for instance, membrane technologies, antibodies, receptors and molecularly imprinted polymers could greatly enhance the development of special fibres and further promote future applications. The combination of SPME with micro-separation techniques also seems very interesting.

11.7 SUPERCRITICAL FLUID EXTRACTION COUPLED WITH SUPERCRITICAL FLUID CHROMATOGRAPHY

The gas-like mass transfer and liquid-like solvating properties of supercritical fluids, together with considerations of automation, speed and cost, make supercritical fluids very attractive for application in coupled systems. Supercritical fluid extraction (SFE) lends itself to either off-line or on-line coupling with various separation techniques (144). For on-line coupling, the general advantages are also true in this case, i.e. high sensitivity, no contamination and easy automation. For the off-line approach, the following advantages can be mentioned: larger sample sizes, different applicable separation methods and operational simplicity. Many applications of the off-line combination of SFE and LC or GC in the bio-pharmaceutical area have been described (145–148).

Prostaglandins have been extracted from drugs (149), as well as from aqueous solutions, by loading the sample on an SPE cartridge and, subsequently, carrying out on-line SFC (150). A glycoside, ouabain, was used as a model compound to study the coupling of SFE and SFC, combined with fraction collection, thus allowing determination of the biological activity of the collected fractions (151). Important information pertaining to solute elution density, efficiency of extraction, solute trapping, and supercritical fluid chromatography (SFC) was obtained. A cytostatic, mitomycin C, has been determined in plasma by SPE–SFC after application on an XAD-2 sorbent. After washing and drying the sorbent, the drug was supercritically desorbed and chromatographed with 12% methanol in CO_2 (152). This phase-system-switching approach prevents the direct injection of a polar matrix or solvent into an SFC system. Up to 1 ml of plasma containing 20 ng of mitomycin C has been analysed (Figure 11.16). The UV chromatogram obtained at 215 nm shows that the isolation step was surprisingly selective. The choice of 360 nm as the detection wavelength has further improved the selectivity.

In principle, the sample transfer from the supercritical state is relatively easily adaptable to other systems, due to the high volatility of the fluid at atmospheric pressure, particularly for carbon dioxide which is the most frequently used fluid.

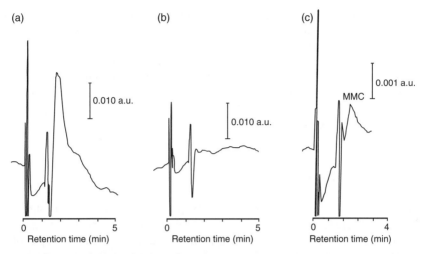

Figure 11.16 Chromatograms of plasma samples obtained by using SPE–SFC with supercritical desorption of the SPE cartridge: (a) blank plasma (20 μl), UV detection at 215 nm; (b) blank plasma (20 μl), UV detection at 360 nm; (c) plasma (1 ml) containing 20 ng mitomycin C (MMC), UV detection at 360 nm. Reprinted from *Journal of Chromatography*, **454**, W. M. A. Niessen *et al.*, 'Phase-system switching as an on-line sample pretreatment in the bioanalysis of mitomycin C using supercritical fluid chromatography,' pp. 243–251, copyright 1988, with permission from Elsevier Science.

However, it seems that the presently small attention being given to SFC has also a negative influence on the interest in SFE, where the latter seems a much more promising technique.

11.8 COUPLED SYSTEMS INVOLVING CAPILLARY ELECTROPHORESIS

In the past decade, capillary electrophoresis (CE) has become a widely accepted tool for the separation of organic and inorganic constituents of complex mixtures. As is the case for LC, with CE various compound classes can be analysed that are not amenable to GC. At the same time, the separation efficiency that can be obtained in CE is much better than in LC. Moreover, CE offers a high speed of analysis and a very low consumption of chemicals. The various modes of CE are useful for diverse compounds of biological interest such as drugs, sugars, peptides, oligonucleotides and proteins. This has revealed the potential for CE in the biomedical laboratory and, in particular, for the analysis of drugs in body fluids (153, 154). However, the use of CE in bioanalysis is often precluded by unfavourable concentration detection limits and by disturbances of the separation process by matrix compounds. The poor sensitivity is due to the very limited sample loadability (typically 1 – 10 nl of sample only) and, when optical detection is used, to the small path length of the capillary (50 – 100 μm). Interferences are often related to salts, proteins and other endogenous substances present in biological samples such as urine and plasma. Hence, in order to effectively apply CE in biomedical analysis, appropriate sample pretreatment procedures which provide both analyte enrichment and sample clean-up, are required. To this end, some multidimensional approaches such as SP(M)E–CE and LC–CE have been explored during the last few years.

11.8.1 SPE–CE AND SPME–CE

Today, off-line SPE is still one of the most frequently used sample preparation techniques for CE (155). In order to overcome the laboriousness of the off-line approach, Veraart *et al.* (156–158) coupled SPE and CE in an at-line fashion by using a robotic SPE device (Figure 11.17(a)). These authors used disposable ODS cartridges for direct extraction of urine and plasma samples, and applied the SPE–CE system to the determination of non-steroidal anti-inflammatory drugs (156), sulfonamides (157) and the anti-coagulant phenprocoumon (158). The automated SPE procedure was performed parallel to the CE analysis, thus yielding desalted, deproteinated and concentrated extracts (Figure 11.17(b) and (c)). Good linearity and detection limits of ca. 100 ng/ml were obtained when UV detection was used, but, as is true for most SPE–CE combinations, only a very small fraction of the final extract was actually analysed by CE.

On-line SPE–CE using a separate SPE column was investigated in the early 1990s, but has never become really successful for biological samples (155). More recently, micro-SPE units have been developed in order to incorporate SPE in a CE capillary (159) (see Figure 11.18). The unit sits between two capillary pieces and contains a small amount of sorbent which is held stationary by micro-frits or by a membrane. With such a module, the introduced sample volume can be increased considerably and up to 1000-fold enrichment of analytes can be achieved. Up until now, a limited number of bioanalytical applications of these so-called in-line SPE–CE systems have been reported and in most of these cases additional sample preparation prior to analysis was still required. These applications include the determination of doxepin (160) and Bench Jones proteins (161) in urine, haloperidol in urine and liver (161, 162), metallothionein in liver (163), 3-phenylamino-1,2-propanediol metabolites in liver cells (164), peptides from cell surfaces (165) and proteins in humor (166). In-line SPE–CE has also been used for protein identification via peptide mapping. Peptides resulting from protein digests were concentrated on an micro-SPE device and subsequently separated and identified by CE–MS–MS (167–169). The in-line SPE–CE approach is still promising and significantly enhances the concentration sensitivity, but it should be added that the CE performance is frequently compromised due to detrimental effects of the packing material, frits, connectors and relatively large volumes of desorbing solvent. Matrix

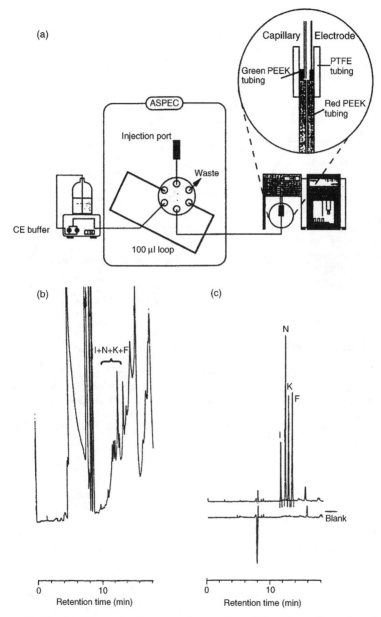

Figure 11.17 (a) Schematic diagram of an at-line SPE–CE configuration. (b and c) Electropherograms of a urine sample (8 ml) spiked with non-steroidal anti-inflammatory drugs (10 µg/ml each) after direct CE analysis (b) and at-line SPE–CE (c). Peak identification is as follows: I, ibuprofen; N, naproxen; K, ketoprofen; F, flurbiprofen. Reprinted from *Journal of Chromatography, B* **719**, J. R. Veraart *et al.*, 'At-line solid-phase extraction for capillary electrophoresis: application to negatively charged solutes,' pp. 199–208, copyright 1998, with permission from Elsevier Science.

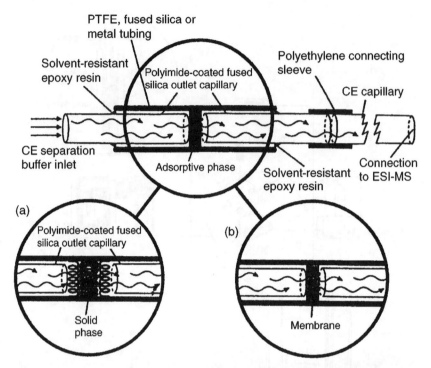

Figure 11.18   Schematic diagram of an in-line SPE unit for CE using (a) polyester wool frits to hold the sorbent, or (b) a particle-loaded membrane. Reprinted from *Journal of Capillary Electrophoresis*, **2**, A. J. Tomlinson and S. Naylor, 'Enhanced performance membrane preconcentration–capillary electrophoresis–mass spectrometry (mPC–CE–MS) in conjunction with transient isotachophoresis for analysis of peptide mixtures,' pp 225–233, 1995, with permission from ISC Technical Publications Inc.

interferences can also be expected when biological fluids are injected directly, since in an in-line set-up the complete sample passes both the micro-unit *and* the CE capillary.

The use of SPME for CE has not (yet) been studied widely. Li and Weber (170) reported an off-line SPME–CE approach for the determination of barbiturates in urine and serum, utilizing a sorbent of plasticized PVC coated around a stainless steel rod. For extraction, the coated rod was inserted for 4 min in a Teflon tube containing 50 μl of sample, and next the rod was repeatedly desorbed in another Teflon tube which each time contained 5 μl of desorption solution. This solution was transferred to an injection vial and an aliquot was injected into the CE system (Figure 11.19). The extraction procedure appeared to be selective and effectively allowed the handling of very small samples.

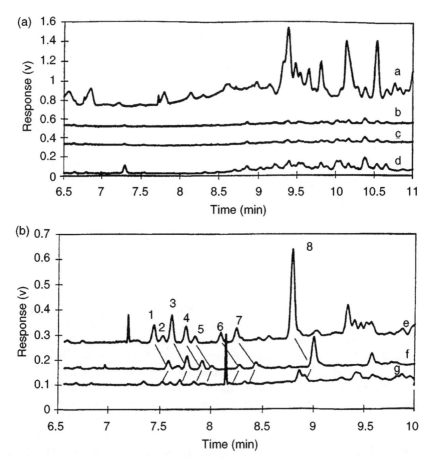

Figure 11.19 SPME–CE analysis of urine samples: (a) blank urine (a) directly injected and extracted for (b) 5 (c) 10 and (d) 30 min; (b) Urine spiked with barbiturates, extracted for (e) 30 and (f, g) 5 min. Peak identification is as follows: 1, pentobartibal; 2, butabarbital; 3, seco-barbital; 4, amobarbital; 5, aprobarbital; 6, mephobarbital; 7, butalbital; 8, thiopental. Concentrations used are 0.15–1.0 ppm (e, f) and 0.05–0.3 ppm (g). Reprinted from *Analytical Chemistry*, **69**, S. Li and S. G. Weber, 'Determination of barbiturates by solid-phase microextraction and capillary electrophoresis,' pp. 1217–1222, copyright 1997, with permission from the American Chemical Society.

11.8.2   LC–CE

Analogous to two-dimensional LC, the on-line coupling of LC and CE has been carried out both in the heart-cut and the comprehensive mode. In a heart-cut LC–CE study, a protein G immunoaffinity LC column was used to selectively preconcentrate insulin from serum (171). A 1 µl elution plug comprising the insulin was switched on-line to the CE system where a part was injected into the capillary for final separation. With CE, efficient separations can be obtained in a short time and, therefore, CE in principle can be a very suitable candidate for the second dimension in a comprehensive multidimensional separation system. In a comprehensive LC–CE system, fractions of the LC effluent are continuously injected into a CE system, thus yielding high peak capacities due to the complementary character of the two dimensions. Comprehensive LC–CE has predominantly been investigated by Jorgenson and co-workers (172–175), with most of their research being devoted to the problem of effectively interfacing the two techniques. Both RPLC–CE and SEC–CE have been reported for the separation of mixtures of (serum) proteins and peptides (tryptic digests). Even a three-dimensional system (combining SEC, RPLC and CE) has been designed in which a SEC separation of peptides, with an analysis time over several hours, is repeatedly sampled into a rapid two-dimensional RPLC–CE system with an analysis time of 7 min (173).

11.9 CONCLUSIONS

The applications described in the preceding sections of this chapter clearly demonstrate that coupled-column chromatography shows very good potential for the selective and sensitive determination of biomedically and pharmaceutically interesting compounds in biological samples. For many bioanalytical queries, a multidimensional chromatographic approach is indispensable to prevent unwanted co-elution of analytes with other sample components, and also to incorporate on-line sample preparation in the analytical procedure. Nowadays, well-developed methodologies exist for multidimensional modes such as LC–LC, SPE–LC, NPLC–GC and SPME–GC, and these have been used routinely for quite a number of real-life applications. Multidimensional systems can yield high selectivities for a single analyte, although during the past few years there has been a growing interest in multi-target analysis systems which can deal with several analytes at the same time. This often means that in the first dimension a larger fraction has to be selected at the expense of a lower attainable selectivity.

Selectivity enhancement can be achieved by coupling a relatively simple chromatographic system to (tandem) MS. Today, easy-to-operate bench-top MS(–MS) systems (single/triple quadrupole or ion-trap) with robust atmospheric-pressure ionization (API) interfaces are widely available, and LC–MS(–MS) has become the method of choice in many stages of drug development, e.g. for quantitative bioanalysis (176). It has even been suggested that the high selectivity (and sensitivity) of MS^n techniques might be sufficient to directly detect and quantify target compounds in complex mixtures by MS alone. However, it is clear now that in most cases some kind of sample preparation and pre-separation remains necessary in order to avoid serious reduction of analyte response (ion-suppression) by matrix effects and/or ion-source contamination (177). Generally, the use of LC–MS(–MS) as a routine technique for the analysis of pharmaceutical and biomedical samples will certainly decrease the importance of multidimensional separation systems. On the other hand, separation techniques, including their combinations, will remain the heart of most analytical methods in the biosciences.

REFERENCES

1. C. F. Poole, 'Planar chromatography at the turn of the century', *J. Chromatogr. A* **856**: 399–427 (1999).

2. C. F. Poole and S. K. Poole, 'Multidimensionality in planar chromatography', *J. Chromatogr. A* **703**: 573–612 (1995).

3. M.-C. Hennion, 'Solid-phase extraction: method development, sorbents and coupling with liquid chromatography', *J. Chromatogr. A* **856**: 3–54 (1999).

4. F. Regnier and G. Huang, 'Future potential of targeted component analysis by multidimensional liquid chromatography–mass spectrometry', *J. Chromatogr. A* **750**: 3–10 (1996).

5. G. J. Opiteck, J. W. Jorgenson, M. A. Moseley III and R. J. Anderegg, 'Two-dimensional microcolumn HPLC coupled to a single-quadrupole mass spectrometer for the elucidation of sequence tags and peptide mapping', *J. Microcolumn Sep.* **10**: 365–375 (1998).

6. G. J. Opiteck, S. M. Ramirez, J. W. Jorgenson and M. A. Moseley-III, 'Comprehensive two-dimensional high-performance liquid chromatography for the isolation of overexpressed proteins and proteome mapping', *Anal. Biochem.* **258**: 349–361 (1998).

7. G. J. Opiteck, J. W. Jorgenson and R. J. Anderegg, 'Two-dimensional SEC/RPLC coupled to mass spectrometry for the analysis of peptides', *Anal. Chem.* **69**: 2283–2291 (1997).

8. G. J. Opiteck, K. C. Lewis, J. W. Jorgenson and R. J. Anderegg, 'Comprehensive on-line LC/LC/MS of proteins', *Anal. Chem.* **69**: 1518–1524 (1997).

9. L. A. Holland and J. W. Jorgenson, 'Separation of nanoliter samples of biological amines by a comprehensive two-dimensional microcolumn liquid chromatography system', *Anal. Chem.* **67**: 3275–3283 (1995).

10. M. M. Bushey and J. W. Jorgenson, 'Automated instrumentation for comprehensive two-dimensional high-performance liquid chromatography of proteins', *Anal. Chem.* **62**: 161–167 (1990).

11. J. B. Wheatley, J. A. Montali and D. E. Schmidt-Jr, 'Coupled affinity-reversed-phase high-performance liquid chromatography systems for the measurement of glutathione *S*-transferases in human tissues', *J. Chromatogr. A* **676**: 65–79 (1994).

12. K. Matsuoka, M. Taoka, T. Isobe, T. Okuyama and Y. Kato, 'Automated high-resolution two-dimensional liquid chromatographic system for the rapid and sensitive separation of complex peptide mixtures', *J. Chromatogr.* **515**: 313–320 (1990).

13. D. Wu, M. Berna, G. Maier and J. Johnson, 'An automated multidimensional screening approach for rapid method development in high-performance liquid chromatography', *J. Pharm. Biomed. Anal.* **16**: 57–68 (1997).

14. E. A. Hogendoorn and P. van Zoonen, 'Coupled-column reversed-phase liquid chromatography in environmental analysis', *J. Chromatogr. A* **703**: 149–166 (1995).

15. N. Lundell and K. Markides, 'Two-dimensional liquid chromatography of peptides: an optimization strategy', *Chromatographia* **34**: 369–375 (1992).

16. E. Eklund, C. Norsten-Höög and T. Arvidsson, 'Determination of free concentration of sameridine in blood plasma by ultrafiltration and coupled-column liquid chromatography', *J. Chromatogr. B* **708**: 195–200 (1998).

17. L. B. Nilsson, 'High sensitivity determination of the remoxipride hydroquinone metabolite NCQ-344 in plasma by coupled column reversed-phase liquid chromatography and electrochemical detection', *Biomed. Chromatogr.* **12**: 65–68 (1998).

18. P. R. Baker, M. A. J. Bayliss and D. Wilkinson, 'Determination of a major metabolite of tipredane in rat urine by high-performance liquid chromatography with column switching', *J. Chromatogr. B* **694**: 193–198 (1997).

19. M. A. J. Bayliss, P. R. Baker and D. Wilkinson, 'Determination of the two major human metabolites of tipredane in human urine by high-performance liquid chromatography with column switching', *J. Chromatogr. B* **694**: 199–209 (1997).

20. E. A. Hogendoorn, U. A. Th Brinkman and P. van Zoonen, 'Coupled-column reversed-phase liquid chromatography-UV analyser for the determination of polar pesticides in water', *J. Chromatogr.* **644**: 307–314 (1993).

21. E. A. Hoogendoorn and P. van Zoonen, 'Coupled-column reversed phase liquid chromatography as a versatile technique for the determination of polar pesticides' in *Environmental Analysis – Techniques, Applications and quality assurance,* Barceló D (Ed.), Vol. 13, Elsevier, Amsterdam, pp. 181–196 (1993).

22. A. Polettini, M. Montagna, E. A. Hogendoorn, E. Dijkman, P. van Zoonen and L. A. van Ginkel, 'Applicability of coupled-column liquid chromatography to the analysis of β-agonists in urine by direct sample injection', *I. Development of a single-residue reversed-phase liquid chromatography–UV method for clenbuterol and selection of chromatographic conditions suitable for multi-residue analysis'*, *J. Chromatogr. A* **695**: 19–31 (1995).

23. E. A. Hoogendoorn, P. van Zoonen, A. Polettini and M. Montagna, 'Hyphenation of coupled-column liquid chromatography and thermospray tandem mass spectrometry for the rapid determination of β2–agonist residues in bovine urine using direct large-volume sample injection. Set-up of single-residue methods for clenbuterol and salbutamol', *J. Mass Spectrom.* **31**: 418–426 (1996).

24. T. Arvidsson and E. Eklund, 'Determination of free concentration of ropivacaine and bupivacaine in blood plasma by ultrafiltration and coupled-column liquid chromatography', *J. Chromatogr. B* **668**: 91–98 (1995).

25. K. Wang, R. W. Blain and A. J. Szuna, 'Multidimensional narrow bore liquid chromatography analysis of Ro 24–0238 in human plasma', *J. Pharm. Biomed. Anal.* **12**: 105–110 (1994).

26. R. M. Mader, B. Rizovski, G. G. Steger, H. Rainer, R. Proprentner and R. Kotz, 'Determination of methotrexate in human urine at nanomolar levels by high-performance liquid chromatography with column switching', *J. Chromatogr.* **613**: 311–316 (1993).

27. A. J. Szuna, T. E. Mulligan, B. A. Mico and R. W. Blain, 'Determination of Ro 23-7637 in dog plasma by multidimensional ion-exchange-reversed-phase high-performance liquid chromatography with ultraviolet detection', *J. Chromatogr.* **616**: 297–303 (1993).

28. K. Lanbeck-Vallén, J. Carlqvist and T. Nordgren, 'Determination of ampicillin in biological fluids by coupled-column liquid chromatography and post-column derivatization', *J. Chromatogr.* **567**: 121–128 (1991).

29. L. G. McLaughlin and J. D. Henion, 'Determination of dexamethasone in bovine tissues by coupled-column normal-phase high-performance liquid chromatography and capillary gas chromatography–mass spectrometry', *J. Chromatogr.* **529**: 1–19 (1990).

30. P. O. Edlund, L. Bowers and J. Henion, 'Determination of methandrostenolone and its metabolites in equine plasma and urine by coupled-column liquid chromatography with ultraviolet detection and confirmation by tandem mass spectrometry', *J. Chromatogr.* **487**: 341–356 (1989).

31. T. M. P. Chichila, P. O. Edlund, J. D. Henion, R. Wilson and R. L. Epstein, 'Determination of melengestrol acetate in bovine tissues by automated coupled-column normal-phase high-performance liquid chromatography', *J. Chromatogr.* **488**: 389–406 (1989).

32. U. R. Tjaden, D. S. Stegehuis, B. J. E. M. Reeuwijk, H. Lingeman and J van der Greef, 'Liquid chromatographic determination of chloramphenicol in kidney tissue homogenates using valve-switching techniques', *Analyst* **113**: 171–174(1988).

33. J. V. Posluszny and R. Weinberger, 'Determination of drug substances in biological fluids by direct injection multidimensional liquid chromatography with a micellar cleanup and reversed-phase chromatography', *Anal. Chem.* **60**: 1953–1958(1988).

34. K. Yamashita, M. Motohashi and T. Yashiki, 'Column-switching techniques for high-performance liquid chromatography of ibuprofen and mefenamic acid in human serum with short-wavelength ultraviolet detection', *J. Chromatogr.* **570**: 329–338 (1991).

35. K. Yamashita, M. Motohashi and T. Yashiki, 'Sensitive high-performance liquid chromatographic determination of propranolol in human plasma with ultraviolet detection using column switching combined with ion-pair chromatography', *J. Chromatogr.* **527**: 196–200 (1990).

36. K. Yamashita, M. Motohashi and T. Yashiki, 'High-performance liquid chromatographic determination of phenylpropanolamine in human plasma and urine, using column switching combined with ion-pair chromatography', *J. Chromatogr.* **527**: 103–114 (1990).

37. K. Yamashita, M. Motohashi and T. Yashiki, 'Sensitive high-performance liquid chromatographic determination of ionic drugs in biological fluids with short-wavelength ultraviolet detection using column switching combined with ion-pair chromatography: application to basic compounds', *J. Chromatogr.* **487**: 357–363 (1989).

38. T. Miyabayashi, K. Yamashita, I. Aoki, M. Motohashi, T. Yashiki and K. Yatani, 'Determination of manidipine and pyridine metabolite in human serum by high-performance liquid chromatography with ultraviolet detection and column switching', *J. Chromatogr.* **494**: 209–217 (1989).

39. G. de Groot, R. Koops, E. A. Hogendoorn, C. E. Goewie, T. J. F. Savelkoul and P. van Vloten, 'Improvement of selectivity and sensitivity by column switching in the determination of glycyrrhizin and glycyrrhetic acid in human plasma by high-performance liquid chromatography', *J. Chromatogr.* **456**: 71–81 (1988).

40. T. A. G. Noctor, 'Bioanalytical applications of enantioselective high-performance liquid chromatography in *A Practical Approach to Chiral Separations by Liquid Chromatography*, Subramanian G (Ed.), VCH, Weinheim, Ch. 12, pp. 357–396 (1994).

41. W. J. Lough and T. A. G. Noctor, 'Multi-column approaches to chiral bioanalysis by liquid chromatography', *Prog. Pharm. Biomed. Anal.* **1**: 241–257 (1994).

42. J. A. Chiarotto and I. W. Wainer, 'Determination of metyrapone and the enantiomers of its chiral metabolite metyrapol in human plasma and urine using coupled achiral–chiral liquid chromatography', *J. Chromatogr.* **665**: 147–154 (1995).

43. N. C. van de Merbel, M. Stenberg, R. Öste, G. Marko-Varga, L. Gorton, H. Lingeman and U. A. Th Brinkman, 'Determination of D- and L-amino acids in biological samples by two-dimensional column liquid chromatography', *Chromatographia* **41**: 6–14 (1995).

44. C. L. Hsu and R. R. Walters, 'Assay of the enantiomers of ibutilide and artilide using solid-phase extraction, derivatization and achiral–chiral column-switching high-performance liquid chromatography', *J. Chromatogr. B* **667**: 115–128 (1995).

45. M. Yamaguchi, K. Yamashita, I. Aoki, T. Tabata, S.-I. Hirai and T. Yashiki, 'Determination of manidipine enantiomers in human serum using chiral chromatography and column-switching liquid chromatography', *J. Chromatogr.* **575**: 123–129 (1992).

46. C.-Y. Hsieh and J.-D. Huang, 'Two-dimensional high-performance liquid chromatographic method to assay *p*-hydroxyphenylphenylhydantoin enantiomers in biological

fluids and stereoselectivity of enzyme induction in phenytoin metabolism', *J. Chromatogr.* **575**: 109–115 (1992).

47. Y. Oda, N. Asakawa, Y. Yoshida and T. Sato, 'On-line determination and resolution of the enantiomers of ketoprofen in plasma using coupled achiral–chiral high-performance liquid chromatography', *J. Pharm. Biomed. Anal.* **10**: 81–87 (1992).

48. K.-M. Chu, S.-M. Shieh, S.-H. Wu and O. Y.-P. Hu, 'Enantiomeric separation of a cardiotonic agent pimobendan and its major active metabolite, UD-CG 212 BS, by coupled achiral–chiral normal-phase high-performance liquid chromatography', *J. Chromatogr. Sci* **30**: 171–176 (1992).

49. H. Fujimoto, I. Nishino, K. Ueno and T. Umeda, 'Determination of the enantiomers of a new 1,4-dihydropyridine calcium antagonist in dog plasma achiral / chiral coupled high performance liquid chromatography with electrochemical detection', *J. Pharm. Sci.* **82**: 319–322 (1993).

50. Y. Oda, N. Asakawa, T. Kajima, Y. Yoshida and T. Sato, 'On-line determination and resolution of verapamil enantiomers by high-performance liquid chromatography with column-switching', *Pharm. Res.* **8**: 997–1001 (1991).

51. A. Walhagen and L.-E. Edholm, 'Chiral separation on achiral stationary phases with different functionalities using β-cyclodextrin in the mobile phase and application to bioanalysis and coupled columns', *Chromatographia* **32**: 215–223 (1991).

52. F. Gimenez, R. Farinotti, A. Thuillier, G. Hazebroucq and I. W. Wainer, 'Determination of the enantiomers of mefloquine in plasma and whole blood using a coupled achiral–chiral high-performance liquid chromatographic system', *J. Chromatogr.* **529**: 339–346 (1990).

53. L. Silan, P. Jadaud, L. R. Whitfield and I. W. Wainer, 'Determination of low levels of the stereoisomers of leucovorin and 5-methyltetrahydrofolate in plasma using a coupled chiral–achiral high-performance liquid chromatographic system with post-chiral column peak compression', *J. Chromatogr.* **532**: 227–236 (1990).

54. A. Walhagen and L.-E. Edholm, 'Coupled-column chromatography of immobilized protein phases for direct separation and determination of drug enantiomers in plasma', *J. Chromatogr.* **473**: 371–379 (1989).

55. D. Masurel and I. W. Wainer, 'Analytical and preparative high-performance liquid chromatographic separation of the enantiomers of ifosfamide, cyclophosphamide and trofosfamide and their determination in plasma', *J. Chromatogr.* **490**: 133–143 (1989).

56. I. W. Wainer and R. M. Stiffin, 'Direct resolution of the stereoisomers of leucovorin and 5-methyltetrahydrofolate using a bovine serum albumin high-performance liquid chromatographic chiral stationary phase coupled to an achiral phenyl column', *J. Chromatogr.* **424**: 158–162 (1988).

57. L.-E. Edholm, C. Lindberg, J. Paulson and A. Walhagen, 'Determination of drug enantiomers in biological samples by coupled column liquid chromatography and liquid chromatography–mass spectrometry', *J. Chromatogr.* **424**: 61–72 (1988).

58. T. Seki, K. Yamaji, Y. Orita, S. Moriguchi and A. Shinoda, 'Simultaneous determination of uric acid and creatinine in biological fluids by column-switching liquid chromatography with ultraviolet detection', *J. Chromatogr. A* **730**: 139–145 (1996).

59. C. Tagesson, M. Kaellberg, C. Klintenberg and H. Starkhammar, 'Determination of urinary 8-hydroxydeoxyguanosine by automated coupled-column high-performance liquid chromatography: a powerful tecnique for assaying in vivo oxidative DNA damage in cancer patients', *Eur. J. Cancer* **31A**: 934–940 (1995).

60. T. K. Nadler, S. K. Paliwal and F. E. Regnier, 'Rapid, automated, two-dimensional high-performance liquid chromatographic analysis of immunoglobulin G and its multimers', *J. Chromatogr. A* **676**: 331–335 (1994).

61. J. S. Patrick and A. L. Lagu, 'Determination of recombinant human proinsulin fusion protein produced in *Escherichia coli* using oxidative sulfitolysis and two-dimensional HPLC', *Anal. Chem.* **64**: 507–511 (1992).

62. J. F. K. Huber and G. Lamprecht, 'Assay of neopterin in serum by means of two-dimensional high-performance liquid chromatography with automated column switching using three retention mechanism', *J. Chromatogr. B* **666**: 223–232 (1995).

63. T. Okuda, Y. Nakagawa and M. Motohashi, 'Complete two-dimensional separation for analysis of acidic compounds in plasma using column-switching reversed-phase liquid chromatography', *J. Chromatogr. B* **726**: 225–236 (1999).

64. T. Miyabayashi, T. Okuda, M. Motohashi, K. Izawa and T. Yashiki, 'Quantitation of a new potent angiotensin II receptor antagonist, TCV-116 and, its metabolites in human serum and urine', *J. Chromatogr. B* **677**: 123–132 (1996).

65. B.-M. Eriksson, B.-A. Persson and M. Wikström, 'Determination of urinary vanillylmandelic acid by direct injection and coupled-column chromatography with electrochemical detection', *J. Chromatogr.* **527**: 11–19 (1990).

66. G. Zürcher and M. Da Prada, 'Simple automated high-performance liquid chromatographic column-switching tecnique for the measurement of dopa and 3-O-methyldopa in plasma', *J. Chromatogr.* **530**: 253–262 (1990).

67. K. Johansen and P. O. Edlund, 'Determination of water-soluble vitamins in blood and plasma by coupled-column liquid chromatography', *J. Chromatogr.* **506**: 471–479 (1990).

68. B.-M. Eriksson and B.-A. Persson, 'Determination of 5-hydroxytryptamine and 5-hydroxyindoleacetic acid in plasma by direct injection in coupled-column liquid chromatography with electrochemical detection', *J. Chromatogr.* **459**: 351–360 (1988).

69. R. A. Coe, L. S. DeCesare and J. W. Lee, 'Quantitation of efletirizine in human plasma and urine using automated solid-phase extraction and column-switching high-performance liquid chromatography', *J. Chromatogr. B* **730**: 239–247 (1999).

70. P. Campíns-Falcó, R. Herráez-Hernández and A. Sevillano-Cabeza, 'Column-switching techniques for high-performance liquid chromatography of drugs in biological samples', *J. Chromatogr.* **619**: 177–190 (1993).

71. R. W. Fedeniuk and P. J. Shand, 'Theory and methodology of antibiotic extraction from biomatrices', *J. Chromatogr. A* **812**: 3–15 (1998).

72. A. K. Sakhi, T. E. Gundersen, S. M. Ulven, R. Blomhoff and E. Lundanes, 'Quantitative determination of endogenous retinoids in mouse embryos by high-performance liquid chromatography with on-line solid-phase extraction, column switching and electrochemical detection', *J. Chromatogr. A* **828**: 451–460 (1998).

73. H. M. M. Arafa, F. M. A. Hamada, M. M. A. Elzamar and H. Nau, 'Fully automated determination of selective retinoic acid receptor ligands in mouse plasma and tissue by reversed-phase liquid chromatography coupled on-line with solid-phase extraction', *J. Chromatogr. A* **729**: 125–136 (1996).

74. S. Emara, H. Askal and T. Masujima, 'Rapid determination of methotrexate in plasma by high-performance liquid chromatography with online solid-phase extraction and automated precolumn derivatization', *Biomed. Chromatogr.* **12**: 338–342 (1998).

75. J. A. Pascual and J. Sanagustín, 'Fully automated analytical method for codeine quantification in human plasma using on-line solid-phase extraction and high-performance liquid chromatography with ultraviolet detection', *J. Chromatogr. B* **724**: 295–302 (1999).

76. G. P. McMahon and M. T. Kelly, 'Determination of aspirin and salicylic acid in human plasma by column-switching liquid chromatography using online solid-phase extraction', *Anal. Chem.* **70**: 409–414 (1998).

77. H. Lindenblatt, E. Krämer, P. Holzmann-Erens, E. Gouzoulis-Mayfrank and K.-A. Kovar, 'Quantitation of psilocin in human plasma by high-performance liquid

chromatography and electrochemical detection: comparison of liquid–liquid extraction with automated on-line solid-phase extraction', *J. Chromatogr. B* **709**: 255–263 (1998).

78. H. Svennberg and P.-O. Lagerström, 'Evaluation of an on-line solid-phase extraction method for determination of almokalant, an antiarrhythmic drug, by liquid chromatography', *J. Chromatogr. B* **689**: 371–377 (1997).

79. D. Barron, J. Barbosa, J. A. Pascual and J. Segura, 'Direct determination of anabolic steroids in human urine by online solid-phase extraction/liquid chromatography/mass spectrometry', *J. Mass Spectrom.* **31**: 309–319 (1996).

80. J. Huwyler, S. Rufer, E. Küsters and J. Drewe, 'Rapid and highly automated determination of morphine and morphine glucuronides in plasma by on-line solid-phase extraction and column liquid chromatography', *J. Chromatogr. B* **674**: 57–63 (1995).

81. S. Bompadre, L. Ferrante, F. P. Alò and L. Leone, 'On-line solid-phase extraction of ceftazidime in serum and determination by high-performance liquid chromatography', *J. Chromatogr. B* **669**: 265–269 (1995).

82. H. Zehender, J. Denouël, M. Roy, L. Le Saux and P. Schaub, 'Simultaneous determination of terbinafine (Lamisil) and five metabolites in human plasma and urine by high-performance liquid chromatography using on-line solid-phase extraction', *J. Chromatogr. B* **664**: 347–355 (1995).

83. O. V. Olesen and B. Poulsen, 'On-line fully automated determination of clozapine and desmethylclozapine in human serum by solid-phase extraction on exchangeable cartridges and liquid chromatography using a methanol buffer mobile phase on unmodified silica', *J. Chromatogr.* **622**: 39–46 (1993).

84. V. Ascalone, P. Guinebault and A. Rouchouse, 'Determination of mizolastine, a new antihistaminic drug, in human plasma by liquid–liquid extraction, solid-phase extraction and column-switching tecniques in combination with high-performance liquid chromatography', *J. Chromatogr.* **619**: 275–284 (1993).

85. D. Chollet and P. Künstner, 'Fast systematic approach for the determination of drugs in biological fluids by fully automated high-performance liquid chromatography with on-line solid-phase extraction and automated cartridge exchange', *J. Chromatogr.* **577**: 335–340 (1992).

86. K. Saeed and M. Becher, 'On-line solid-phase extraction of piroxicam prior to its determination by high-performance liquid chromatography', *J. Chromatogr.* **567**: 185–193 (1991).

87. G. García-Encina, R. Farrán, S. Puig and L. Martínez, 'Validation of an automated liquid chromatographic method for omeprazole in human plasma using on-line solid-phase extraction', *J. Pharm. Biomed. Anal.* **21**: 371–382 (1999).

88. K. Kronkvist, M. Gustavsson, A.-K. Wendel and H. Jaegfeldt, 'Automated sample preparation for the determination of budesonide in plasma samples by liquid chromatography and tandem mass spectrometry', *J. Chromatogr. A* **823**: 401–409 (1998).

89. B. Kaye, W. J. Herron, P. V. Macrae, S. Robinson, D. A. Stopher, R. F. Venn and W. Wild, 'Rapid, solid phase extraction tecnique for the high-throughput assay of darifenacin in human plasma', *Anal. Chem.* **68**: 1658–1660 (1996).

90. J. Hempenius, J. Wieling, J. P. G. Brakenhoff, F. A. Maris and J. H. G. Jonkman, 'High-throughput solid-phase extraction for the determination of cimetidine in human plasma', *J. Chromatogr. B* **714**: 361–368 (1998).

91. M. Jemal, D. Teitz, Z. Ouyang and S. Khan, 'Comparison of plasma sample purification by manual liquid-liquid extraction, automated 96-well liquid–liquid extraction and automated 96-well solid-phase extraction for analysis by high-performance liquid chromatography with tandem mass spectrometry', *J. Chromatogr. B* **732**: 501–508 (1999).

92. R. Rühl and H. Nau, 'Determination of adapalene (CD271/Differin®) and retinol in plasma and tissue by on-line solid-phase extraction and HPLC analysis', *Chromatographia* **45**: 269–274 (1997).

93. Z. Yu and D. Westerlund, 'Direct injection of large volumes of plasma in a column-switching system for the analysis of local anaesthetics', II. Determination of bupivacaine in human plasma with an alkyl-diol silica precolumn', *J. Chromatogr. A* **725**: 149–155 (1996).

94. R. Oertel, K. Richter, T. Gramatté and W. Kirch, 'Determination of drugs in biological fluids by high-performance liquid chromatography with on-line sample processing', *J. Chromatogr. A* **797**: 203–209 (1998).

95. R. A. M. van der Hoeven, A. J. P. Hofte, M. Frenay, H. Irth, U. R. Tjaden, J. van der Greef A. Rudolphi, K.-S. Boos, G. Marko Varga and L. E. Edholm, 'Liquid chromatography–mass spectrometry with on-line solid-phase extraction by a restricted-access C_{18} precolumn for direct plasma and urine injection', *J. Chromatogr. A* **762**: 193–200 (1997).

96. J. Hermansson, A. Grahn and I. Hermansson, 'Direct injection of large volumes of plasma/serum of a new biocompatible extraction column for the determination of atenolol, propanolol and ibuprofen', Mechanisms for the improvement of chromatographic performance', *J. Chromatogr. A* **797**: 251–263 (1998).

97. F. Mangani, G. Luck, C. Fraudeau and E. Vérette, 'On-line column-switching high-performance liquid chromatography analysis of cardiovascular drugs in serum with automated sample clean-up and zone-cutting technique to perform chiral separation', *J. Chromatogr. A* **762**: 235–241 (1997).

98. K. Benkestock, 'Determination of propiomazine in rat plasma by direct injection on coupled liquid chromatography columns with electrochemical detection', *J. Chromatogr. B* **700**: 201–207 (1997).

99. B. Nilsson, 'Extraction and quantitation of cortisol by use of high-performance liquid affinity chromatography', *J. Chromatogr.* **276**: 413–417 (1983).

100. B. Johansson, 'Simplified quantitative determination of plasma phenytoin: on-line precolumn high-performance liquid immunoaffinity chromatography with sample prepurification', *J. Chromatogr.* **381**: 107–113 (1986).

101. A. Farjam, G. J. de Jong, R. W. Frei, U. A. Th Brinkman, W., Haasnoot, A. R. M. Hamers, R. Schilt and F. A. Huf, 'Immunoaffinity pre-column for selective on-line sample pre-treatment in high-performance liquid chromatography determination of 19-nortestosterone', *J. Chromatogr.* **452**: 419–433 (1988).

102. W. Haasnoot, R. Schilt, A. R. M. Hamers, F. A. Huf, A. Farjam, R. W. Frei and U. A. Th Brinkman, 'Determination of β-19-nortestosterone and its metabolite α-19-nortestosterone in biological samples at the sub parts per billion level by high-performance liquid chromatography with on-line immunoaffinity sample pretreatment', *J. Chromatogr.* **489**: 157–171 (1989).

103. A. Farjam, R. de Vries, H. Lingeman and U. A. Th Brinkman, 'Immuno precolumns for selective on-line sample pretreatment of aflatoxins in milk prior to column liquid chromatography', *Int. J. Environ. Anal. Chem.* **44**: 175–184 (1991).

104. A. Farjam, N. C. van de Merbel, H. Lingeman, R. W. Frei and U. A. Th Brinkman, 'Non-selective desorption of immuno precolumns coupled on-line with column liquid chromatography: determination of aflatoxins', *Int. J. Environ. Anal. Chem.* **45**: 73–87(1991).

105. A. Farjam, N. C. van de Merbel, A. A. Nieman, H. Lingeman and U. A. Th Brinkman, 'Determination of aflatoxin M1 using a dialysis-based immunoaffinity sample pretreatment system coupled on-line to liquid chromatography', *J. Chromatogr.* **589**: 141–149(1992).

106. W. Haasnoot, M. E. Ploum, R. J. A. Paulussen, R. Schilt and F. A. Huf, 'Rapid determination of clenbuterol residues in urine by high-performance liquid chromatography with on-line automated sample processing using immunoaffinity chromatography', *J. Chromatogr.* **519**: 323–335 (1990).

107. V. Kircher and H. Parlar, 'Determination of Δ^9 tetrahydrocannabinol from human saliva by tandem immunoaffinity chromatography–high-performance liquid chromatography', *J. Chromatogr. B* **677**: 245–255 (1996).

108. C. S. Creaser, S. J. Feely, E. Houghton and M. Seymour, 'Immunoaffinity chromatography combined on-line with high-performance liquid chromatography–mass spectrometry for the determination of corticosteroids', *J. Chromatogr. A* **794**: 37–43 (1998).

109. E. Davoli, R. Fanelli and R. Bagnati, 'Purification and analysis of drug residues in urine samples by on-line immunoaffinity chromatography/high-performance liquid chromatography/continuos-flow fast atom bombardment mass spectrometry', *Anal. Chem.* **65**: 2679–2685 (1993).

110. G. S. Rule and J. D. Henion, 'Determination of drugs from urine by on-line immunoaffinity chromatography–high-performance liquid chromatography–mass spectrometry', *J. Chromatogr.* **582**: 103–112 (1992).

111. J. Cai and J. Henion, 'On-line immunoaffinity extraction–coupled column capillary liquid chromatography/tandem mass spectrometry: trace analysis of LSD analogs and metabolites in human urine', *Anal. Chem.* **68**: 72–78 (1996).

112. J. Cai and J. Henion, 'Quantitative multi-residue determination of β-agonists in bovine urine using on-line immunoaffinity extraction–coupled column packed capillary liquid chromatography–tandem mass spectrometry', *J. Chromatogr. A* **691**: 357–370 (1997).

113. M. W. F. Nielen, H. E. van Ingen, A. J. Valk, R. W. Frei and U. A. Th Brinkman, 'Metal-loaded sorbents for selective on-line sample handling and trace enrichment in liquid chromatography', *J. Liq. Chromatogr.* **10**: 617–633, (1987).

114. C. Lipschitz, H. Irth, G. J. de Jong, U. A. Th Brinkman and R. W. Frei, 'Trace enrichment of pyrimidine nucleobases, 5-fluoro-uracil and bromacil on a silver-loaded thiol stationary phase with on-line reversed-phase high-performance liquid chromatography', *J. Chromatogr.* **471**: 321–334 (1989).

115. H. Irth, R. Tocklu, K. Welten, G. J. de Jong, U. A. Th Brinkman and R. W. Frei (1989), 'Trace-level determination of 3′-azido-3′-deoxythymidine in human plasma by preconcentration on a silver (I)-thiol stationary phase with on-line reversed-phase high-performance liquid chromatography', *J. Chromatogr.* **491**: 321–330 (1989).

116. H. Irth, R. Tocklu, K. Welten, G. J. de Jong, R. W. Frei and U. A. Th Brinkman, 'Trace enrichment on a metal-loaded thiol stationary phase in liquid chromatography: effect of analyte structure and pH value on the (de)sorption behaviour', *J. Pharm. Biomed. Anal.* **7**: 1679–1690(1989).

117. B. Sellergren, 'Noncovalent molecular imprinting: antibody-like molecular recognition in polymeric network materials', *Trends. Anal. Chem.* **16**: 310–320 (1997).

118. K. S. Boos and C. H. Grimm, 'High performance liquid chromatography integated with solid-phase extraction in bioanalysis using restricted access precolumn packings', *Trends Anal. Chem.* **18**: 175–180 (1999).

119. K. Grob-Jr, H. P. Neukom and R. Etter, 'Coupled high-performance liquid chromatography–capillary gas chromatography as a replacement for gas chromatography–mass spectrometry in the determination of diethylstilbestrol in bovine urine', *J. Chromatogr.* **357**: 416–422 (1986).

120. V. Gianesello, L. Bolzani, E. Brenn and A. Gazzaniga, 'Determination of broxaterol in plasma by coupled HPLC–GC', *J. High Resolut. Chromatogr. Chromatogr. Commun.* **11**: 99–102 (1988).

121. F. Munari and K. Grob, 'Automated on-line HPLC–HRGC: instrumental aspects and application for the determination of heroin metabolites in urine', *J. High. Resolut. Chromatogr. Chromatogr. Commun.* **11**: 172–176 (1988).

122. M. Ghys, J. van Dijck, C. Dewaele, M. Verstappe, M. Verzele and P. Sandra, 'On-line micro size-exclusion chromatography–capillary gas chromatography', *Proceedings of the 10th Symposium on capillary Chromatography*, Riva del Garda, Italy, P. Sandra and G. Redant (Eds), Hüthig, Heidelberg pp. 726–735 (1989).

123. E. C. Goosens, D. de Jong, J. H. M. van den Berg, G. J. de Jong and U. A. Th Brinkman, 'Reversed-phase liquid chromatography coupled on-line with capillary gas chromatography', I. Introduction of large volumes of acqueous mixtures through an on-column interface', *J. Chromatogr.* **552**: 489–500 (1991).

124. E. C. Goosens, D. de Jong, G. J. de Jong and U. A. Th Brinkman, 'Reversed-phase liquid chromatography coupled on-line with capillary gas chromatography. II. Use of a solvent vapor exit to increase introduction volumes and introduction rates into the gas chromatograph', *J. Microcolumn Sep* **6**: 207–215 (1994).

125. E. C. Goosens, I. M. Beerthuizen, D. de Jong, G. J. de Jong and U. A. Th Brinkman, 'Reversed-phase liquid chromatography coupled on-line with capillary gas chromatography', Use of an anion-exchange membrane to remove an ion-pair reagent from the eluent', *Chromatographia* **40**: 267–271(1995).

126. E. C. Goosens, K. H. Stegman, D. de Jong, G. J. de Jong and U. A. Th Brinkman, 'Investigation of online reversed-phase liquid chromatography–gas chromatography–mass spectrometry as a tool for the identification of impurities in drug substances', *Analyst* **121**: 61–66 (1996).

127. J. Ogorka, G. Schwinger, G. Bruat and V. Seidel, 'On-line coupled reversed-phase high-performance liquid chromatography–gas chromatography–mass spectrometry', A powerful tool for the identification of unknown impurities in pharmaceutical products', *J. Chromatogr.* **626**: 87–96 (1992).

128. P. Wessels, J. Ogorka, G. Schwinger and M. Ulmer, 'Elucidation of the structure of drug degradation products by on-line coupled reversed phase HPLC–GC–MS and on-line derivatization', *J. High Resolut. Chromatogr.* **16**: 708–712 (1993).

129. T. Hyötyläinen, T. Andersson and M. L. Riekkola, 'Liquid chromatographic sample cleanup coupled on-line with gas chromatography in the analysis of beta-blockers in human serum and urine', *J. Chromatogr. Sci.* **35**: 280–286 (1997).

130. T. Hyötyläinen, H. Keski-Hynnilä and M. L. Riekkola, 'Determination of morphine and its analogues in urine by on-line coupled reversed-phase liquid chromatography–gas chromatography with on-line derivatization', *J. Chromatogr. A* **771**: 360–365 (1997).

131. R. Herráez-Hernández, A. J. H. Louter, N. C. van de Merbel and U. A. Th Brinkman, 'Automated on-line dialysis for sample preparation for gas chromatography: determination of benzodiazepines in human plasma', *J. Pharm. Biomed. Anal.* **14**: 1077–1087 (1996).

132. A. Farjam, J. J. Vreuls, W. J. G. M. Cuppen, U. A. Th Brinkman and G. J. de Jong, 'Direct introduction of large-volume urine samples into an on-line immunoaffinity sample pretreatment-capillary gas chromatography system', *Anal. Chem.* **63**: 2481–2487 (1991).

133. J. J. Vreuls, A. J. H. Louter and U. A. Th Brinkman, 'On-line combination of aqueous-sample preparation and capillary gas chromatography', *J. Chromatogr. A* **856**: 279–314 (1999).

134. C. L. Arthur and J. Pawliszyn, 'Solid phase microextraction with thermal desorption using fused silica optical fibers', *Anal. Chem.* **62**: 2145–2148 (1990).

135. C. L. Arthur, L. M. Killam, K. D. Buchholz and J. Pawliszyn, 'Automation and optimization of solid-phase microextraction', *Anal. Chem.* **64**: 1960–1966 (1992).

136. S. Motlagh and J. Pawliszyn, 'Online monitoring of flowing samples using solid phase microextraction–gas chromatography', *Anal. Chim. Acta.* **284**: 265–273 (1993).

137. Z. Zhang and J. Pawlyszin, 'Headspace solid-phase microextraction', *Anal. Chem.* **65**: 1843–1852 (1993).

138. E. H. M. Koster, C. Wemes, J. B. Morsink and G. J. de Jong, 'Determination of lidocaine in plasma by direct solid-phase microextraction combined with gas chromatography', *J. Chromatogr. B* **739**: 175–182 (2000).

139. S. Ulrich and J. Martens, 'Solid-phase microextraction with capillary gas–liquid chromatography and nitrogen-phosphorus selective detection for the assay of antidepressant drugs in human plasma', *J. Chromatogr. B* **696**: 217–234 (1997).

140. M. Krogh, K. Johansen, F. Tonnesen and K. E. Rasmusen, 'Solid-phase microextraction for the determination of the free concentration of valproic acid in human plasma by capillary gas chromatography', *J. Chromatogr. B* **673**: 299–305 (1997).

141. W. H. J. Vaes, E. Urrestarazu Ramos, H. J. M. Verhaar, W. Seinen and J. L. M. Hermens, 'Measurement of the free concentration using solid-phase microextraction: binding to protein', *Anal. Chem.* **68**: 4463–4467 (1996).

142. I. Koide, O. Noguchi, K. Okada, A. Yokoyama, H. Oda, S. Yamamoto and H. Kataoka, 'Determination of amphetamine and methamphetamine in human hair by headspace solid-phase microextraction and gas chromatography with nitrogen-phosphorus detection', *J. Chromatogr. B* **707**: 99–104(1998).

143. J.-L. Liao, C-M. Zeng, S. Hjertén and J. Pawliszyn, 'Solid phase micro extraction of biopolymers, exemplified with adsorption of basic proteins onto a fiber coated with polyacrylic acid', *J. Microcolumn Sep.* **8**: 1–4. (1996)

144. T. Greibrokk, 'Applications of supercritical fluid extraction in multidimensional systems', *J. Chromatogr. B* **703**: 523–536 (1995).

145. K. Hartonen and M. L. Riekkola, 'Detection of β-blockers in urine by solid-phase extraction–supercritical fluid extraction and gas chromatography–mass spectrometry', *J. Chromatogr. B* **676**: 45–52 (1996).

146. D. L. Allen, K. S. Scott and J. S. Oliver, 'Comparison of solid-phase extraction and supercritical fluid extraction for the analysis of morphine in whole blood', *J. Anal. Toxicol.* **23**: 216–218 (1999).

147. L. J. Mulcahey and L. T. Taylor, 'Supercritical fluid extraction of active components in a drug formulation', *Anal. Chem.* **64**: 981–984 (1992).

148. B. R. Simmons and J. T. Stewart, 'Supercritical fluid extraction of selected pharmaceuticals from water and serum', *J. Chromatogr. B* **688**: 291–302 (1997).

149. D. A. Roston, 'Supercritical fluid extraction-supercritical fluid chromatography for analysis of a prostaglandin: HPMC dispersion', *Drug Dev. Ind. Pharm.* **18**: 245–255 (1992).

150. I. J. Koski, B. A. Jansson, K. E. Markides and M. L. Lee, 'Analysis of prostaglandins in aqueous solutions by supercritical fluid extraction and chromatography', *J. Pharm. Biomed. Anal.* **9**: 281–290 (1991).

151. Q. L. Xie, K. E. Markides and M. L. Lee, 'Supercritical fluid extraction–supercritical fluid chromatography with fraction collection for sensitive analytes', *J. Chromatogr. Sci.* **27**: 365–370 (1989).

152. W. M. A. Niessen, P. J. M. Bergers, U. R. Tjaden and J. van der Greef, 'Phase-system switching as an on-line sample pretreatment in the bioanalysis of mitomycin C using supercritical fluid chromatography', *J. Chromatogr.* **454**: 243–251 (1988).

153. C. M. Boone, J. C. M. Waterval, H. Lingeman, K. Ensing and W. J. M. Underberg, 'Capillary electrophoresis as a versatile tool for the bioanalysis of drugs', A review', *J. Pharm. Biomed. Anal.* **20**: 831–863 (1999).

154. W. Thormann, C. X. Zhang and A. Schmutz, 'Capillary electrophoresis for drug analysis in body fluids', *Ther. Drug Monit.* **18**: 506–520 (1996).

155. J. R. Veraart, H. Lingeman and U. A. Th Brinkman, 'Coupling of biological sample handling and capillary electrophoresis', *J. Chromatogr. A* **856**: 483–514 (1999).

156. J. R. Veraart, C. Gooijer, H. Lingeman, N. H. Velthorst and U. A. Th Brinkman, 'At-line solid-phase extraction for capillary electrophoresis: application to negatively charged solutes', *J. Chromatogr. B* **719**: 199–208 (1998).

157. J. R. Veraart, C. Gooijer, H. Lingeman, N. H. Velthorst and U. A. Th Brinkman, 'At-line solid-phase extraction coupled to capillary electrophoresis: determination of amphoteric compounds in biological samples', *J. High Resolut. Chromatogr.* **22**: 183–187(1999).

158. J. R. Veraart, C. Gooijer, H. Lingeman, N. H. Velthorst and U. A. Th Brinkman, 'Determination of phenprocoumon in plasma and urine by at-line solid-phase extraction-capillary electrophoresis', *J. Pharm. Biomed. Anal.* **17**: 1161–1166 (1998).

159. A. J. Tomlinson and S. Naylor, 'Enhanced performance membrane preconcentration–capillary electrophoresis–mass spectrometry (mPC-CE–MS) in conjunction with transient isotachophoresis for analysis of peptide mixtures', *J. Capillary Electrophor.* **2**: 225–233 (1995).

160. M. E. Swartz and M. Merion, 'On-line sample preconcentration on a packed-inlet capillary for improving the sensitivity of capillary electrophoretic analysis of pharmaceuticals', *J. Chromatogr.* **632**: 209–213 (1993).

161. A. J. Tomlinson, L. M. Benson, R. P. Oda, W. D. Braddock, B. L. Riggs, J. A. Katzmann and S. Naylor, 'Novel modifications and clinical applications of preconcentration-capillary electrophoresis–mass spectrometry', *J. Capillary Electrophor.* **2**: 97–104 (1995).

162. L. M. Benson, A. J. Tomlinson and S. Naylor, 'Time course analysis of a microsomal incubation of a therapeutic drug using preconcentration capillary electrophoresis (Pc-CE)', *J. High Resolut. Chromatogr.* **17**: 671–673 (1994).

163. J. H. Beattie, R. Self and M. P. Richards, 'The use of solid phase concentrators for online pre-concentration of metallothionein prior to isoform separation by capillary zone electrophoresis', *Electrophoresis* **16**: 322–328 (1995).

164. L. M. Benson, A. J. Tomlinson, A. N. Mayeno, G. J. Gleich, D. Wells and S. Naylor, 'Membrane preconcentration-capillary electrophoresis–mass spectrometry (mPC-CE–MS) analysis of 3–phenylamino-1,2–propanediol (PAP) metabolites', *J. High Resolut. Chromatogr.* **19**: 291–293(1996).

165. A. J. Tomlinson, S. Jameson and S. Naylor, 'Strategy for isolating and sequencing biologically derived MHC class I peptides', *J. Chromatogr. A* **744**: 273–278 (1996).

166. E. Rohde, A. J. Tomlinson, D. H. Johnson and S. Naylor, 'Protein analysis by membrane preconcentration-capillary electrophoresis: systematic evaluation of parameters affecting preconcentration and separation', *J. Chromatogr. B* **713**: 301–311 (1998).

167. D. Figeys, Y. Zhang and R. Aebersold, 'Optimization of solid phase microextraction-capillary zone electrophoresis–mass spectrometry for high sensitivity protein identification', *Electrophoresis* **19**: 2338–2347 (1998).

168. D. Figeys, A. Ducret and R. Aebersold, 'Identification of proteins by capillary electrophoresis–tandem mass spectrometry', *Evaluation of an on-line solid-phase extraction device', J. Chromatogr. A* **763**: 295–306 (1997).

169. D. Figeys, A. Ducret, J. R. Yates-III and R. Aebersold, 'Protein identification by solid phase microextraction-capillary zone electrophoresis–microelectrospray-tandem mass spectrometry', *Nat. Biotechnol.* **14**: 1579–1583 (1996).

170. S. Li and S. G. Weber, 'Determination of barbiturates by solid-phase microextraction and capillary electrophoresis', *Anal. Chem.* **69**: 1217–1222 (1997).

171. L. J. Cole and R. T. Kennedy, 'Selective preconcentration for capillary zone electrophoresis using protein G immunoaffinity capillary chromatography', *Electrophoresis* **16**: 549–556 (1995).

172. A. W. Moore-Jr, J. P. Larmann-Jr, A. V. Lemmo and J. W. Jorgenson, 'Two-dimensional liquid chromatography–capillary electrophoresis techniques for analysis of proteins and peptides', *Methods Enzymol.* **270**: 401–419 (1996).

173. A. W. Moore-Jr and J. W. Jorgenson, 'Comprehensive three-dimensional separation of peptides using size exclusion chromatography/reversed phase liquid chromatography/ optically gated capillary zone electrophoresis', *Anal. Chem.* **67**: 3456–3463 (1995).

174. A. W. Moore-Jr and J. W. Jorgenson, 'Rapid comprehensive two-dimensional separations of peptides via RPLC-optically gated capillary zone electrophoresis', *Anal. Chem.* **67**: 3448–3455 (1995).

175. J. P. Larmann-Jr, A. V. Lemmo, A. W. Moore-Jr and J. W. Jorgenson, 'Two-dimensional separations of peptides and proteins by comprehensive liquid chromatography-capillary electrophoresis', *Electrophoresis* **14**: 439–447 (1993).

176. W. M. A. Niessen, 'State-of-the-art in liquid chromatography–mass spectrometry, *J. Chromatogr. A* **856**: 179–197 (1999).

177. J. Henion, E. Brewer and G. Rule, 'Sample preparation for LC/MS/MS: analyzing biological and environmental samples', *Anal. Chem.* **70**: 650A–656A (1998).

12 Multidimensional Chromatography: Industrial and Polymer Applications

Y. V. KAZAKEVICH and R. LOBRUTTO

Seton Hall University, South Orange, NJ, USA

12.1 INTRODUCTION

In this present chapter, applications of multidimensional chromatography to industrial and polymer samples are described, together with general principles and details of the interfacing setups. The main focus is on complex analyte mixtures or samples that cannot be analyzed solely by using a single mode of chromatography. Multidimensional chromatography offers the ability to analyze certain components in a mixture, which is otherwise very difficult, when employing one type of analytical technique. The advantages of using multidimensional chromatography are increases in the selectivity and efficiency, plus identification of certain components in a multicomponent mixture.

The use of coupled column technology allows preseparation of complex samples of industrial chemicals and polymers and on-line transferring of selected fractions from a primary column to a secondary column for further separation. The resolution of individual components in a complex matrix and increased peak capacity may be obtained when using coupled column chromatography. Gas chromatography (GC) is a very efficient separation technique, but it is not always effective for the resolution of all components in complex mixtures due to either coelution or a lack of volatility. The preseparation of the complex sample may be carried out by using liquid chromatography (LC), supercritical fluid chromatogaphy (SFC), and even another GC column of different polarity. The first column may be used to isolate specific components and fractionate the chemicals by class or group before chromatographic analysis on the subsequent column. The optimal coupled column chromatography procedure includes the on-line sample pretreatment and cleanup prior to the final analytical technique, and results in faster analysis of the components. Coupled separation techniques could significantly minimize and often exclude off-line sample pretreatment and preseparation procedures, which usually include filtration through prepacked sample tubes, preparative thin layer chromatography, liquid–liquid partitioning, Soxhlet extraction, and supercritical fluid extraction. These off-line techniques could result in solute loss, contamination, long workup times and introduction of human error.

Multidimensional Chromatography, edited by L. Mondello, A. C. Lewis and K. D. Bartle
©2002 John Wiley & Sons Ltd.

12.2 GENERAL

Multidimensional techniques have been applied for the analysis of polymer additives and polymer samples employed in polymer chemistry. Such additives, which include antioxidants (mainly sterically hindered phenols), are present in order to enhance the performance of the polymers and to ensure processing and long-term stability. Antioxidants, such as 2,6-di-*tert*-butyl-*p*-cresol and butylated hydroxytoluene (BHT), are added since many polymers are often subject to thermal and oxidative degradation. However, the additives themselves are also subject to oxidation, especially during processing or UV irradiation. The analysis of these additives and their transformation products has become very important for routine quality control, especially in the medical plastic and food packaging industries, where the identity and levels of potentially toxic substances must be accurately controlled and known (1). The use of multidimensional chromatography for the analysis of polymer additives in food products, including edible oils, are described in this contribution.

The usual means of identifying and quantifying the level of these additives in polymer samples is performed by dissolution of the polymer in a solvent, followed by precipitation of the material. The additives in turn remain in the supernatant liquid. The different solubilites of the additives, high reactivity, low stability, low concentrations and possible co-precipitation with the polymer may pose problems and lead to inconclusive results. Another sample pretreatment method is the use of Soxhlet extraction and reconcentration before analysis, although this method is very time consuming, and is still limited by solubility dependence. Other approaches include the use of supercritical fluids to extract the additives from the polymer and subsequent analysis of the extracts by microcolumn LC (2).

Multidimensional chromatography has also been applied for the analysis of industrial chemicals and related samples. Industrial samples which have been analyzed by multidimensional chromatography include coal tar, antiknock additives in gasoline (3), light hydrocarbons (4, 5), trihaloalkanes and trihaloalkenes in industrial solvents (6–8), soot and particulate extracts, and various industrial chemicals that might be present in gasoline and oil samples.

In this present chapter, the applications of multidimensional chromatography using various types of coupled techniques for the analysis of industrial and polymer samples, and polymer additives, are described in detail. The specific applications are organized by technique and a limited amount of detail is given for the various instrumental setups, since these are described elsewhere in other chapters of this volume.

12.3 LC–GC

High performance liquid chromatography (HPLC) is an excellent technique for sample preseparation prior to GC injection since the separation efficiency is high, analysis time is short, and method development is easy. An LC–GC system could be fully automated and the selectivity characteristics of both the mobile and stationary phases can be modified to effectively clean up unwanted components in sophisticated matrices. This, in turn, reduces the number of components actually present in the final analytical step (GC) by allowing heart-cut fractions of the selected peaks of interest. HPLC can effectively separate compounds based on chemical classes and can be used to enrich very dilute samples. The preseparation of components by HPLC prior to introduction into the GC system also increases the sensitivity. This can be achieved because of the efficient transfer of the whole HPLC fraction containing the compounds of interest and the efficient removal of interfering materials that may suppress the detection limits obtained by analysis during the GC step. The pretreatment of samples by HPLC offers three main advantages: preseparation into chemical classes by their different polarities, cleanup of the components of interest from a 'dirty' matrix, and sample enrichment (9).

Normally, most LC–GC applications use normal phase LC as the first separation method. Therefore, those low-boiling, non-polar solvents which are normally used in normal phase LC, plus small LC fraction volumes, are recommended. In addition, the volitization of the eluent which is being transferred is another important consideration. The volume of the transferred LC eluent may pose a problem due to column and detector overload and loss of resolution of the transferred fractions due to solvent overlap of the desired peaks. For these reasons, low eluent flow rates and relatively short LC retention times are normally used. The introduction of large volumes has an effect on the efficiency of the solute resolution and therefore the inlet temperature must be maintained so that it is close to the LC eluent boiling point. Ensuring the proper temperature avoids overload of the GC detector and column and thus prevents distorted, broad and even split peaks. These effects have been minimized by the introduction of retention gaps prior to the GC column of uncoated, deactivated capillary columns that allow the focusing and concentration of the solutes and have low retention capabilities. Solvent evaporation can be allowed in these retention gaps by two techniques, namely concurrent solvent evaporation (the GC initial temperature is higher than the boiling point of the eluent) and partial concurrent solvent evaporation (the GC initial temperature is lower than the boiling point of the eluent).

12.4 LC–GC APPLICATIONS

12.4.1 NORMAL PHASE LC–GC APPLICATIONS

The methods of analysis of polymer additives and chemicals, such as hydrocarbons, alcohols, etc., are not only restricted to the field of polymer chemistry but can also be applied for the analysis of such materials in the field of food chemistry. In addition, the analysis of polyaromatic hydrocarbons in edible oils has been of extreme importance. Polymeric packaging materials that are intended for food-contact use may contain certain additives that can migrate into the food products which are actually packaged in such products. The amounts of the additives that are permitted to migrate into food samples are controlled by government agencies in order to show that the levels of migrated additives from polymeric materials are within the specified migration limits. The highest level of migration of these additives, on account of their lipopphilic nature, is usually seen in edible oils or fats. Many references to the analysis of oils by using HPLC–GC are given in the book by Grob (10). These food products are usually very difficult to analyze due to the complex nature of the oils and fats and usually many sample preparation steps are needed before chromatographic separation is applied. Therefore, the use of multidimensional chromatography has allowed for less sample handling, better quantitative results and efficient analysis of such samples. Usually, normal phase HPLC is used for sample preparation, including the separation of the additive from the oil and fats. Then, the eluent fraction containing the additive is transferred to either the GC system in the second dimension or to another LC column, followed by GC analysis.

Baner and Guggenberger (11) have reported the analysis of a Tinuvin 1577 polymer-additive-rectified olive oil, virgin olive oil, Miglyol 812 S, corn and sunflower oils by using on-line coupled normal phase HPLC–GC. These authors have reported the quantitative determination of Tinuvin 1577 from poly (ethylene terepthalate) (PET) and polycarbonate (PC) polymers in these oil samples. The HPLC column used for the analysis of virgin olive oil was a 125 mm × 4.6 mm Spherisorb Si-5 μm column with a guard column, with the eluent being 30% dichloromethane in hexane. A 50 μL sample injection loop and a flow rate of 400 μL/min were used in this work. The LC/GC interface was a 300–500 μL transfer loop and was dependent upon the HPLC mobile phase flow rate and width of the Tinuvin 1577 HPLC peak. The GC column was a DB–5HT (15 m × 0.25 mm, 0.1 μm coating thickness) high-temperature fused-silica capillary. The GC oven temperature program was 160 °C for 8 min (the sample transfer and solvent evaporation temperature), a 10 °C/min ramp to 260°C, a 5 °C/min ramp to 320 °C, and finally a 10 °C/min ramp to 360 °C for 15 min. The GC conditions included a He flow rate of 0.9 ml/min at 160°C, nitrogen make-up gas and flame-ionization detection.

The Tinuvin 1577 (*MW*, 425.5) eluted on the silica column before the olive oil triglycerides. Then, the HPLC eluent fraction containing the Tinuvin 1577 was transferred into the GC unit by using a loop-type interface. The detection limit was found to be 0.19 ± 0.07 mg/L and the quantifiable limit was determined as being three times the limit of detection (LOD). Figure 12.1 shows the HPLC and GC chromatograms of the blank oil versus the detection limit concentration of Tinuvin 1577 in virgin olive oil. This method has also been applied to separate other higher-molecular-weight polymer additives such as Irganox 245 (*MW*, 586.8) and Irganox 1010 (*MW*, >1200), an antioxidant polymer additive.

12.5 SEC–GC APPLICATIONS

The premise of size exclusion chromatography (SEC) is that solute molecules are separated according to their effective molecular size in solution. SEC allows the separation of fractions according to their molecular weight and eliminates the

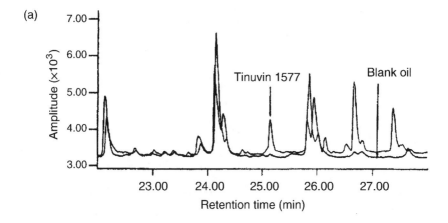

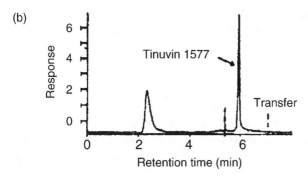

Figure 12.1 Analysis of Tinuvin 1577 in 30% virgin olive oil (in hexane), showing (a) the gas chromatogram comparing the pure oil with a sample at the Tinuvin 1577 detection limit concentration, and (b) the corresponding liquid chromatogram. Reprinted from *Journal of High Resolution Chromatography*, **20**, A. L. Baner and A. Guggenberger, 'Analysis of Tinuvin 1577 polymer additive in edible oils using on-line coupled HPLC–GC', pp. 669–673, 1997, with permission from Wiley-VCH.

high-molecular-weight material within the sample, thus preventing its transfer into the GC capillary column. The combination of SEC with capillary GC is very useful in the analysis of volatile compounds from complex mixtures.

A multidimensional system using capillary SEC–GC–MS was used for the rapid identification of various polymer additives, including antioxidants, plasticizers, lubricants, flame retardants, waxes and UV stabilizers (12). This technique could be used for additives having broad functionalities and wide volatility ranges. The determination of the additives in polymers was carried out without performing any extensive manual sample pretreatment. In the first step, microcolumn SEC excludes the polymer matrix from the smaller-molecular-size additives. There is a minimal introduction of the polymer into the capillary GC column. Optimization of the pore sizes of the SEC packings was used to enhance the resolution between the polymer and its additives, and smaller pore sizes could be used to exclude more of the polymer

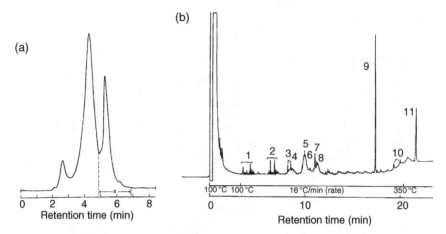

Figure 12.2 Chromatograms of an ABS copolymer sample: (a) microcolumn SEC trace; (b) capillary GC trace of peak 'x'. Peak identification is as follows: 1, C_{14} alkanes; 2, C_{15} alkanes; 3, C_{18} alkanes; 4, nonylphenol; 5, palmitic acid; 6, styrene–acetonitrile trimer; 7, stearic acid; 8, styrene–acetonitrile trimer; 9, Irganox 1076; 10, trinonylphenyl phosphate; 11, Ethanox 330. Reprinted with permission from Ref. (12).

material in the matrix. This is an effective technique to prevent the introduction of non-volatile components into the capillary GC–MS system.

As representative examples of the use of this technique, Figures 12.2, 12.3 and 12.4 show, respectively, the SEC–GC traces obtained in the analysis of samples

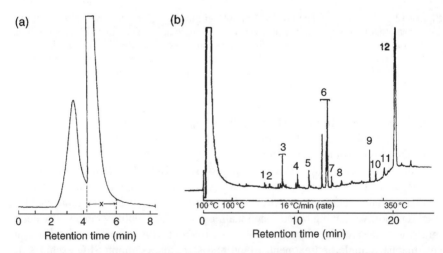

Figure 12.3 Chromatograms of an ignition-resistant high-impact polystyrene sample: (a) Microcolumn SEC trace; (b) capillary GC trace of peak 'x'. Peak identification is as follows 1, ionol; 2, benzophenone; 3, styrene dimer; 4, palmitic acid; 5, stearic acid; 6, styrene trimers; 7, styrene trimer; 8, styrene oligomer; 9, Irganox 1076 and Irganox 168; 10, styrene oligomer; 11, nonabromodiphenyl oxide; and 12, decabromodiphenyl oxide. Reprinted with permission from Ref. (12).

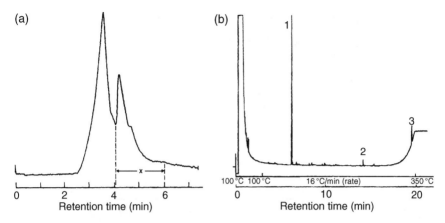

Figure 12.4   Chromatograms of a styrene–isoprene–styrene triblock copolymer sample: (a) microcolumn SEC trace; (b) capillary GC trace of the introduced section 'x'. Peak identification is as follows: 1, ionol; 2, not identified; 3, Irganox 565. Reprinted with permission from Ref. (12).

of emulsion polymerized acrylonitrile–butadiene–styrene–(ABS) copolymer, ignition-resistant high-impact polystyrene, and styrene–isoprene–styrene triblock copolymer. The following conditions were used in these three analyses. LC: UV detection at 254 nm; injection volume of 200 nL; fused-silica capillary column (30 cm × 250 μm i.d.) with an Ultrastyragel 10 000 (styrene/divinylbenzene) polymeric packing; mobile phase, THF at a flow rate of 3 μL/min. GC: DB-1 column (15 m × 0.32 i.d., 0.25 μm film thickness); He carrier gas; column connected to an uncoated deactivated inlet (5m × 0.32 mm i.d.), with the latter being connected to a 10-port valve, which was used to transfer components from the SEC system to the capillary GC–MS system.

The coupling of SEC with GC has also been used for the analysis of polymer additives from a polystyrene matrix (13). The transfer technique in this case includes concurrent solvent evaporation using a loop-type interface, early vapor exit and co-solvent trapping. The latter allowed recoveries of almost 100% for solutes as volatile as n-tridecane. The adaptation made to the standard loop-type interface by the addition of an extra valve between the LC detector and the LC–GC transfer valve led to improved quantitative results by avoiding the problem of mixing within the injection loop and also improved the recovery of n-alkanes from the mixture. Since the sample pretreatment incorporated dissolving the sample instead of extracting it, quantitative results were obtained. In addition, the effects of shifting the retention time window for the transfer were investigated and demonstrated that recoverabilities of C_{13}–C_{38} compounds of up to almost 100% could be obtained. The fraction obtained from 4.25–5.25 min, shows the greatest recovery, as can be seen in Figure 12.5.

The coupling of SEC to GC is not an easy process and in order to avoid additional LC interactions that could effect the predominate size exclusion separation relatively polar solvents such as THF are usually employed. The drawback is that polar

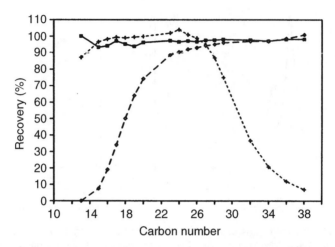

Figure 12.5 Effect of shifting the time window for the transfer. Operation in the SEC–GC analysis of polymer additives in a poly styene matrix, shown for the following fractions: ◇◇, 4.00–5.00 min; ■■, 4.25–5.25 min; − + − + − + −, 4.45–5.45 min. Reprinted from *Proceedings of the 15th International Symposium on Capillary Chromatography*, J. Blomberg *et al.*, 'Automated sample clean-up using on-line coupling of size-exclusion chromatography to high resolution gas chromatography', pp. 837–847, 1993, with permission from Wiley-VCH.

solvents show poor wettability properties when they are introduced in to the retention gaps and the conventional on-column techniques cannot be used for the introduction of large fractions (such as 500 μL). Therefore, a more applicable transfer technique, such as concurrent solvent evaporation of THF (ca. 100 °C) with a loop-type interface, was used, employing *n*-decane as the co-solvent. For solutes that eluted within 120–150 °C above the transfer temperature, quantitative results could not be obtained by solely using concurrent solvent evaporation, so here the problem was circumvented by using co-solvent trapping. The optimization of the LC–GC transfer within a larger temperature range and co-solvent concentration was achieved by modification of the transfer temperature, and increasing the length of the retention gap.

The applicability of this experimental setup and the repeatability of the data obtained were checked by analyzing polystyrene (PS706) after the addition of several well-known polymer additives. The SEC trace is presented in Figure 12.6(a), with the capillary GC trace of the additive fractions being shown in Figure 12.6(b). A relative standard deviation of less than 2% was obtained from 10 consecutive analyses of several additives, including Cyasorb UV531, Tinuvin 120 and Irgfos 168, added to the polymer material.

The quantitative analysis of additives in a polycarbonate homopolymer has been carried out by micro-SEC–capillary GC and by a conventional precipitation technique (14). The validity of the on-line technique was demonstrated and equivalent

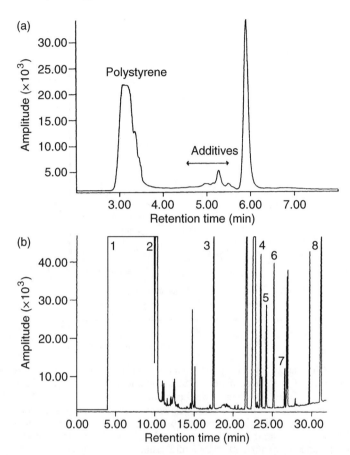

Figure 12.6 (a) SEC trace of polystyrene PS706 plus additives. (b) Capillary GC trace of the additive fraction: 1, THF (main solvent); 2, *n*-decane (co-solvent); 3, *n*-eicosane (internal standard); 4, Tinuvin 327; 5, Cyasorb UV531; 6, Tinuvin 120; 7, Tinuvin 770; 8, Irgafos 168. Reprinted from *Proceedings of the 15th International Symposium on Capillary Chromatography*, J. Blomberg *et al.*, 'Automated sample clean-up using on-line coupling of size-exclusion chromatography to high resolution gas chromatography', pp. 837–847, 1993, with permission from Wiley-VCH.

concentrations were obtained from both techniques for most of the components, with the exception of Tinuvin 329. A greater concentration was obtained for this additive by using the coupled technique, which may be attributed to additive loss during the precipitation technique. In addition, quantitative results were obtained from ten separate analyses of the sample, with the coupled technique showing a 2–3 times lower %RSD for most of the polycarbonate additives. Typical chromatograms obtained by using this coupled technique are shown in Figure 12.7.

Cortes *et al.* (15) have demonstrated the use of multidimensional chromatography employing on-line coupled microcolumn size exclusion chromatography and

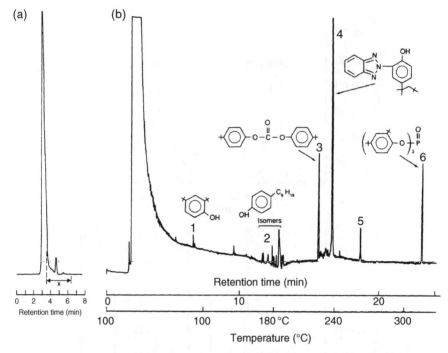

Figure 12.7 Chromatograms of a polycarbonate sample: (a) microcolumn SEC trace; (b) capillary GC trace of introduced fractions. SEC conditions: fused-silica (30 cm × 250 mm i.d.) packed with PL-GEL (50 Å pore size, 5 mm particle diameter); eluent, THF at a Flow rate of 2.0ml/min; injection size, 200 NL; UV detection at 254 nm; 'x' represents the polymer additive fraction transferred to LC system (ca. 6 μL). GC conditions: DB-1 column (15m × 0.25 mm i.d., 0.25 μm film thickness); deactivated fused-silica uncoated inlet (5 m × 0.32 mm i.d.); temperature program, 100 °C for 8 min, rising to 350 °C at a rate of 12°C/min; flame ionization detection. Peak identification is as follows: 1, 2,4-*tert*-butylphenol; 2, nonylphenol isomers; 3, di(4-*tert*-butylphenyl) carbonate; 4, Tinuvin 329; 5, solvent impurity; 6, Irgaphos 168 (oxidized). Reprinted with permission from Ref. (14).

pyrolysis gas chromatography for characterization of a styrene–acrylonitrile copolymer. This system used a 10-port valve and a glass chamber interface for the analysis of non-volatile compounds. A more complete description of the microcolumn liquid chromatography–pyrolysis gas chromatography interface is given in the original reference (15). It is advantageous to use a coupled mode since fractions do not have to be collected manually, evaporated, redissolved in a certain solvent and then manually transferred to a pyrolysis probe by using a syringe. For the characterization of polymers, the coupled techniques of SEC and pyrolysis GC allow the determination of average polymer composition as a function of molecular size. This may also elucidate a more comprehensive understanding of the polymer properties and the polymerization chemistry of particular systems.

Solutions of the styrene–acrylonitrile copolymer were separated on a micro-size exclusion column (50 cm × 250 μm i.d., packed with Zorbax PSM-1000), using THF as the mobile phase, and with UV detection at 220 nm, and specific fractions were then transferred into an on-line pyrolysis chamber. Within this chamber, the non-volatile polymer fractions were pyrolyzed and the products separated on a capillary GC column (50 m × 0.25 mm i.d., coated with 5% phenylmethylsilicone). The variability in composition was studied by obtaining the area ratios of the styrene and acrylonitrile peaks that were formed in each molecular size fraction. The relative composition of this polymer in the six fractions over the molecular-weight range from 1.1×10^4 to 1.8×10^4 was determined to be 5.96, with a relative standard deviation of 2.4%. The relative composition of the styrene–acrylonitrile copolymer was independent of molecular size. Figure 12.8 presents the micro SEC trace obtained for this polymer, and also shows the molecular-size fractions that were transferred to the pyrolysis-GC system. A typical pyrolysis-GC trace is given in Figure 12.9, for a fraction with a molecular-weight range of 1 800 000–450 000.

The analysis of industrial samples such as the pyrolysis products of Turkish lignites has been carried out by using GC, SEC and coupled HPLC/GC (16). The combustion products of lignites result in atmospheric pollution. The pyrolysis of

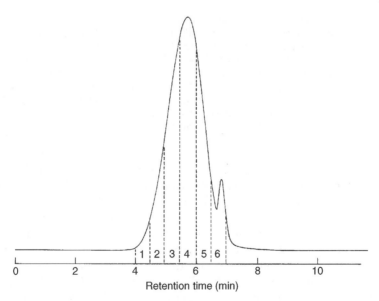

Figure 12.8 Microcolumn size exclusion chromatogram of a styrene–acrylonitrile copolymer sample; fractions transferred to the pyrolysis system are indicated 1–6. Conditions: fused-silica column (50 cm × 250 μm i.d.) packed with Zorbax PSM-1000 (7μm d_f); eluent, THF; flow rate, 2.0 μL/min; detector, Jasco Uvidec V at 220 nm; injection size, 20 nL. Reprinted from *Analytical Chemistry*, **61**, H. J. Cortes *et al.*, 'Multidimensional chromatography using on-line microcolumn liquid chromatography and pyrolysis gas chromatography for polymer characterization', pp. 961–965, copyright 1989, with permission from the American Chemical Society.

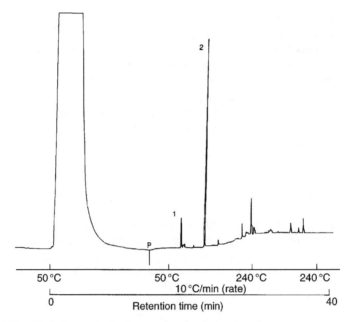

Figure 12.9 Typical pyrolysis chromatogram of fraction from a styrene–acrylonitrile copolymer sample obtained from a microcolumn SEC system: 1, acrylonitrile; 2, styrene. Conditions: 5 % Phenylmethylsilicone (0.33 μm d_f) column (50 m × 0.2 mm i.d.); oven temperature, 50 to 240 °C at 10 °C/min; carrier, gas, helium at 60 cm/s; flame-ionization detection at 320 °C; make-up gas, nitrogen at a rate of 20 mL/min.; 'P' indicates the point at which pyrolysis was made. Reprinted from *Analytical Chemistry*, **61**, H. J. Cortes *et al.*, 'Multidimensional chromatography using on-line microcolumn liquid chromatography and pyrolysis gas chromatography for polymer characterization', pp. 961–965, copyright 1989, with permission from the American Chemical Society.

Beypazari, Can and Goynuk lignites was carried out in a conventional Heinze retort. The pyrolysis product of the Goynuk lignite was analyzed by using coupled HPLC/GC, with the condensable total oil and tar products from pyrolysis being dissolved in dichloromethane. A silica–aminosilane bonded silica column, in series with a UV detector, a 10-port valve interface with a 150 μl sample loop and a Carlo Erba gas chromatograph equipped with a 25 m × 0.33 mm i.d. retention gap, connected to a 25 m × 0.33 mm i.d. BP-5 0.5 μm analytical fused column, with flame-ionization detection, was used. The HPLC eluent consisted of 10% dichloromethane in pentane, with helium being used as the GC carrier gas. After elution of the aliphatic fraction which contained saturates and olefins, backflusing was carried out and then analysis of the aromatic fractions was performed. On-line GC of these fractions allowed separation of the components and identification by their retention indices. It was found that alkanes and alkenes were present in the aliphatic fraction, and the aromatic fraction contained quantities of polycyclic aromatic compounds (PACs). In addition, fractionation of the sample by ring size was carried out, with this being achieved by a forward flow of HPLC eluent and then backflushing after the elution of the three-ring PAC. Analysis of the backflush fraction revealed the presence of some low-molecular-mass polar PACs, including nitrogen-containing PACs as well as higher-molecular-mass PAC species, e.g. containing 4–5 rings.

12.6 SEC–REVERSED PHASE LC APPLICATIONS

Quantitative determination of the polymer additives in an acrylonitrile–butadiene–styrene (ABS) terpolymer by using microcolumn size exclusion chromatography with THF as the eluent, coupled on-line to a reversed phase column employing typical solvent systems such as acetonitrile–water has been demonstrated (16). A pre-separation was carried out by size exclusion chromatography using a narrow-pore-size packing (50 Å). This packing completely resolved the smaller-molecular-size additives from the polymer fraction, as can be seen in Figure 12.10 (a). For the transfer from the SEC system to the reversed phase LC system, a 6-port Valco valve with a 10 μL external loop was used. The transferred fraction was then analyzed by reversed phase HPLC for determination of the polymer additives (Figure 12.10(b)). The introduction of 6 μl of the THF fraction from the SEC system into the reversed phase system allowed good resolution between the components, and gave peak shapes which were not broad or distorted. The use of microcolumn SEC permitted a small THF volume containing the additive fraction to be transferred. However, if a larger i.d. column had been used, a greater volume of THF would need to be transferred, thus causing broad and distorted peaks in the second dimension.

Johnson *et al.* (17) have coupled an SEC system, operating in the normal phase mode, using Micropak TSK gels with THF as the eluent, to a gradient LC system in the reversed phase mode, using MicroPak-MCH (monolayer octadecylsilane phase) with acetonitrile–water as the eluent, for the analysis of various additives in rubber stocks. These additives include carbon black, processing oils, antioxidants, vulcanizing accelerators and sulfur. The rubber stocks that were analyzed were butadiene–acrylonitrile (Chemigum N-615) and styrene–butadiene (Plioflex 1502) copolymers. The compositions of the compound rubber stocks is presented in Table 12.1.

The system used in this analysis consisted of four columns, with the first three being size exclusion columns (50 cm 3000H, 50cm 2000H and 80 cm 1000H Micropak TSK gels), followed by a C18 analytical column (Micropak MCH). The chromatogram of the Chemigum rubber stock obtained from the three coupled SEC columns can be seen in Figure 12.11(a). After this stage, a 10 μL fraction of each peak in the SEC chromatogram was transferred to the RPLC system via an injection valve. In this mode, a gradient was applied to elute the more hydrophobic components. Standards were run in order to identify the retention times of the components in the SEC coupled system, as well as in the SEC–RPLC system. RPLC traces are shown for two the fractions, i.e. dibutylphthalate and elemental sulfur,

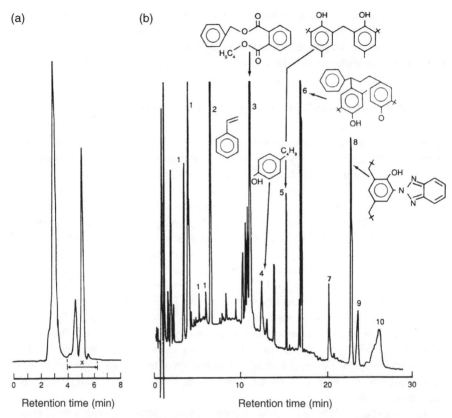

Figure 12.10 Microcolumn SEC–LC analysis of an acrylonitrile–butadiene–styrene (ABS) terpolymer sample: (a) SEC trace; (b) LC trace. SEC conditions: fused-silica column (30 cm × 250 mm i.d.) packed with PL-GEL (50 Å pore size, 5 mm particle diameter); eluent, THF at a flow rate of 2.0 mL/min; injection size, 200 nL; UV detection at 254 nm; 'x' represents the polymer additive fraction (6 μL) transferred to LC system. LC conditions: NovaPak C18 Column (15 cm × 4.6 mm i.d.); eluent, acetonitrile–water (60:40) to (95:5) in 15 min gradient; flow rate of 1.5 mL/min; detection at 214 nm. Peaks identification is follows: 1, styrene–acrylonitrile; 2, styrene; 3, benzylbutyl phthalate; 4, nonylphenol isomers; 5, Vanox 2246; 6, Topanol; 7, unknown; 8, Tinuvin 328; 9, Irganox 1076; 10, unknown. Reprinted with permission from Ref. (14).

transferred from the SEC coupled system to the reversed system in Figures 12.11 (b) and 12.11(c), respectively.

The same system was used for analysis of the Plioflex stock. The SEC trace is seen in Figure 12.12(a), with the RPLC traces for two of the fractions, (1) Wingtay 100 and (2) mixed disulfide and 2,2′ thiobis (benzothiazole) (MBTS), transferred from the SEC coupled system to the reversed system presented in Figures 12.12(b) and 12.12(c), respectively. Plioflex is a more complex sample than Chemigum since it contains processing oil (see Table 12.1). Therefore, the analysis was more complicated due to the presence of a wide molecular weight range of aromatic compounds in this oil (eluted between 30 and 50 min) which could be seen to interfere

Table 12.1   Compositions of the compound rubber stocks

Component	Content [a]	
	Chemigum N-615	Plioflex 1502
Polymer	100	100
HAF carbon black	50	50
Aromatic processing oil	–	5
Dibutyl phthalate	5	–
Stearic acid	1	1
Wingstay 100 [b]	–	1
Wingstay 300 [c]	1	–
Zinc oxide	5	5
MBTS [d]	1	1
TMTD [e]	0.3	0.3
Sulfur	1.5	1.5
Total	164.8	164.8

[a] Expressed as parts per hundred of 'pure' polymer.
[b] Mixed diarylphenylenediamine.
[c] Alkylarylphenylenediamine
[d] 2,2^1-thiobis (benzothiazole).
[e] Tetramethyltrhiuradisulfide.

with other components in the SEC trace (see Figure 12.12(a)). Hence, analysis in the second dimension by RPLC of the transferred fractions allowed the identification of these components.

12.7   GC–GC APPLICATIONS

When columns of the same polarity are used, the elution order of components in GC are not changed and there is no need for trapping. However, when columns of different polarities are used trapping or heart-cutting must be employed. Trapping can be used in trace analysis for enrichment of samples by repetitive preseparation before the main separation is initiated and the total amount or part of a mixture can then be effectively and quantitatively transferred to a second column. The main considerations for a trap are that it should attain either very high or very low temperatures over a short period of time and be chemically inactive. The enrichment is usually carried out with a cold trap, plus an open vent after this, where the trace components are held within the trap and the excess carrier gas is vented. Then, in the re-injection mode the vent behind the trap is closed, the trap is heated and the trapped compounds can be rapidly flushed from the trap and introduced into the second column. Peak broadening and peak distortion, which could occur in the preseparation, are suppressed or eliminated by this re-injection procedure (18).

Heart-cutting, with trapping and re-injection, using isothermal dual capillary column chromatography for separation of the UV photolysis products of methyl

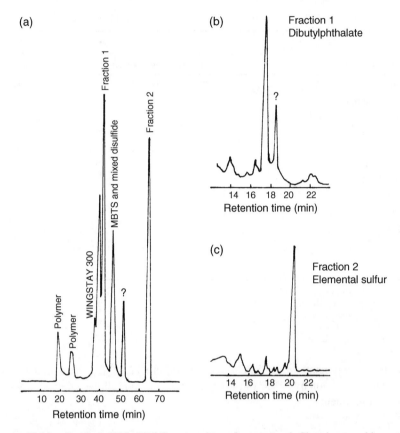

Figure 12.11 Coupled SEC–RPLC separation of compound Chemigum rubber stock: (a) SEC trace; (b) RPLC trace of fraction 1, dibutylphthalate; (c) RPLC trace of fraction 2, elemental sulfur. Coupled SEC conditions: MicroPak TSK 3000H (50 cm) × 2000H (50 cm) × 1000 H (80 cm) columns (8 mm i.d.); eluent, THF at a flow rate of 1 mL/min; UV detection at 215 nm (1.0 a.u.f.s.); injection volume, 200 μL. RPLC conditions: MicroPak MCH (25 cm × 2.2 mm i.d.) column; flow rate, 0.5 mL/min; injection volume, 10 μL; gradient, acetonitrile–water (20 : 80 v/v) to 100% acetonitrile at 3% acetonitrile/min; UV detection at 254 nm (0.05 a.u.f.s.). Reprinted from *Journal of Chromatography,* **149**, E. L. Johnson *et al.*, 'Coupled column chromatography employing exclusion and a reversed phase. A potential general approach to sequential analysis', pp. 571–585, copyright 1978, with permission from Elsevier Science.

isopropyl ether has been carried out, as well as heart-cutting without trapping of the selected fractions (19). The reaction mixture was preseparated on a 20 m poly (propylene glycol) column and then re-injected on a Marlophen column (100 m × 0.25 mm i.d.). The separation with trapping demonstrated the effectiveness of the trapping and re-injection procedure since more of the solvent peak that results from the fronting of the tail is eliminated and does not interfere by overlapping with peaks of higher retention on the second column. This allows well resolved chromatograms

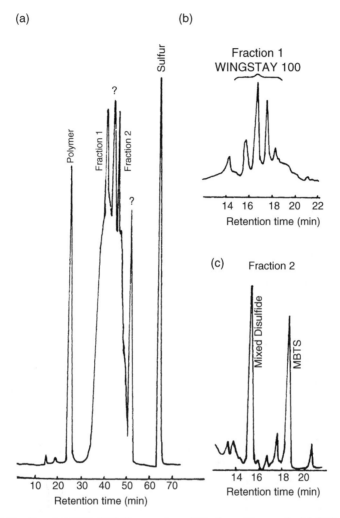

Figure 12.12 Coupled SEC–RPLC separation of Plioflex rubber stock: (a) SEC; (b) RPLC trace of fraction 1, Wingstay 100 (Five-peak pattern is representative of diarylphenylenediamine isomers); (c) RPLC trace of fraction 2, mixed disulfide and MBTS (2,2´-thiobis (benzothiazole)). Obtained under the same conditions as given for Figure 12.11. Reprinted from *Journal of Chromatography*, **149**, E. L. Johnson *et al.*, 'Coupled column chromatography employing exclusion and a reversed phase. A potential general approach to sequential analysis', pp. 571–585, copyright 1978, with permission from Elsevier Science.

and higher peak intensities for trace components to be obtained, as can be seen in Figures 12.13(a) (with trapping) and 12.13(b) (without trapping).

Multi-column switching can be an effective approach for the determination of high and low concentrations of sample components in complex mixtures. This is a very powerful technique for the analytical and preparative separation of components

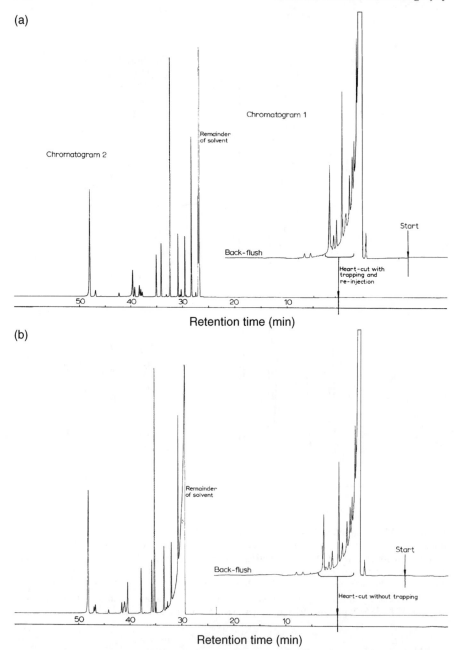

Figure 12.13 Illustration of isothermal dual capillary column chromatography used for separation of UV photolysis products of methyl isopropyl ether. (a) Heart-cut and back-flushing at preseparation: chromatogram 1, PPG pre-column (20 m × 0.25 mm i.d.); 55 °C, 0.2 bar N$_2$; 3 μL. Chromatogram 2, Marlophen main column (100 m × 0.25 mm i.d.); 1.5 bar N$_2$; sample, heart-cut from chromatogram 1. (b) Obtained under the same conditions as (a), but without trapping of the heart-cut. Reprinted with permission from Ref. (19).

that are not completely resolved, or not resolved at all, when using a single GC column.

The determination of trace impurities in various major industrial materials such as gasoline, styrene and aniline by using a combination of programmed-temperature sample introduction, mass-flow-controlled multi-column dual-oven capillary gas chromatography and on-line mass spectrometry has been demonstrated (20). The cryotrapping at the inlet of the second column resulted in significant improvement in the resolution of critical peak pairs and reproducibility of the analysis. The multi-column analysis of aniline was performed by using an OV-17 (25 m × 0.32 mm i.d., $d_f = 1.0$ μm precolumn (Figure 12.14(a)) and an HP(1–50 m × 0.32 mm i.d., $d_f = 1.05$ μm) second column. The shaded compound (peak) is solvent flushed via a splitline to permit detection by mass spectrometry of impurities in the aniline and also to prevent overloading of the MS detector. The analysis of aniline was carried out both with (Figure 12.14(c)) and without cryotrapping (Figure 12.14(b)), where it is found that the use of cryotrapping leads to increased resolution between the phenol and aniline present in the sample.

Another example of multi-column analysis has been demonstrated for the determination of impurities in styrene. The marked compounds in the styrene sample (Figure 12.15(a)) were solvent flushed via a splitline, with the analysis being carried out with a cryotrapping separation (CTS) (see Figure 12.15(b)). The first column, was an Ultra-2 (25 m × 0.32 mm i.d., $d_f = 0.25$ μm) precolumn, while the main column was a DB-WAX (30 m × 0.32 mm, $d_f = 0.25$ μm) with an FID being employed as the detection system.

This multi-column swithching (GC–GC) technique has also been shown to be a powerful method for the separation of benzene and 1-methyl-cyclopentane in gasoline, as well as for the analysis of m-and p-xylenes in ethylbenzene.

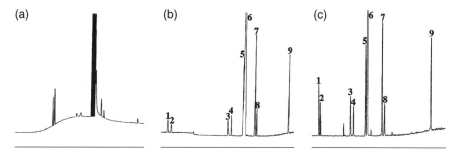

Figure 12.14 Chromatographic analysis of aniline: (a) Precolumn chromatogram (the compound represented by the shaded peak is solvent flushed); (b) main column chromatogram without cryotrapping; (c) main column chromatogram with cryotrapping. Conditions: DCS, two columns and two ovens, with and without cryotrapping facilities: columns OV-17 (25 m × 0.32 mm i.d., 1.0 μm d.f.) and HP-1 (50 m × 0.32 mm, 1.05 μm d_f). Peak identification is as follows: 1, benzene; 2, cyclohexane; 3, cyclohexylamine; 4, cyclohexanol; 5, phenol; 6, aniline; 7, toluidine; 8, nitrobenzene; 9, dicyclohexylamine. Reprinted with permission from Ref. (20).

(a)   (b)

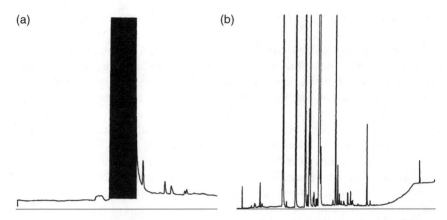

Figure 12.15   Chromatographic analysis of styrene and its impurities: (a) precolumn chromatogram (the compound represented by the shaded peak is solvent flushed prior to transfer to
the second column); (b) main column chromatogram with cryotrapping. Conditions: DCS,
two columns and two ovens: CTS, columns, Ultra-2 (25 m × 0.32 mm i.d., 0.52 μm d.f.) and
DB-WAX (30 m × 0.32 mm i.d., 0.25 μm d_f). Reprinted with permission from Ref. (20).

The system used to carry out such an analysis is shown in Figure 12.16 with
the schematic diagrams of the dual-column switching (DCS) and multi-column
switching (MCS) modules being presented in Figure 12.17. The solvent flush via the
splitline is carried out when the MFC3 valve is open (Figure 12.17 (b)).

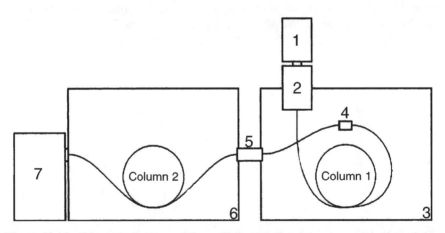

Figure 12.16   Schematic diagrams of the applied system involving cryotrapping, consisting
of a manual or automated injector (1), a temperature-programmable cold-injection system
with septumless sampling head (2), a gas chromatograph (3) configured with an FID monitor,
column-switching-device (4) and MCS or DCS pneumatics, connected via a heated transfer
line with an included cryotrap (5) to a second gas chromatograph (6) with a mass-selective
detector (7) (MCS, multi-column switching; DCS, dual-column switching). Reprinted with
permission from Ref. (20).

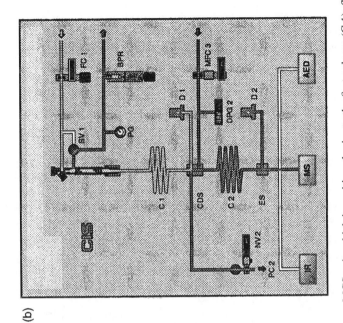

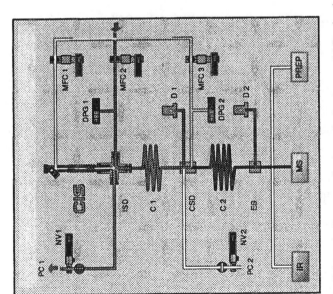

Figure 12.17 Schematic diagrams of solvent-flush systems. (a) Dual-column MCS unit, which is positioned prior to the first column (C 1); flushing is carried out with both NV 1 and MFC 2 open. (b) Dual-column DCS unit; solvent flushing is carried out via the splitline, with MFC 3 open (NV, needle valve; MFC, mass flow control). Reprinted with permission from Ref. (20).

12.8   SFC–GC AND NORMAL PHASE-LC–SFC APPLICATIONS

Supercritical fluid chromatography (SFC) is an intermediate mode between gas and liquid chromatography which combines the best features of each of these techniques. This is a chromatographic technique in which the mobile phase is neither a liquid or a gas but a supercritical fluid, with physical properties that are intermediate between those of gases and liquids. Supercritical chromatography is of importance either in the first or second dimension since it permits the separation and determination of compounds that cannot be analyzed by HPLC or GC. Such compounds could be either non-volatile, of high molecular mass, reactive or thermally labile, thus making them unsuitable for GC. Thermally labile compounds could include polymer additives, with the low operating temperatures of SFC being important for the analysis of such compounds. In addition, one of the advantages that SFC has over HPLC is that when using CO_2 as a mobile phase it can be easily interfaced to a multitude of detection systems, including UV, FID, MS and FTIR. SFC could also be an attractive alternative to normal phase LC and SEC for the analysis of high-molecular-mass species such as polymers and fossil fuels since SEC analyses are hindered by poor efficiency, while normal phase LC analysis lacks a universal detection system. Hence, SFC with FID detection allows the analysis of high-boiling petroleum fractions, group-type analysis of saturates, olefins and aromatics in petroleum fuels (21).

12.9   NORMAL PHASE-LC–SFC APPLICATIONS

The use of microcolumn LC generates small peak volumes and allows the coupling of SFC by using a solvent-venting injection procedure. The preseparation of polymer additives from various polymers was carried out on a 320 μm i.d. fused-silica column packed with 5 μm polystyrene particles, using THF as the eluent (22). The low-molecular-weight additives ($MW < 1200$) were eluted at the total permeation volume since they were more retained than their excluded polymer counterparts. The evaporation of the solvent from the collected fraction was carried out by flushing the loop with nitrogen gas, in turn leaving the polymer additives coated on the wall of the retention gap. The additives were then transferred to the SFC column by switching the liquid CO_2 flow via the retention gap to the SFC system. The additives were focused by the effect of the higher temperature, while their separation was achieved by gradient elution by systematically changing the density of the supercritical fluid.

An application of an LC–SFC system has been demonstrated by the separation of non-ionic surfactants consisting of mono- and di-laurates of poly (ethyleneglycol) (23). Without fractionation in the precolumn by normal phase HPLC (Figure 12.18 (a)) and transfer of the whole sample into the SFC system, the different homologues coeluted with each other. (Figure 12.18(b)). In contrast with prior fractionation by HPLC into two fractions and consequent analysis by SFC, the homologues in the two fractions were well resolved (Figures 12.18(c) and 12.18(d)).

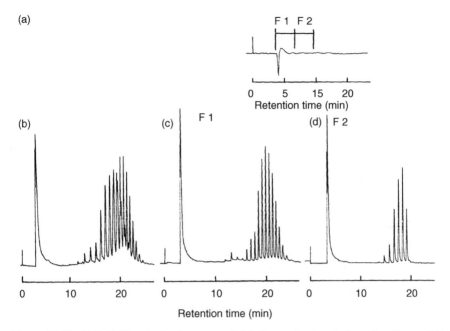

Figure 12.18   LC–SFC analysis of mono- and di-laurates of poly(ethylene glycol) ($n = 10$) in a surfactant sample: (a) normal phase HPLC trace; (b) chromatogram obtained without prior fractionation; (c) chromatogram of fraction 1 (F1); (d) chromatogram of fraction 2 (F2). LC conditions: column (20 cm × 0.25 cm i.d.) packed with Shimpak diol; mobile phase, n-hexane/methylene chloride/ethanol (75/25/1); flow rate, 4 μL/min; UV detection at 220 nm. SFC conditions: fused-silica capillary column (15 m × 0.1 mm i.d.) with OV-17 (0.25 μm film thickness); Pressure-programmed at a rate of 10 atm/min from 80 atm to 150 atm, and then at a rate of 5 atm/min; FID detection. Reprinted with permission from Ref. (23).

12.10   SFC–GC APPLICATIONS

An on-line supercritical fluid chromatography–capillary gas chromatography (SFC–GC) technique has been demonstrated for the direct transfer of SFC fractions from a packed column SFC system to a GC system. This technique has been applied in the analysis of industrial samples such as aviation fuel (24). This type of coupled technique is sometimes more advantageous than the traditional LC–GC coupled technique since SFC is compatible with GC, because most supercritical fluids decompress into gases at GC conditions and are not detected by flame-ionization detection. The use of solvent evaporation techniques are not necessary. SFC, in the same way as LC, can be used to preseparate a sample into classes of compounds where the individual components can then be analyzed and quantified by GC. The supercritical fluid sample effluent is decompressed through a restrictor directly into a capillary GC injection port. In addition, this technique allows selective or multi-step heart-cutting of various sample peaks as they elute from the supercritical fluid chromatograph. The determination of the heart-cut times are achieved by monitoring the response from the on-line ultraviolet absorbance and flame ionization detectors. A schematic diagram of the instrumental setup used for these analyses is given in Figure 12.19.

In addition, solute focusing is possible by maintaining a low initial temperature (e.g. 40 °C) for a long period of time (8–12 min) to allow the mixture of decompressed carbon dioxide, helium gas and the solutes to focus on the GC column. The optimization of the GC inlet temperature can also lead to increased solute focusing. After supercritical fluid analysis, the SF fluid effluent is decompressed through a heated capillary restrictor from a packed column (4.6 mm i.d.) directly into a hot GC split vaporization injector.

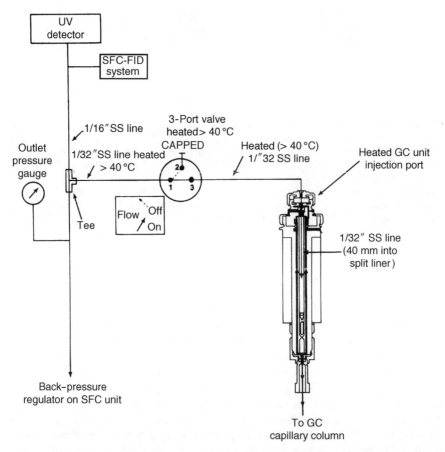

Figure 12.19 Schematic diagram of the interface system used for supercritical fluid chromatography–gas chromatography. Reprinted from *Journal of High Resolution Chromatography*, **10**, J. M. Levy *et al.*, 'On-line multidimensional supercritical fluid chromatography/capillary gas chromatography', pp. 337–341, 1987, with permission from Wiley-VCH.

The preseparation utilized a 5 μm cyano column (250 cm × 4.6 mm i.d.) and a 5 μm silica column (250 cm × 4.6 mm i.d.) in series, followed by GC analysis on an SE-54 column (25 m × 0.2 mm i.d., 0.33 μm film thickness). The SFC system separated the aviation sample into two peaks, including saturates and single-ring aromatics as the first peak, and two-ring aromatic fractions as the second peak. These fractions were selectively cut and then transferred to the GC unit for further analysis. (Figure 12.20).

Another application of SFC–GC was for the isolation of chrysene, a polyaromatic hydrocarbon, from a complex liquid hydrocarbon industrial sample (24). A 5 μm octadecyl column (200 cm × 4.6 mm i.d.) was used for the preseparation, followed by GC analysis on an SE-54 column (25 m × 0.2 mm i.d., 0.33 μm film thickness). The direct analysis of whole samples transferred from the supercritical fluid chromatograph and selective and multi-heart-cutting of a particular region as it elutes from the SFC system was demonstrated. The heart-cutting technique allows the possibility of separating a trace component from a complex mixture (Figure 12.21).

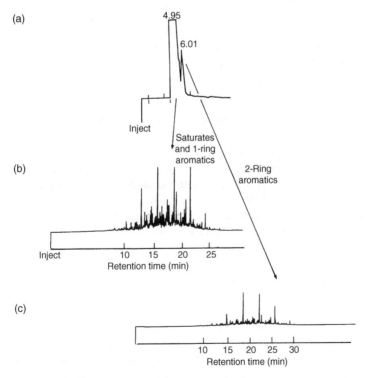

Figure 12.20 SFC–GC analysis of a sample of aviation fuel: (a) SFC separation into two peaks; (b and c) corresponding GC traces of the respective peaks (flame-ionization detection used throughout). Reprinted from *Journal of High Resolution Chromatography*, **10**, J. M. Levy *et al.*, 'On-line multidimensional supercritical fluid chromatography/capillary gas chromatography', pp. 337–341, 1987, with permission from Wiley-VCH.

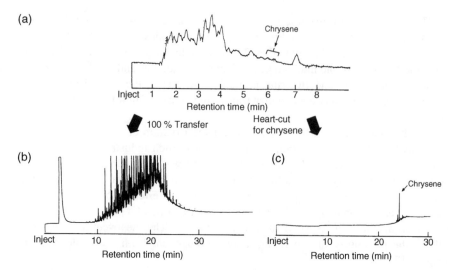

Figure 12.21   SFC–GC heart-cut analysis of chrysene from a complex hydrocarbon mixture: (a) SFC trace (UV detection); (b) GC trace without heart-cut (100% transfer); (c) GC trace of heart-cut fraction (flame-ionization detection used for GC experiments). Reprinted from *Journal of High Resolution Chromatography*, **10**, J. M. Levy *et al.*, 'On-line multidimensional supercritical fluid chromatography/capillary gas chromatography', pp. 337–341, 1987, with permission from Wiley-VCH.

SFC–GC has also been used for group-type separations of high-olefin gasoline fuels, including saturates, olefins, and aromatics (25). The SFC–GC characterization of the aromatic fraction of gasoline fuel was carried out by using CO_2 on four packed columns in series, i.e. silica, Ag+-loaded silica, cation-exchange silica and NH_2 silica. The heart-cut fractions were transferred into a capillary column coated with a methyl polysiloxane stationary phase, with cryofocusing at 50 °C being used for focusing during the transfer into the GC system. The SFC–FID and GC–FID chromatographs are shown in Figure 12.22.

12.11   SFC–SFC APPLICATIONS

An on-line SFC–SFC coupled technique involving a rotary valve interface was used to provide an efficient separation of coal tar (see in Figure 12.23) (26). A schematic diagram of the multidimensional packed capillary to open tubular column SFC–SFC system is shown in Figure 12.24. The rotary valve interface was used to provide the flexibility of using two independently controlled pumps, which gave an increased performance of the system when compared to the traditional one-pump system. In addition, an on-column cryogenic trap was used to suppress efficiency losses due to the first packed column. This cryogenic unit efficiently traps the selected fractions and focuses the sample, then allowing the transfer of the fractions into a narrow band

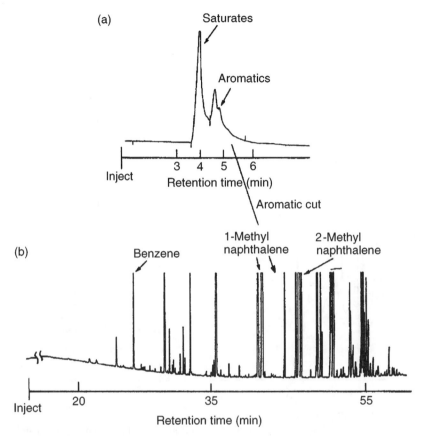

Figure 12.22 SFC–GC analysis of aromatic fraction of a gasoline fuel. (a) SFC trace; (b) GC trace of the aromatic cut. SFC conditions: four columns (4.6 mm i.d.) in series (silica, silver-loaded silica, cation-exchange silica, amino-silica); 50 °C; 2850 psi; CO_2 mobile phase at 2.5 mL/min; FID detection. GC conditions: methyl silicone column (50 m × 0.2 mm i.d.); injector split ratio, 80:1; injector temperature, 250 °C; carrier gas helium; temperature programmed, −50 °C (8 min) to 320 °C at a rate of 5 °C/min; FID detection. Reprinted from *Journal of Liquid Chromatography*, **5**, P. A. Peaden and M. L. Lee, 'Supercritical fluid chromatography: methods and principles', pp. 179–221, 1987, by courtesy of Marcel Dekker Inc.

onto the second column. The first column was an aminosilane stationary phase packed column and gave good resolution of the coal tar extract by chemical class separation (i.e. of aromatic rings) due to high selectivity. The second column was an open tubular fused-silica column (10.5 m × 50 µm i.d.) coated with a liquid crystalline polysiloxane stationary phase. The compounds were separated on the second column according to shape-selective separation of the isomeric compounds, thus leading to increased efficiency. By using a packed first column, this increased the loading capacity and increased sample concentration, which may not be obtained with an open tubular column.

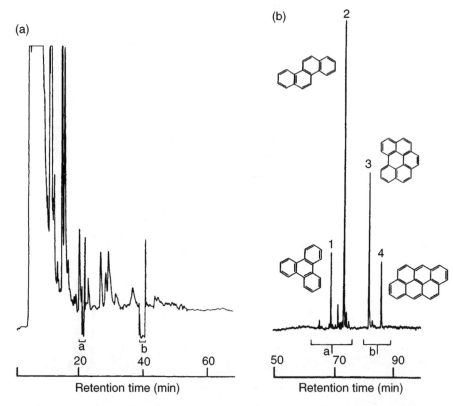

Figure 12.23 SFC–SFC analysis, involving a rotary valve interface, of a standard coal tar sample (SRM 1597). Two fractions were collected from the first SFC separation (a) and then analyzed simultaneously in the second SFC system (b); cuts 'a' and 'b' are taken between 20.2 and 21.2 min, and 38.7 and 40.2 min, respectively. Peak identification is as follows: 1, tri-phenylene 2, chrysene 3, benzo[*ghi*]perylene; 4, anthracene. Reprinted from *Analytical Chemistry*, **62**, Z. Juvancz *et al.*, 'Multidimensional packed capillary coupled to open tubular column supercritical fluid chromatography using a valve-switching interface', pp. 1384–1388, copyright 1990, with permission from the American Chemical Society.

12.12   CONCLUSIONS

Multidimensional chromatography has proven to be useful for the analysis of complex samples such as polymer or industrial mixtures. All of the separation techniques available today have definite limitations in terms of their selectivity and separation range, which leads to the necessity of multi-stage separation procedures for samples which contain a wide variety of different components. Current technological processes require fast and rugged analytical methods which can provide comprehensive information about the process stages and products. This dictates the necessity of development of automated complex separation procedures with minimal sample pre-treatment, and the use of on-line multidimensional chromatographic techniques is a logical solution to these requirements.

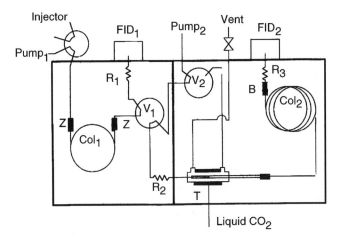

| Liquid CO_2

Figure 12.24 Schematic diagram of the multidimensional packed capillary to open tubular column SFC–SFC system. Reprinted from *Analytical Chemistry*, **62**, Z. Juvancz *et al.*, 'Multidimensional packed capillary coupled to open tubular column supercritical fluid chromatography using a valve-switching interface', pp. 1384–1388, copyright 1990, with permission from the American Chemical Society.

One of the significant drawbacks of multidimensional analytical methods is the specificity of the conditions of each separation mode for a particular sample type, together with restrictive requirements for the type and operational conditions of the interface between them. Therefore, extensive work in the method development stage, along with the availability of highly skilled personnel for operating such systems, are required.

Methods developed for on-line technological control have to be tested for the variation of the product composition due to process variations. However, if rugged analytical procedures are developed these multidimensional methods may only require minimal attention during on-line operation. Multidimensional chromatography for the analysis of complex polymer and industrial samples offers chromatographers high productivity and efficiency and is an excellent alternative to off-line methods.

REFERENCES

1. R. G. Lichtenthaler and F. Ranfelt, 'Determination of antioxidants and their transformation products in polyethylene by high-performance liquid chromatography', *J. Chromatogr.* **149**: 553–560 (1978).

2. Y. Hirata and Y. Okamoto, 'Supercritical fluid extraction combined with microcolumn liquid chromatography for the analysis of polymer additives', *J. Microcolumn Sep.* **1**: 46–50 (1989).

3. G. Castello, 'Gas-chromatographic determination of alkyllead compounds in aromatic-based fuels', *Chim. Ind.* **51**: 700–704 (1969).

4. R. J. Laub and J. H. Purnell, 'Quantitative approach to the use of multicomponent solvents in gas-liquid chromatography', *Anal. Chem.* **48**: 799–803 (1976).

5. G. Castello and G. D'Amato, 'Mixed porapak columns for the gas-chromatographic separation of light hydrocarbons', *Ann. Chim.* **69**: 541–549 (1979).

6. G. Castello and G. D'Amato, 'Comparison of the polarity of porous polymer-bead stationary phases with that of some liquid phases', *J. Chromatogr.* **366**: 51–57 (1986).

7. G. Castello, A. Timossi and T. C. Gerbino, 'Analysis of haloalkanes on wide-bore capillary columns of different polarity connected in series', *J. Chromatogr.* **522**: 329–343 (1990).

8. G. Castello, A. Timossi and T. C. Gerbino, 'Gas chromatographic separation of halogenated compounds on non-polar and polar wide bore capillary columns', *J. Chromatogr.* **454**: 129–143 (1988).

9. F. Andreolini and F. Munari, 'Chromatographic separations through on-line coupled techniques: HPLC–HRGC', in *Proceedings of the 11th International Symposium on Capillary Chromatography*, Monterey, CA, USA, May 14–17, Sandra (Ed.), Hüthig, Heidelberg, Germany, pp. 602–610 (1990).

10. K. Grob (Ed.), *On-Line Coupled LC–GC*, Hüthig, Heidelberg, Germany (1991).

11. A. L. Baner and A. Guggenberger, 'Analysis of Tinuvin 1577 polymer additive in edible oils using on-line coupled HPLC–GC', *J. High Resolut. Chromatogr.* **20**: 669–673 (1997).

12. H. J. Cortes, B. M. Bell, C. D. Pfeiffer and J. D. Graham, 'Multidimensional chromatography using on-line coupled microcolumn size exclusion chromatography–capillary gas chromatography–mass spectrometry for determination of polymer additives', *J. Microcolumn Sep.* **1**: 278–288. (1989)

13. J. Blomberg, P. J. Schoenmakers and N. van den Hoed, 'Automated sample clean-up using on-line coupling of size-exclusion chromatography to high resolution gas chromatography', in *Proceedings of the 15th International Symposium on Capillary Chromatography*, Vol. I, Riva del Garda, Italy, May 24–27, Sandra P. (Ed.), Hüthig, Heidelberg, Germany, pp. 837–847 (1993).

14. H. J. Cortes, G. E. Bormett and J. D. Graham, 'Quantitative polymer additive analysis by multidimensional chromatography using online coupled microcolumn size exclusion chromatography as a preliminary separation', *J. Microcolumn Sep.* **4**: 51–57 (1992).

15. H. J. Cortes, G. L. Jewett, C. D. Pfeiffer, S. Martin and C. Smith, 'Multidimensional chromatography using on-line microcolumn liquid chromatography and pyrolysis gas chromatography for polymer characterization', *Anal. Chem.* **61**: 961–965 (1989).

16. M. Citiroglu, E. Ekinci, M. Tolay and B. Frere, 'Analysis of Turkish lignites pyrolysis products of GC, SEC and coupled HPLC/GC', in *Proceedings of the 13th International Symposium on Capillary Chromatography*, Vol. II, Riva del Garda, Italy, May 13–16, P. Sandra (Ed.), Hüthig, Heidelberg, Germany, pp. 1358–1359 (1991).

17. E. L. Johnson, R. Gloor and R. E. Majors, 'Coupled column chromatography employing exclusion and a reversed phase. A potential general approach to sequential analysis', *J. Chromatogr.* **149**: 571–585 (1978).

18. J. Sevcik, in *Advances in Chromatography 1979*, Proceedings of the 14th International Symposium, Lausanne, Switzerland, September 28–28, A. Zlatkis (Ed.).

19. G. Schomburg, H. Husmann and F. Weeke, 'Aspects of double-column gas chromatography with glass capillaries involving intermediate trapping', in *Advances in Chromatography*, Proceedings of the 10th International Symposium, Münich Germany, November 3–6, A. Zlatkis (Ed.), *J. Chromatogr.* **112**, 205–217 (1975).

20. A. Hoffmann, J. Staniewski, R. Bremer, F. Rogies and J. A. Rijks, 'Applications of mass flow controlled multicolumn switching in on-line capillary GC–MS', in *Proceedings of*

14th International Symposium on Capillary Chromatography, Baltimore, MA, USA, May 25–29, P. Sandra (Ed.), Hüthig, Heidelberg, Germany, pp. 624–631 (1992).

21. H. J. Cortes (Ed.), *Multidimensional Chromatography. Techniques and Applications*, Marcel Dekker, New York, pp. 304–(1990).

22. S. Ashraf, K. D. Bartle, N. J. Cotton, I. L. Davies, R. Moulder and M. W. Raynor, 'Hyphenated techniques in capillary chromatography', in *Proceedings of 12th International Symposium on Capillary Chromatography*, Kobe, Japan, September 11–14, P. Sandra (Ed.), Hüthig, Heidelberg, Germany, pp. 318–322 (1990).

23. H. Daimon, Y. Katsura, T. Kondo, Y. He and Y. Hiratapp, 'Directly coupled techniques using microcolumn and supercritical fluid: SFE/SFC, SFE/LC and LC, SFC', in *Proceedings of 12th International Symposium on Capillary Chromatography*, Kobe, Japan, September 11–14, P. Sandra (Ed.), Hüthig, Heidelberg, Germany, pp. 456–465 (1990).

24. J. M. Levy, J. P. Guzowski and W. E. Huhak, 'On-line multidimensional supercritical fluid chromatography/capillary gas chromatography. *J. High Resolut. Chromatogr.* **10**: 337–341 (1987).

25. P. A. Peaden and M. L. Lee, 'Supercritical fluid chromatography: methods and principles. *J. Liq. Chromatogr.* **5**: 179–221 (1982).

26. Z. Juvancz, K. M. Payne, K. E. Markides and M. L. Lee, 'Multidimensional packed capillary coupled to open tubular column supercritical fluid chromatography using a valve-switching interface', *Anal. Chem.* **62**: 1384–1388 (1990).

13 Multidimensional Chromatography in Environmental Analysis

R. M. MARCÉ

Universitat Rovira i Virgili, Tarragona, Spain

13.1 INTRODUCTION

Multidimensional chromatography has important applications in environmental analysis. Environmental samples may be very complex, and the fact that the range of polarity of the components is very wide, and that there are a good many isomers or congeners with similar or identical retention characteristics, does not allow their separation by using just one chromatographic method.

The main aims in environmental analysis are sensitivity (due to the low concentration of microcontaminants to be determined), selectivity (due to the complexity of the sample) and automation of analysis (to increase the throughput in control analysis). These three aims are achieved by multidimensional chromatography: sensitivity is enhanced by large-volume injection techniques combined with peak compression, selectivity is obviously enhanced if one uses two separations with different selectivities instead of one, while on-line techniques reduce the number of manual operations in the analytical procedure.

For analytical purposes, environmental analysis can be divided into the control of pollution and the analysis of target compounds. For the control of pollution, it is important to monitor both well-known priority pollutants and all of the other non-priority pollutants. The selectivity of the analytical column may therefore not be sufficient. In most cases, however, mass spectrometry (MS) detection can solve the problem and this is why gas chromatography GC–MS is widely used in routine analysis. Sometimes, however, MS, and even MS/MS (which requires complex instrumentation) may not solve the problem and multidimensional chromatography is then a suitable technique. The low levels at which the micropollutants are to be determined is another drawback and multidimensional techniques are a good solution to this problem. In pollution control, high throughput is required and this may be obtained by automating the analysis via multidimensional chromatography, which also reduces the possible sources of error.

In target-compound analysis, a particular compound, usually present in a complex matrix and at trace levels, needs to be quantified. Here, selectivity and

Multidimensional Chromatography, edited by L. Mondello, A. C. Lewis and K. D. Bartle
©2002 John Wiley & Sons Ltd.

sensitivity are the most important requirements, and, as well as the use of a suitable selective detector, multidimensional chromatography has also played an important role.

Soil extracts are usually very complex. In water samples, humic and fulvic acids make analysis difficult, especially when polar substances are to be determined. Multidimensional chromatography can also make a significant contribution here to this type of analysis.

The use of multidimensional chromatography in environmental analysis has been reviewed in the literature (1–6). Of the multidimensional systems described in previous chapters, GC–GC liquid chromatography LC–LC and LC–GC, whose applications to environmental analysis will be detailed in this chapter, are the ones most often used in environmental analysis.

Other multidimensional systems, such as supercritical fluid chromatography (SFC–GC or LC–SFC), will not be described here because, although some applications to environmental analysis have been described (4, 7–9), they have not been very widely used in this field.

13.2 MULTIDIMENSIONAL GAS CHROMATOGRAPHY

13.2.1 INTRODUCTION

Gas chromatography, because of its high resolution, is widely used in environmental analysis to determine a wide range of pollutants. This technique is applied to both volatile and, after a derivatization step, to nonvolatile analytes. Environmental samples are usually quite complex because of the different pollutants which may be present. Multidimensional chromatography (MDGC) or GC–GC coupling would therefore be expected to be widely applied in environmental analysis. Despite its many advantages, however, a major drawback with MDGC is that, in principle, many heart-cuts from the first column should be subjected to a second separation. In other words, the method can become extremely time consuming. One remedy is to combine several heart-cuts and analyse these in one second run. However, the risk of co-elution then markedly increases and this is particularly dangerous when detection with no identification power is used. This is what happens, for example, with chlorinated analytes, for which electron-capture detection is the most widely used, due to its high sensitivity (5).

In general, capillary gas chromatography provides enough resolution for most determinations in environmental analysis. Multidimensional gas chromatography has been applied to environmental analysis mainly to solve separation problems for complex groups of compounds. Important applications of GC–GC can therefore be found in the analysis of organic micropollutants, where compounds such as polychlorinated dibenzodioxins (PCDDs) (10), polychlorinated dibenzofurans (PCDFs) (10) and polychlorinated biphenyls (PCBs) (11–15), on account of their similar properties, present serious separation problems. MDGC has also been used to analyse other pollutants in environmental samples (10, 16, 17).

MDGC has also been used in the air analysis field. For instance, it has been applied to the analysis of volatile organic compounds (VOCs) in air, thus enabling a wider range of these compounds to be analysed (18).

Today, however, GC–GC coupling is seldom used to determine pesticides in environmental samples (2), although comprehensive MDGC has been applied to determine pesticides in more complex samples, such as human serum (19). On the other hand, new trends in the pesticide market, which is now moving towards the production of optically active enantiomers and away from racemic mixtures, may make this area suitable for GC–GC application. The coupling of non-chiral columns to chiral columns appears to be a suitable solution to the separation problems that such a trend might cause.

Multidimensional gas chromatography has also been used in the qualitative analysis of contaminated environmental extracts by using spectral detection techniques such as infrared (IR) spectroscopy and mass spectrometry (MS) (20). These techniques produce the most reliable identification only when they are dealing with pure substances; this means that the chromatographic process should avoid overlapping of the peaks.

Most applications in environmental analysis involve heart-cut GC–GC, while comprehensive multidimensional gas chromatography is the most widely used technique for analysing extremely complex mixtures such as those found in the petroleum industry (21).

13.2.2 EXAMPLES OF MULTIDIMENSIONAL GAS CHROMATOGRAPHY APPLIED TO ENVIRONMENTAL ANALYSIS

A typical example of MDGC in environmental analysis is the determination of PCBs. These are ubiquitous contaminants of the environment in which they occur as complex mixtures of many of the 209 theoretically possible congeners. The compositions of environmental mixtures vary according to sample type.

Attempts to optimize the capillary GC separation conditions of 209 PCBs on a single column of either single or mixed phases have had only limited success. MDGC has therefore been very important. In some cases, mass spectrometry and, in particular high-resolution mass spectrometry, may be enough to determine different isomers which co-elute in a single column, and sensitivity may be enhanced by selected ion monitoring (SIM) or negative chemical ionization (NCI).

In MDGC, the usual configuration normally has a non-polar phase (such as SE 54 or CPSil 8) on the first column to make the initial, well-characterized separation. The sample is chromatographed on this column to a point just before the elution of the unresolved peaks. The column flow is then switched into a second column of a different, usually more polar, phase such as CPSil 19 or CPSil88, for the duration of the elution of these resolved peaks only. The column is again isolated and the small group of unresolved peaks is separated on the second column (15). Other columns which have been used include BPX5 (22), OV1(23) or Ultra 2 (11) as the first column, and HT8 (23), OV-210 (12, 24) or FFAP (14) as the second column.

Kinghorn *et al.* used MDGC and electron-capture detection (ECD) to determine seven specific chlorobiphenyl congeners (key congeners) which co-eluted with other components in the mixture of an Arochlor standard (22). Figure 13.1 shows the separation from the first column (BPX5) with flame-ionization detection (FID) and the separation of the different cuts in the second column (HT8) with ECD. The separation of most congeners is good.

The chirality of the PCB congeners has also been taken into account in this separation. From a total of 209 congeners, 78 are axially chiral in their nonplanar

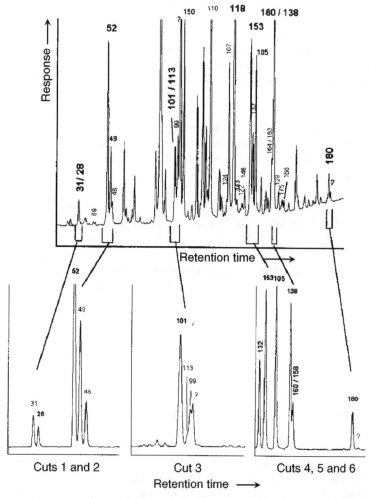

Figure 13.1   Monitor (FID) (a) and analytical (ECD) (b) channel responses for PCB congeners in Aroclor 1254, showing selection of the six heart-cut events: First columns, HT8; second columns, BPX5. Reprinted from *Journal of High Resolution Chromatography*, **19**, R. M. Kinghorn *et al.*, 'Multidimensional capillary gas chromatography of polychlorinated biphenyl marker compounds', pp. 622–626, 1996, with permission from Wiley-VCH.

conformations and 19 form stable enantiomers due to the restricted rotation around the central carbon–carbon bond at ambient temperatures (14). There are at least nine of the conformationally stable chiral PCBs which are present in commercial formulations and these are expected to accumulate in the environment. In order to analyse them, an achiral column was coupled to a chiral column (14, 25). The gas chromatographic separation of most of the chiral PCB congeners was achieved on different cyclodextrin phases (14).

MDGC has been used for separating commercial formulations of PCBs (11, 12, 22, 23, 26) although it is not widely used on real samples. In some examples, MDGC has been applied to determine PCBs in sediment samples (13, 14, 27) and water samples (14, 24).

For PCB analysis, Glausch *et al.* used an achiral column coated with DB-5 and a chiral column coated with immobilized Chirasil-Dex (14). The column was switched with a pneumatically controlled six-port valve and peak broadening was minimized by cooling the first part of the second column with air precooled with liquid nitrogen, thus focussing the cut fraction. These authors determined the chiral polychlorinated biphenyls 95, 132 and 149 in river sediments, using microsimultaneous steam distillation–solvent extraction as the sample treatment technique. Figure 13.2 presents the MDGC-ECD chromatograms of PCB fractions from sediment samples, where it can be seen that the separation of the PCB enantiomers is good.

In many environmental extracts, the analytes of interest overlap with other analytes or matrix components. MDGC is therefore essential for improving accuracy when identifying non-target methods. In a typical example (20), contaminated water, clay and soil samples have been analysed. While water and clay extracts can be analysed by GC-IR–MS, soil samples, because of their greater complexity, need MDGC-IR–MS to identify the pollutants present in the sample. In such cases, an MDGC-IR–MS system with multiple parallel cryogenic traps and sample recycling is used (see Figure 13.3).

The chromatogram from the first column was divided into five areas of five heart-cuts each. Some peaks identified in the chromatogram from the first column were used as heart-cut markers. This method has some limitations which mainly concern contamination of the system, and also with the determination of less volatile pollutants. However, such a system is able to detect and accurately identify about 40 pollutants.

Another interesting application of MDGC is in the rapid determination of isoprene (the most reactive hydrocarbon species) and dimethyl sulfide (DMS) (the major source of sulfur in the marine troposphere and a precursor to cloud formation) in the atmosphere (16). The detection limits were 5 and 25 ng l^{-1}, respectively.

A programmed temperature-vaporization (PTV) injector (with a sorbent-packed liner) was used to preconcentrate and inject the sample. Thermal desorption was performed and the analytes were passed to a primary column (16 m × 0.32 mm i.d., film thickness 5 μm, 100% methyl polysiloxane) and separated according to analyte vapour pressure. Selected heart-cuts were transferred to a second column (15 m × 0.53 mm i.d., Al_2O_3/Na_2SO_4 layer, open tubular column with 10 μm stationary phase) where final separation was performed according to chemical functionality.

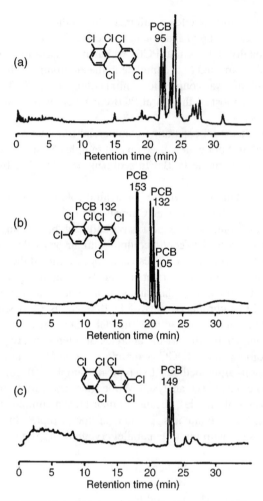

Figure 13.2 MDGC-ECD chromatograms of PCB fractions from sediment samples, demonstrating the separation of the enantiomers of (a) PCB 95, (b) PCB 132, and (c) PCB 149; non-labelled peaks were not identified. Reprinted from *Journal of Chromatography*, A **723**, A. Glausch *et al.*, 'Enantioselective analysis of chiral polychlorinated biphenyls in sediment samples by multidimensional gas chromatography–electron-capture detection after steam distillation–solvent extraction and sulfur removal', pp. 399–404, copyright 1996, with permission from Elsevier Science.

Figure 13.4 shows the standard chromatogram obtained by using this method. In this sample, isoprene and major interference compounds, i.e. 2-methyl pentane and hexane, were spiked at $1 \mu g \, l^{-1}$, plus DMS at $10 \mu g \, l^{-1}$.

When using one-dimensional GC (1D-GC), the analysis took about 60 min (shown in the inset to Figure 13.4), while with the use of two-dimensional (2D-GC) it took only about 12 min.

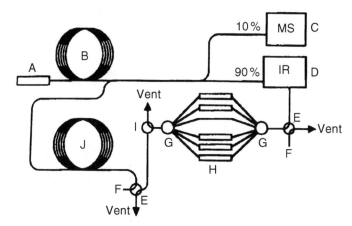

Figure 13.3 Schematic diagram of the parallel cryogenic trap MDGC-IR–MS system: A, splitless injection port; B, Rt$_x$-5 non-polar first-stage separation column; C, HP 5970B MSD; D, HP 5965B IRD; E, four-port two-way valve (300 °C maximum temperature); F, external auxiliary carrier gas; G, six-port selection valve (300 °C maximum temperature); H, stainless-steel cryogenic traps; I, three-port two- way valve (300 °C maximum temperature); J, Rt$_x$-5 intermediate polarity column. Reprinted from *Journal of Chromatography A, 726,* K. A. Krock and C. L. Wilkins, 'Qualitative analysis of contaminated environmental extracts by multidimensional gas chromatography with infrared and mass spectral detection (MDGC–IR–MS)', pp. 167–178, copyright 1996, with permission from Elsevier Science.

13.3 MULTIDIMENSIONAL LIQUID CHROMATOGRAPHY

13.3.1 INTRODUCTION

Liquid chromatography (LC) is a good alternative to GC for polar or thermolabile compounds. While polar compounds need to be derivatized for GC analysis, this is therefore not necessary for LC analysis.

When environmental samples are analysed by reverse-phase liquid chromatography, the most widely used technique, polar interferences usually appear (ions, plus humic and fulvic acids). This makes it difficult to determine more polar compounds that elute in the first part of the chromatogram. This is specially important when detection is not selective, e.g. UV detection, which is one of the most common techniques in routine analysis. In such cases, multidimensional chromatography plays an important role.

The application range of coupled-column technology is determined by the separation power of the first column. In general, it can be said that low resolution favours multiresidue methods (MRMs), while high resolution leads to methods for a single analyte or for a group of analytes with similar properties.

Multidimensional LC–LC, using two high-resolution columns with orthogonal separation mechanisms, has only a few applications in environmental analysis. The limitations that such a multidimensional system has with regard to selectivity must

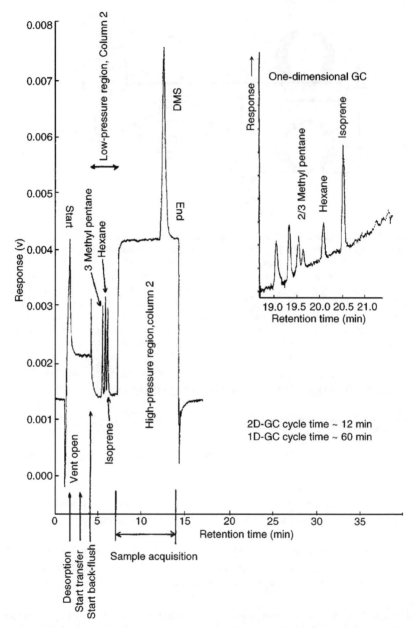

Figure 13.4   Two-dimensional GC separation of isoprene, interference hydrocarbons, and DMS, with the inset showing the one-dimensional separation of isoprene and interference hydrocarbons for comparison. Reprinted from *Environmental Science and Technology,* **31**, A. C. Lewis *et al.*, 'High-speed isothermal analysis of atmospheric isoprene and DMS using online two-dimensional gas chromatography', pp. 3209–3217, copyright 1997, with permission from the American Chemical Society.

be taken into account. For example, the differences in the physico-chemical bases of the separation processes involved may lead to poorly compatible mobile phase systems, thus requiring complex interfaces. Moreover, the separation obtained in the first column can, at least partly, be decreased in the second column.

In spite of such limitations, some examples can be found in literature. For example, a reversed-phase C_{18} column has also been coupled to a weak ion-exchange column to determine gluphosinate, glyphosate and aminomethylphosphonic acid (AMPA) in environmental water (28). This method will be described further below.

Zebühr et al. (29) developed an automated system for determining PAHs, PCBs and PCDD/Fs by using an aminopropyl silica column coupled to a porous graphitic carbon column. This method gives five fractions, i.e. aliphatic and monoaromatic hydrocarbons, polycyclic aromatic hydrocarbons, PCBs with two or more ortho-chlorines, mono-ortho PCBs, and non-ortho PCBs and PCDD/Fs. This method employed five switching valves and was successfully used with extracts of sediments, biological samples and electrostatic filter precipitates.

As mentioned above, the most commonly used liquid chromatographic technique is reverse-phase liquid chromatography (RPLC), which is also the most often used coupled technique. When two RPLC systems are coupled to analyse aqueous samples, there is an additional advantage because large sample volumes can be injected without causing extensive band broadening. This means that there is on-line enrichment. In general, therefore, the less polar the analyte, then the more the sample volume can be enriched without causing band broadening or breakthrough. However, when highly polar analytes have to be determined, the enrichment and clean-up needed to eliminate the interference become more limited. However, results have been good for some analytes, as we will see later.

Most work on LC–LC in environmental analysis has been developed by the Van Zoonen group (30, 31) and the Hernández group (32–34).

A commonly used system in environmental analysis is the heart-cutting technique which uses the separation power of the first column to obtain a higher selectivity than with the previously described precolumn enrichment. The two columns are coupled via a switching valve, as shown in Figure 13.5.

Separation in column 1 (C-1) removes early-eluting interference compounds, and so considerably increases the selectivity. The fraction of interest separated in C-1 is then transferred to column 2 (C-2) where the analytes of the fraction are separated. These transfers can be carried out either in forward mode or backflush mode. The forward mode is preferred because the backflush mode has two disadvantages for polar to moderately polar analytes. For most polar compounds, it leads to additional band broadening, while for more retained analytes there is a decrease in the separation obtained earlier in the process (31).

An important parameter in LC–LC is the transfer volume, i.e. the time that C-1 is coupled to C-2, since the selectivity is highly dependent on this. In environmental samples, it is important to remove early-eluting interference in order to ensure selective analysis. A short analysis time is important for routine analysis of environmental samples.

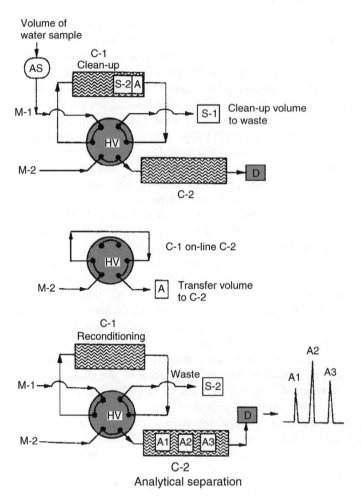

Figure 13.5 Schematic presentation of the procedure involved in coupled-column RPLC: AS, autosampler; C-1 and C-2, first and second separation columns, respectively; M-1 and M-2, mobile phases; S-1 and S2, interferences; A, target analytes; HV, high-pressure valve; D, detector. Reprinted from *Journal of Chromatography, A* **703**, E. A. Hogendoorn and P. van Zoonen, 'Coupled-column reversed-phase liquid chromatography in environmental analysis', pp. 149–166, copyright 1995, with permission from Elsevier Science.

When a first column of a very short length (and therefore a low selectivity) is used (this is especially suitable for multiresidue methods), we talk about an on-line pre-column (PC) switching technique coupled to LC (PC–LC or solid-phase extraction (SPE)-LC). This is particulary useful for the enrichment of analytes, and enables a higher sample volume to be injected into the analytical column and a higher sensitivity to be reached. The sample is passed through the precolumn and analytes are retained, while water is eliminated; then, by switching the valve, the analytes retained in the precolumn are transferred to the analytical column by the mobile phase, and with not just a fraction, as in the previous cases.

Depending on the kind of sorbent in the precolumn connected on-line to the analytical column, the retention of analytes in the precolumn may be more or less selective.

The sorbents that are most frequently used in environmental analysis are C_{18}-silica based sorbents, polymeric sorbents (usually styrenedivinilbenzene) and graphitized carbon. In order to increase the selectivity of these sorbents, immunosorbents (35, 36) have been developed and used with good results, while recently, molecularly imprinted polymers have started be to used (35, 36).

Polar compounds present the most problems because of their low breakthrough volumes with common sorbents. In the last few years, highly crosslinked polymers have become commercially available which involve higher retention capacities for the more polar analytes (37, 38). Polymers have also been chemically modified with polar groups in order to increase the retention of the compounds previously mentioned (35, 37).

Instead of a sorbent contained in a precolumn, discs can also be used with a special device (38, 39) which enables the number of discs to be changed easily, although this technique is currently limited to the kind of discs that are commercially available.

The characteristics of the sorbent in the precolumn may lead to problems when coupling the two systems. Therefore, when the analytes are more retained in the precolumn than in the analytical column, peak broadening may appear, even when the analytes are eluted in the backflush mode (40). This has been solved with a special design in which the analytes retained in the precolumn are eluted with only the organic solvent of the mobile phase and the corresponding mobile phase is subsequently formed (40, 41).

On-line SPE–LC has been widely used in environmental analysis to solve the problems caused by the low concentrations of the analytes to be detected and also to automate the analysis (42–44).

13.3.2 EXAMPLES OF MULTIDIMENSIONAL LIQUID CHROMATOGRAPHY IN ENVIRONMENTAL ANALYSIS

LC–LC is applied to environmental samples with two major aims, i.e. to determine a single analyte and to determine a group of analytes (by the multiresidue methods) at the low levels required by legislation in both cases. Some examples of these are discussed below. In addition, some applications for the particular case of SPE–LC, will also be described.

The single-residue methodology has been used to determine analytes with different characteristics. The main advantage of this technique is the short analysis time. LC–LC methodology has been applied to various polar analytes, such as bentazone (46), and less polar compounds such as isoproturon (46) or pentaclorphenol (47). Most applications refer to water samples (46, 48), although solid samples have also been studied (31, 49, 50). Soil samples contain more interfering compounds so clean-up is even more important when analysing soil extracts. Coupled RPLC is therefore a suitable technique for efficient on-line clean-up procedures. Some examples of various applications are shown in Table 13.1.

Some of the different methods which demonstrate the suitability of this technique for determining single analytes will now be described in greater detail.

Ethylenethiourea (ETU) is a highly polar metabolite of the ethylenebisdithiocarbamate (EBDC) fungicides. This is a relatively stable degradation product which is also present in EBDC formulations (at concentrations 0.02–5%) (48). ETU is mainly determined by GC using a precolumn derivatization, which is time consuming, or by LC using UV or electrochemical detection with a clean-up step. Both techniques are very laborious and are also not sensitive enough. LC–LC is therefore an attractive alternative to these methods and has been used to determine ETU in aqueous samples (48). The experimental conditions are given in Table 13.1. The authors, after studying different packing materials of the alkylbonded columns, decided to work with a 5 μm Hypersyl ODS for better separation of the interfering compounds and peak compression. Ammonium acetate (pH 7.5) was also added to the mobile phase since this gave a better peak profile than with pure water at pH 3; the flow rates were 1.0 (M-1) and 1.1 (M-2) ml min^{-1}. Experiments showed that the injection volumes on C-1 should not exceed 200 μl and these were eluted with 2.6 ml of mobile phase. C-1 was switched on-line with C-2, and the ETU-containing fraction transferred to C-2 by using 0.44 ml of mobile phase (40 s). The retention time of ETU on C-2 was about 3 min, while the total analysis took less than 10 min.

From Figure 13.6, which compares the direct LC analysis and the LC–LC analysis, we can see that there is a significant increase in selectivity.

With this method, levels of 0.1 μg l^{-1} can be detected in ground water and, if an off-line liquid–liquid extraction step is added, levels of 0.1 μg l^{-1} can be detected (48).

Another example is the determination of bentazone in aqueous samples. Bentazone is a common medium-polar pesticide, and is an acidic compound which co-elutes with humic and/or fulvic acids. In this application, two additional boundary conditions are important. First, the pH of the M-1 mobile phase should be as low as possible for processing large sample volumes, with a pH of 2.3 being about the best that one can achieve when working with alkyl-modified silicas. Secondly, modifier gradients should be avoided in order to prevent interferences caused by the continuous release of humic and/or fulvic acids from the column during the gradient (46).

With bentazone, small changes in the composition of the mobile phase have a dramatic effect on the final results (see Figure 13.7).

The methanol gradient from 50 to 60% releases quite a lot of interfering components. Omitting the step gradient does not provide enough selectivity and so the best conditions were obtained with a pH gradient. The experimental conditions are shown in Table 13.1.

This method can quantify levels of 0.1 μg l^{-1} in real samples and reproducibility values are good; the total analysis time was 8 min. Figure 13.8 compares the chromatogram obtained by using this method with one obtained without column switching.

Table 13.1 Examples of the application of LC–LC in environmental analysis

Analyte	Matrix	Detection[a] (nm)	Injection volume (µl)	C-1 / C-2[b]	Clean-up with M-1 / Transfer with M-2[c]	LOD[d] (µg l^{-1})	Reference
Bromacil Diuron 3,4-dichloroaniline	Water	UV(254) UV(254) UV(254)	100	15 × 3.2 C_{18}, 7 µm 150 × 4.6 C_{18}, 5 µm	M-1: 2.5 ml MeOH (10%) M-2: 0.5 ml MeOH (65%)	0.2 0.01 0.02	50
DNOC Dinoterb Dinoseb	Soil	UV(365) UV(365) UV(365)	75	15 × 3.2 C_{18}, 7 µm 50 × 4.6 C_{18}, 3 µm	M-1: 1.1 ml MeOH (23%) in buffer (pH 2.9) M-2: 0.8 ml MeOH (60%) in buffer (pH 2.9)	10	49
Ethylenthiourea	Water	UV(233)	200	150 × 4.6 C_{18}, 5 µm 150 × 4.6 C_{18}, 5 µm	M-1: 2.6 ml CH_3CN (1%) and NH_3 (0.2%) M-2: 0.44 ml CH_3CN (1%) and NH_3 (0.2%)	1 0.1[e]	48
Methylisocyanate	Water	UV(237)	770	50 × 4.6 C_{18}, 3 µm 100 × 4.6 C_{18}, 3 µm	M-1: 1.7 ml CH_3CN (40%) M-2: 0.4 ml CH_3CN (50%)	1 0.1	51
Bentazone	Water	UV(220)	2000	50 × 4.6 C_{18}, 3 µm 100 × 4.6 C_{18}, 3 µm	M-1: 4.7 ml MeOH (50%) in phosphate buffer M-2: 0.5 ml MeOH (60%) in phosphate buffer	0.1	46
Chlorophenoxy acids	Water	UV(228)	400	10 × 3 GFF, 5 µm 100 × 4.6 C_{18}, 3 µm	M-1: 1 ml MeOH (5%) in TFA (0.05%) M-2: 0.5 ml MeOH (50%) in TFA (0.05%)	0.1	52
Isoproturon	Water	UV(244)	4000	50 × 4.6 C_{18}, 3 mm 100 × 4.6 C_{18}, 3 µm	M-1: 5.85 ml CH_3CN (47.5%) M-2: 0.4 ml CH_3CN (47.5%)	0.1	46
Simazine Atrazine	Water	UV(223)	20 000	30 × 4 C_{18}, 5 µm 100 × 4.6 C_{18}, 3 µm	M-1: 2.9 ml CH_3CN/H_2O (40/60, v/v)	0.1–0.2	33

Table 13.1 (continued)

Analyte	Matrix	Detection[a] (nm)	Injection volume (μl)	C-1 C-2[b]	Clean-up with M-1 Transfer with M-2[c]	LOD[a] ($\mu g\,l^{-1}$)	Reference
Terbuthylazine Terbutryn					M-2: 0.7 ml CH_3CN/H_2O (70/30, v/v)	<0.2[e]	
Aldehydes and ketones	Air	UV (360)	2000	$100 \times 4.6\ C_{18}$, 3 μm $100 \times 4.6\ C_{18}$, 3 μm	M-1, M-2: 4 ml MeOH/CH_3CN/H_2O gradient	0.5 [e,f]	53
Gluphosinate Glyphosate AMPA	Water	FD (263/317)	2000	$30 \times 4.6\ C_{18}$, 5 μm $250 \times 4.6\ NH_2$, 5 μm	M-1: 2.21 ml CH_3CN (35 %) in buffer (pH 5.5) M-2: 0.53 ml CH_3CN (35 %) in buffer (pH 5.5)	1	28
Glyphosate AMPA	Soil	FD (263/317)	2000	$30 \times 4.6\ C_{18}$, 5 μm $250 \times 4.6\ NH_2$, 5 μm	M-1: 2.12 ml CH_3CN (35 %) in buffer (pH 5.5) M-2: 0.41 ml CH_3CN (35 %) in buffer (pH 5.5)	1	50
Atrazine DIA DEA HA	Water	UV (220)	2000	$30 \times 4\ C_{18}$, 5 μm $125 \times 4\ C_{18}$, 5 μm	M-1: 2.6 ml CH_3CN/H_2O (20/80, v/v) M-2: 4.2 ml CH_3CN/H_2O gradient	0.2–0.5 0.02–0.1[e]	34

[a] UV, ultraviolet; FD, fluorescence detection.
[b] C-1, first column; C-2, second column; dimensions given in mm.
[c] M-1, first mobile phase; M-2, second mobile phase.
[d] Limit of detection.
[e] After solid-phase extraction.
[f] LOD in pmd l^{-1}

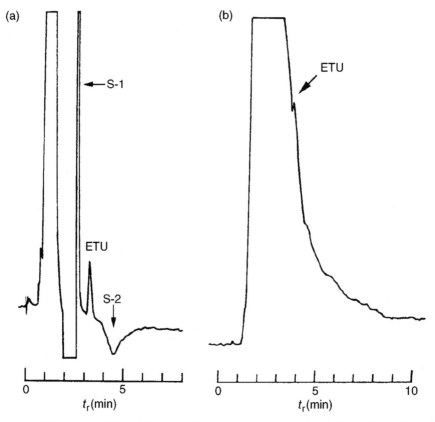

Figure 13.6 Direct RPLC analysis of a blank ground water sample spiked with 4.5 (μg l^{-1} ETU, (a) with and (b) without column-switching. A 60 $\times$ 4.6 mm i.d. column and a 150 $\times$ 4.6 mm i.d. column were used for C-1 and C-2, respectively, with pure water as M-1 and methanol–0.025 M ammonium acetate (pH, 7.5) (5:95, v/v) as M-2; S-1 and S-2 are the interfering peaks. Reprinted from *Chromatographia*, **31**, E. A. Hogendoorn *et al.*, 'Column-switching RPLC for the trace-level determination of ethylenethiourea in aqueous samples', pp. 285–292, 1991, with permission from Vieweg Publishing.

Since only a small fraction of the interfering material reaches the second column and subsequently the detector, the next analysis can start after the analyte has been transferred to column 2. This provides a high throughput (about 7 samples per hour).

This method can also be used to analyse soil samples. For instance, fenpropimorph, which is a non-polar pesticide with good UV sensitivity but poor selectivity, has, after treatment, been determined in soil samples (31). In this example, an amount of soil was extracted overnight with acetonitrile; this was then poured into a Buchner filter and rinsed with the same solvent. The acetonitrile solution was concentrated and, prior to LC analysis, the extract was diluted with water and 100 μl were then injected into the LC system.

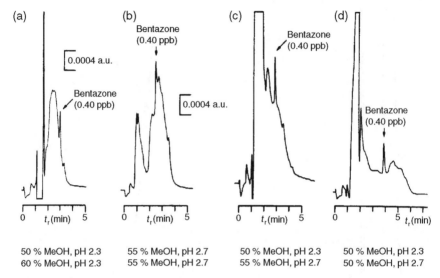

Figure 13.7 Selectivity effected by employing different step gradients in the coupled-column RPLC analysis of a surface water containing 0.40 μg l^{-1} bentazone, by using direct sample injection (2.00 ml). Clean-up volumes, (a), (c) and (d) 4.65 ml of M-1, and (b) 3.75 ml of M-1: transfer volumes, (a), (c) and (d), 0.50 ml of M-1, and (b), 0.40 ml of M-1. The displayed chromatograms start after clean-up on the first column. Reprinted from *Journal of Chromatography*, *A* **644**, E. A. Hogendoorn *et al.*, 'Coupled-column reversed-phase liquid chromatography-UV analyser for the determination of polar pesticides in water', pp. 307–314, copyright 1993, with permission from Elsevier Science.

The experimental conditions are shown in Table 13.1, while Figure 13.9 shows the chromatogram of a soil sample extract spiked with fenpropimorph obtained by this method, plus the chromatogram from the two columns connected in series without column switching.

In multiresidue analysis, where more analytes with a wide polarity range need to be determined, large transfer volumes are required, and consequently, the selectivity is lower. However, since the major interferences in water analysis are the polar humic and fulvic acids, removing this early eluting interference in coupled-column RPLC will also be feasible in multiresidue methodology.

Several examples have been described in the literature, and some of these are included in Table 13.1.

For instance, a group of triazine herbicides has been determined in environmental water samples. In this example (33), a group of four triazines (simazine, atrazine, terbuthylazine and terbutryn) were determined by LC–LC by using the experimental conditions specified in Table 13.1. The flow-rate was 1 ml min^{-1} in both cases. The total analysis time was only 7 min, which enables a sample throughput of up to 60 samples per day. However, the limits of detection from this system were only about 0.1–0.15 μg l^{-1}, which is not enough for drinking water according to EEC regulations. These authors proposed a solid-phase extraction step

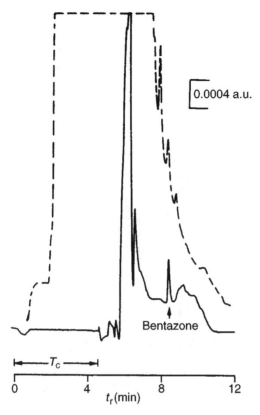

Figure 13.8 RPLC-UV (220 nm) chromatogram of a surface water sample containing 0.40 μg l^{-1} bentazone (injection volume, 2.00 ml): solid line, chromatogram obtained with the coupled-column procedure; dashed line, chromatogram obtained with the same two columns coupled on-line without column switching, using a mobile phase of methanol -0.02 M phosphate buffer (pH 2.7 (50:50, v/v)) at 1 ml min^{-1}; T_c, clean-up time on the first column using the coupled-column procedure. Reprinted from *Journal of Chromatography, A* **644**, E. A. Hogendoorn *et al.*, 'Coupled-column reversed-phase liquid chromatography-UV analyser for the determination of polar pesticides in water', pp. 307–314, copyright 1993, with permission from Elsevier Science.

for determining the pesticides at levels of 0.1 μg l^{-1}, which requires limits of detection of 0.02 μg l^{-1}.

Atrazine is a widely used triazine which can degrade to several products, including deisopropylatrazine (DIA), deethylatrazine (DEA) and hydroxyatrazine (HA). These species are highly polar and their determination by GC requires a derivatization step. LC methods combined with SPE (off-line or on-line) are therefore the ones which are most commonly used. The LC–LC method proposed in the literature (34) can allow low levels to be detected with a small sample volume. The experimental conditions are shown in Table 13.1. Due to the different polarity between the most

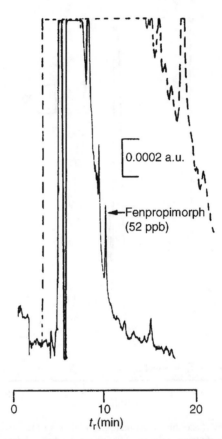

Figure 13.9 Coupled-column RPLC-UV (215 nm) analysis of 100 μl of an extract of a spiked soil sample (fenpropimorph, 0.052 mg Kg^{-1}). LC conditions: C-1, 5 μm Hypersil SAS (60 m × 4.6 mm i.d.); C-2, 5 μm Hypersil ODS (150 m × 4.6 mm i.d.); M-1, acetonitrile-0.5 % ammonia in water (50 : 50, v/v); M-2, acetonitrile – 0.5 % ammonia in water (90:10, v/v); flow-rate, 1 ml min^{-1}; clean-up volume, 5.9 ml; transfer volume, 0.45 ml. The dashed line represents the chromatogram obtained when using the two columns connected in series without column switching. Reprinted from *Journal of Chromatography A*, **703**, E. A. Hogendoorn and P. van Zoonen, 'Coupled-column reversed-phase liquid chromatography in environmental analysis', pp. 149 – 166, copyright 1995, with permission from Elsevier Science.

polar metabolite (DIA) and atrazine, a large transfer volume is required and this leads to a decrease in the selectivity. However, since the more polar compounds are the major source of interference, separation on C-1 allows the large excess of early eluting polar interference to be removed and thus improves the selectivity. The analytes are separated in C-2 by gradient elution because of the difference in polarity. The limits of detection are between 0.2 (atrazine) and 0.5 (HA) μg l^{-1}, but it should be pointed out that only 2 ml of sample are needed. When an off-line C$_{18}$ SPE stage is applied, limits of detection down to 0.02 – 0.1 μg l^{-1} were reached. Figure 13.10

shows the chromatogram of a surface water spiked at 2 μgl^{-1} of atrazine, and its metabolites, obtained under the conditions shown in Table 13.1.

Another interesting group of pesticides contains gluphosinate, glyphosate and aminomethylphosphonic acid (AMPA). Gluphosinate and glyphosate are widely used as non-selective contact herbicides, while aminomethylphosphonic acid is the main metabolite of glyphosate. All are very polar and detection by LC thus requires a derivatization step to enhance their fluorescence since they do not exhibit UV absorption. The method (28) includes precolumn derivatization with a 9-fluorenyl-methylchloroformate (FMOC-Cl) reagent, followed by LC–LC separation. The experimental conditions are shown in Table 13.1.

Since there is a high percentage of acetonitrile in the derivatization solution, the latter must be diluted in order to decrease the retention of the derivative in the first column and so that the derivatives can be separated from the excess reagent. The three compounds can be determined at a level of 1 μg l^{-1}, with the sample through-put being at least 40 samples per day.

When AMPA or gluphosinate are determined alone, the sensitivity is higher because a higher dilution is not required. For glyphosate, when the transfer volume is precisely adjusted to 280 μl for the FMOC-glyphosate-containing fraction, a limit of detection of about 0.2 μg l^{-1} can be reached (28).

Chlorophenoxy acids are relatively polar pesticides which are usually determined by LC because volatile derivatives have to be prepared for GC analysis. This group of herbicides can be detected by multiresidue methods combined with automated procedures for sample clean-up, although selectivity and sensitivity can be enhanced by coupled-column chromatographic techniques (52). The experimental conditions for such analyses are shown in Table 13.1.

The first attempts employing two C_{18} columns showed that the selectivity was not high enough, although this improved when the first column was substituted by a 5 μm GFF II internal surface reversed-phase material. This is known as a restricted-access-material (RAM) column which, since it restricts some compounds because of their size and includes reversed-phase interaction and ionic exchange, is very useful for analysing herbicides in samples with high contents of humic and fulvic acids (54).

Figure 13.11 shows the chromatogram obtained for a surface water sample spiked with various chorophenoxy acids at a level of 0.5 μg l^{-1}, under the same conditions as previously and after enrichment on a C_{18} column and clean-up on silica SPE cartridges.

With dicamba, a more polar chlorobenzoic acid herbicide, a gradient step is needed to elute all of the compounds in one chromatographic run. Depending on the buffer and selectivity of the detector, the baseline can be severely disturbed. If this happens, a step-gradient elution is recommended (52), and in this way the method can detect all of the compounds at very low levels.

Another interesting application of LC–LC is the determination of low-molecular-mass carbonyl compounds in air. Carbonyl compounds, such as aldehydes and ketones, are now being given more and more attention, both as pollutants and as

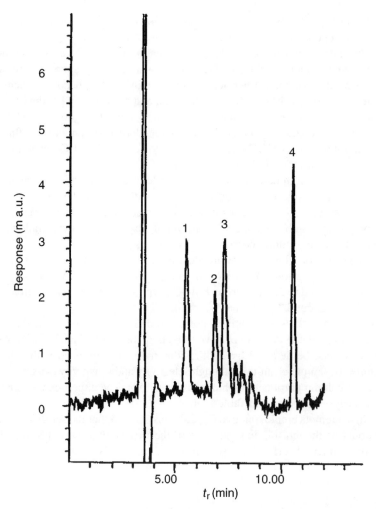

Figure 13.10 LC–LC chromatogram of a surface water sample spiked at 2 μg l⁻¹ with atrazine, and its metabolites (registered at 220 nm). Conditions: volume of sample injected, 2 ml; clean-up time, 2.60 min; transfer time, 4.2 min; The blank was subtracted. Peak identification is as follows: 1, DIA; 2, HA; 3, DEA; 4, atrazine. Reprinted from *Journal of Chromatography, A* **778**, F. Hernández *et al.*, 'New method for the rapid determination of triazine herbicides and some of their main metabolites in water by using coupled-column liquid chromatography and large volume injection', pp. 171–181, copyright 1997, with permission from Elsevier Science.

possible key compounds in photochemical reactions. In this example (53), 13 different aldehydes and ketones were determined in air and water samples by using coupled-column RPLC. These compounds were determined by UV detection of the derivative formed with 2,3-dinitrophenylhydrazine in C₁₈ SPE cartridges. With two

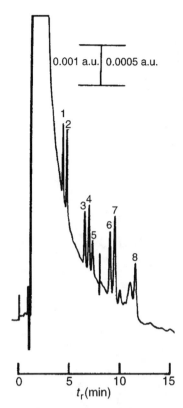

Figure 13.11 Column-switching RPLC trace of a surface water sample spiked with eight chlorophenoxyacid herbicides at the $0.5\,\mu g\ l^{-1}$ level: 1, 2,4-dichlorophenoxyacetic acid; 2, 4-chloro-2-methylphenoxyacetic acid; 3, 2-(2,4-dichlorophenoxy) propanoic acid; 4, 2-(4-chloro-2-methylphenoxy) propanoic acid; 5, 2,4,5-trichlorophenoxyacetic acid; 6, 4-(2,4-dichlorophenoxy) butanoic acid; 7, 4-(4-chloro-2-methylphenoxy) butanoic acid; 8, 2-(2,4,5-trichlorophenoxy) propionic acid. Reprinted from *Analytica Chimica Acta*, **283**, J. V. Sancho-Llopis *et al.*, 'Rapid method for the determination of eight chlorophenoxy acid residues in environmental water samples using off-line solid-phase extraction and on-line selective precolumn switching', pp. 287–296, copyright 1993, with permission from Elsevier Science.

coupled-columns, the sensitivity and selectivity are both higher because interfering peaks do not enter the second column. This is demonstrated in Figure 13.12, which shows the chromatograms obtained both with and without column switching. The experimental conditions are reported in Table 13.1. As can be seen from this figure, there are significant differences in the selectivity and sensitivity for the two chromatograms, which again demonstrates the suitability of RPLC–RPLC methodology.

The coupling of low-resolution liquid chromatography (or SPE) to liquid chromatography has been widely applied to environmental analysis because of the improvement in sensitivity.

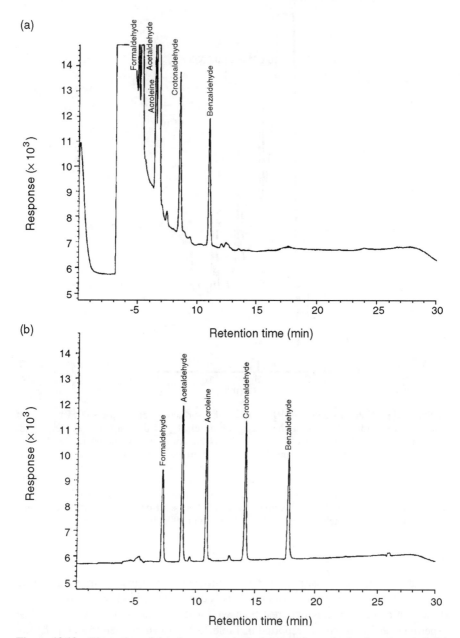

Figure 13.12   Illustration of the clean-up method, showing the analysis of an air sample (a) with and (b) without column switching. Details of the analytical conditions are given in the text. Reprinted from *Journal of Chromatography*, *A* **697**, P. R. Kootstra and H. A. Herbold, 'Automated solid-phase extraction and coupled-column reversed-phase liquid chromatography for the trace-level determination of low-molecular-mass carbonyl compounds in air', pp. 203–211, copyright 1995, with permission from Elsevier Science.

A general step is the coupling of a 10 mm long precolumn (2–3 mm i.d.), filled with C_{18} or polystyrene–divinylbenzene, to a 10–50 μm RPLC column. A PLRP-s precolumn achieved higher breakthrough volume for the more polar compounds and is thus highly recommended. This set-up has been used to determine several groups of pesticides (40, 55–62), phenols (41, 63–65), PAHs (66, 67), naphthalenesulfonates (68, 69), etc. The sample volume depends on the breakthrough volumes of the compounds to be determined, with typical values being between 10 and 200 ml.

For example, Figure 13.13 shows the chromatogram obtained when 200 ml of tap water was spiked at levels of $1\,\mu g\; l^{-1}$ of such pesticides. The limits of detection achieved by using this method were between 0.05 and 0.5 $\mu g\, l^{-1}$, although more polar compounds such as vamidothion or 4-nitrophenol could not be determined (61).

Various highly crosslinked polymers, with slightly different properties, such as Envi-Chrom P, Lichrolut EN, Isolute ENV or HYSphere-1, have been applied in environmental analysis, mainly for polar compounds. For phenol, for instance, which is a polar compound, the recoveries (%) when 100 ml of sample was analysed were 5, 16 and 6 for PLRP-s, Envi-Chrom P and Lichrolut EN, respectively (70).

Chemically modified polymers have been used to determine polar compounds in water samples (37, 71). Chemical modification involves introducing a polar group into polymeric resins. These give higher recoveries than their unmodified analogues for polar analytes. This is due to an increase in surface polarity which enables the aqueous sample to make better contact with the surface of the resin (35).

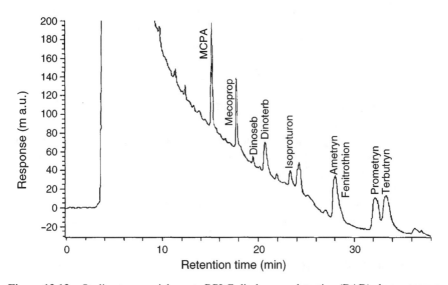

Figure 13.13 On-line trace enrichment–RPLC-diode-array detection (DAD) chromatogram (at 230 nm) obtained from 200 ml of tap water spiked with various pesticides at levels of 1 μg l^{-1}. Reprinted from *Chromatographia*, **43**, C. Aguilar *et al.*, 'Determination of pesticides by on-line trace enrichment–reversed-phase liquid chromatography–diode-array detection and confirmation by particle-beam mass spectrometry', pp. 592–598, 1996, with permission from Vieweg Publishing.

One problem with SPE–LC is the low selectivity of the precolumn. Immunosorbents, which have been developed for a few compounds (72), can considerably improve selectivity. The selectivity was high for different pollutants (73, 74). Figure 13.14 shows that selectivity is higher with an anti-isoproturon cartridge than with a PLRP-S precolumn.

Some groups of pollutants also have specific problems. For instance, PAHs tend to adsorb on the walls of the system with which they come into contact and so an organic solvent or surfactant must be added to the sample. Several solvents have been tested (66, 67): isopropanol or acetonitrile are the most often used solvents, while Brij is the most recommended surfactant (66). A very critical parameter in these cases is their concentration.

Another problem is the determination of ionic substances. Although these can be analysed by an ionic exchange precolumn, there are some significant limitations and, in most cases, common sorbents are preferred. In this case, an ion-pair reagent must be added to the sample to form an ion-pair and so reduce the polarity. This is used, for instance, to determine naphthalenesulfonic acids in environmental samples. Tetrabuthylamonium (TBA) is the most common ion-pair reagent (68), although when LC is coupled to mass spectrometry, it must be replaced by a volatile ion-pair reagent, e.g. triethylamine (69).

One problem with environmental samples that has already been mentioned concerns humic and fulvic acids which may be retained in the precolumn and co-elute with the more polar compounds. Of course, this depends on the selectivity of the sorbent in the precolumn. A simple solution is to add sodium sulphite to the solution prior to preconcentration. This approach has led to good results (37, 71).

Figure 13.15 shows the influence of adding sodium sulphite on the chromatogram of a river water sample.

Another solution to this particular problem is to use a restricted-access-material (RAM) column prior to the precolumn (54).

For the determination of a wide range of neutral, acidic and basic pollutants, two precolumns can be coupled in series (75). A PLRP-S precolum is used to trap the neutral and non-ionized acidic pollutants and a precolumn packed with the same sorbent and loaded with a sodium dodecylsulfate (SDS) (to form the ion-pair) is used to trap the positively charged (basic) pollutants. Each precolumn is coupled to one analytical column.

Sensitivity is usually higher with SPE–LC than with LC–LC, although both selectivity and sample throughput are lower (76).

13.4 LIQUID CHROMATOGRAPHY–GAS CHROMATOGRAPHY

13.4.1 INTRODUCTION

The main drawback of GC is sample introduction and this is especially important when analytes are to be determined at trace levels. Today, however, there is no problem with introducing 10–100 μl of organic solvents such as ethyl acetate or alkanes

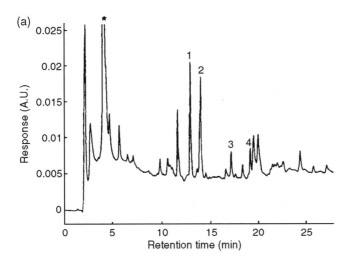

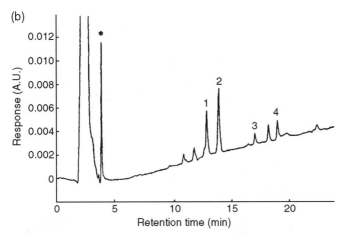

Figure 13.14 LC-diode-array detection (DAD) chromatogram (at 220 nm) obtained after preconcentration of 50 ml of ground water sample spiked with various pollutants at levels of 3 μg l^{-1} passed through (a) a PLRP-S cartridge and (b) an anti-isoproturon cartridge. Peak identification is as follows: 1, chlortoluron; 2, isoproturon plus diuron; 3, linuron; 4, dibenzuron; *, water matrix. Reprinted from *Journal of Chromatography, A* **777**, I. Ferrer *et al.* 'Automated sample preparation with extraction columns by means of anti-isoproturon immunosorbents for the determination of phenylurea herbicides in water followed by liquid chromatography diode array detection and liquid chromatography–atmospheric pressure chemical ionization mass spectrometry', pp. 91–98, copyright 1997, with permission from Elsevier Science.

if a retention gap is used. The techniques developed for this can also be used to transfer relevant fractions from an LC–column and this has been extremely important in developing LC–GC methods for environmental trace analysis.

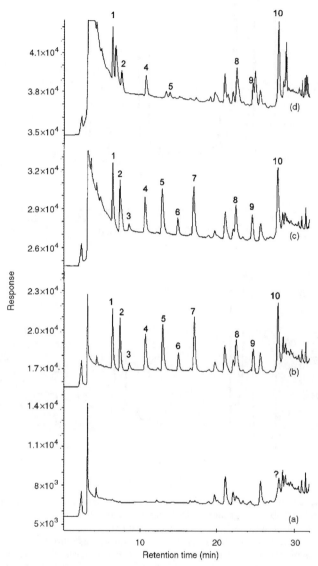

Figure 13.15 Chromatograms obtained by on-line trace enrichment of 50 ml of Ebro river water with and without the addition of different volumes of 10% Na$_2$SO$_3$ solution for every 100 ml of sample: (a) blank with the addition of 1000 µl of sulfite; (b) spiked with 4 µg l^{-1} of the analytes and 1000 µl of sulfite; (c) spiked with 4 µg l^{-1} of the analytes and 500 µl of sulfite; (d) spiked with 4 µg l^{-1} of the analytes without sulfite. Peak identification is as follows: 1, oxamyl; 2, methomyl; 3, phenol; 4, 4-nitrophenol; 5, 2,4-dinitrophenol; 6, 2-chlorophenol; 7, bentazone; 8, simazine; 9, MCPA; 10, atrazine. Reprinted from *Journal of Chromatography, A* **803**, N. Masqué *et al.*, 'New chemically modified polymeric resin for solid-phase extraction of pesticides and phenolic compounds from water', pp. 147–155, copyright 1998, with permission from Elsevier Science.

LC–GC is a very powerful analytical technique because of its selectivity and sensitivity in analysing complex mixtures and therefore it has been used extensively to determine trace components in environmental samples (2, 5, 77). LC allows preseparation and concentration of the components into compound types, with GC being used to analyse the fractions. The advantages of on-line LC–GC over the off-line system are, first, the less sample which is required and, secondly, that there is less need for laborious sample pretreatment because the method is automated (78).

Important developments in LC–GC have been made by Grob and co-workers (79–81) and by the Brinkman group (82–87), who have mainly studied the application of this technique to environmental analysis. This coupled technique has usually been applied to water, although air and soil extracts have also been analysed.

On-line coupling of normal-phase liquid chromatography (NPLC) and gas chromatography is today a well developed and robust procedure and has been regularly applied to environmental analysis. When a fraction of the NPLC sample is introduced in to the GC unit, a large-volume interface (LVI) is needed but, due to the volatility of the organic solvent used in NPLC, this does not present such a great problem.

However, for water analysis, reverse-phase liquid chromatography is more suitable but its coupling with GC has some drawbacks because of the partly aqueous effluent. Several systems have been developed (88, 89) and applied to determine pollutants in water.

An alternative way of eliminating water is to pass the aqueous sample through a small LC column, also called a precolumn or an SPE column, where the analytes are retained and the water is mostly eliminated. Subsequent elution of these analytes with an organic solvent (usually ethyl or methyl acetate) and more compatible handling with GC systems enables the sample to be significantly concentrated and low levels of analytes to be determined. However, an additional step is needed prior to the elution of the retained analytes in order to eliminate the small amounts of water in the LC column. This is usually carried out by drying the LC column with nitrogen (89) or by adding a small drying column containing sodium sulfate or silica (89, 90). The former approach is normally used and we can find several applications to environmental analysis in the literature (91–93).

These small columns,(usually 10 mm × 1–4.6 mm i.d.) are normally packed with 10–40 μm sorbents such as C_{18}-bonded silica, C_8-bonded silica or styrene–divinylbenzene copolymer. These sorbents are not very selective and more selective sorbents, such as the immunosorbent (94), have also been used with good results. Coupling of SPE–gas chromatography is in fact the one most often used in environmental analysis because it reaches a high level of trace enrichment, eliminates water and elutes retained compounds easily with an organic solvent that can be injected into the gas chromatograph.

In order to achieve the widest application range, partially concurrent solvent evaporation (PCSE) with an on-column interface is normally used during the transfer of analytes from the LC-type precolumn to the GC system. Fully concurrent solvent evaporation (FSCE), with a loop-type interface, is used in some cases, although the more volatile analytes can be lost. A co-solvent (an organic solvent with a higher boiling point than the transfer solvent) can be added before the transfer step to minimize this effect (95).

A programmed-temperature vaporizer (PTV) has also been used as an interface for introducing the LC fraction to the GC unit (84,96) and to desorb the analytes retained in the SPE sorbent contained in the PTV liner. Water samples can then be injected directly in to the PTV injector.

13.4.2 EXAMPLES OF LIQUID CHROMATOGRAPHY–GAS CHROMATOGRAPHY APPLIED TO ENVIRONMENTAL ANALYSIS

One example of normal-phase liquid chromatography coupled to gas chromatography is the determination of alkylated, oxygenated and nitrated polycyclic aromatic compounds (PACs) in urban air particulate extracts (97). Since such extracts are very complex, LC–GC is the best possible separation technique. A quartz microfibre filter retains the particulate material and supercritical fluid extraction (SFE) with CO_2 and a toluene modifier extracts the organic components from the dust particles. The final extract is then dissolved in n-hexane and analysed by NPLC. The transfer at 100 μl min^{-1} of different fractions to the GC system by an on-column interface enabled many PACs to be detected by an ion-trap detector. A flame ionization detector (FID) and a 350 μl loop interface was used to quantify the identified compounds. The experimental conditions employed are shown in Table 13.2.

Figure 13.16 shows the LC separation of this extract and the GC/FID chromatogram of the oxy-PAC fraction. The different oxy-PACs can be quantified without interfering compounds with a non-selective detector such as an FID.

One of the first examples of the application of reverse-phase liquid chromatography–gas chromatography for this type of analysis was applied to atrazine (98). This method used a loop-type interface. The mobile phase was the most important parameter because retention in the LC column must be sufficient (there must be a high percentage of water), although a low percentage of water is only possible when the loop-type interface is used to transfer the LC fraction. The authors solved this problem by using methanol/water (60:40) with 5% 1-propanol and a precolumn. The experimental conditions employed are shown in Table 13.2.

An alternative way of eliminating water in the RPLC eluent is to introduce an SPE trapping column after the LC column (88, 99). After a post-column addition of water (to prevent breakthrough of the less retained compounds), the fraction that elutes from the RPLC column is trapped on to a short-column which is usually packed with polymeric sorbent. This system can use mobile phases containing salts, buffers or ion-pair reagents which can not be introduced directly into the GC unit. This system has been successfully applied, for example, to the analysis of polycyclic aromatic hydrocarbons (PAHs) in water samples (99).

Another interface for RPLC–GC is the programmed-temperature-vaporization (PTV) system, an interesting application of which is the determination of phthalates

Table 13.2 Examples of the application LC–GC in environmental analysis

Analyte	Sample	Sample volume (ml)	LC column (mm × mm)	Interface/ Transfer volume/solvent	GC system (m × mm)	Detection [a]	LOD [b] (ng l^{-1})	Reference
Atrazine	Water	10	5 µm Spherisorb ODS (100 × 2)	Loop type 150 µl methanol/water (60 : 40) with 5% 1-propanol	RG: Deactivated (2 × 0.53) PC: Carbowax 20M 0.4 µm (3 × 0.32) AC: 0.4 µm Carbowax 20M (40 × 0.32)	NPD	3	98
Polycyclic aromatic hydrocarbons	Water	0.098	5 µm Krosil C$_{18}$ (100 × 3.1) SPE: 10 µm PLRPS (10 × 2)	On-column 75 µl ethyl acetate	RG: Deactivated (5 × 0.53) AC: 0.17 µm HP1 (25 × 0.32)	FID	n.s.	99
Organo-phosphorus pesticides	Water	2.5	Thick extraction (15 × 4.2) discs 90% C$_{18}$, 10% XAD	On-column 70 µl ethyl acetate	RG: Deactivated (5 × 0.32) PC: 0.5 µm DB-5 (3 × 0.32) AC: 0.5 µm DB-5 (25 × 0.32)	NPD	10–100	91
Organo-phosphorus and organo-sulfur pesticides	Water	10	10 µm PLRP-S (10 × 2) Drying cartridge	On-column 100 µl ethyl acetate	RG: Deactivated (5 × 0.32) AC: 0.14 µm DB-1 (15 × 0.32)	FID NPD FPD	100 0.4–6.0 1–10	90
Organo-phosphorus pesticides	Water	2	15–25 µm PLRP-S (10 × 2)	Loop type 500 µl methyl-t-butyl ether with 10% ethyl acetate; co-solvent; 100 µl n-decane	RG: Deactivated (10 × 0.53) AC: 0.2 µm CP Sil–19 CB (50 × 0.32)	NPD	10	95
Alkylated, oxygenated and nitrated polycyclic aromatic compounds	Urban air particulate extracts	20 µl of hexane extract from SFE	5 µm Silica (100 × 2)	On-column loop type 350 µl pentane/ dichloromethane (50-50)	RG: Deactivated (10 × 0.53) PC: not given AC: 0.5 µm DMS (25 × 0.32)	MS FID	n.q. n.s.	97

Table 13.2 (*continued*)

Analyte	Sample	Sample volume (ml)	LC column (mm × mm)	Interface/Transfer volume/solvent	GC system (m × mm)	Detection[a]	LOD[b] (ng l^{-1})	Reference
Pesticides	Water	10	15–25 µm PLRPS (10 × 2)	Loop type 100 µl ethyl acetate; co-solvent, 50 µl ethyl acetate	RG: Deactivated (5 × 0.53) PC: 0.25 µm HP-5MS (1 × 0.25) AC: 0.25 µm HP-5MS (30 × 0.25)	MS/MS	0.01–0.50	100
Phthalates	Water	10	5 µm C$_{18}$ (10 × 3)	PTV 2cm bed Carbofrit 720 µl methanol/water (85 : 15)	PC: 1 µm PS-255 (1 × 0.53) AC: 0.5 µm PS-255 (20 × 0.25)	MS	5–10	96
Pesticides	Water	10	20 µm PLRP-S (10 × 2)	On-column 100 µl ethyl acetate	RG: Deactivated (5 × 0.53) PC: 0.25 µm HP-5MS (2 × 0.25) AC: 0.25 µm HP5MS (3 × 0.25)	MS	2–20	92
Alkyl-, chloro-, and mononitro-phenols	Water	10	15–25 µm PLRP-S (10 × 2)	Loop type 100 µl ethyl acetate	RG: Deactivated (5 × 0.53) PC: 0.52 µm HPUltra (3 × 0.32) AC: 0.52 µm HPUltra 2 (25 × 0.32)	MS	1–25	101
s-triazines	Water	10	Immunoaffinity (10 × 3) sorbent and 20 µm PLRP-S (10 × 2)	On-column 100 µl ethyl acetate	RG: Deactivated (0.2 × 0.1) PC: 0.25 µm HP-5MS (1.5 × 0.25) AC: 0.25 µm HP-5MS (25 × 0.25)	FID NPD	15 1.5	94

[a] NPD, nitrogen–phosphorus detector; FID, flame-ionization detector; MS, mass spectrometer.
[b] n.q., not quantified; n.s., not specified.

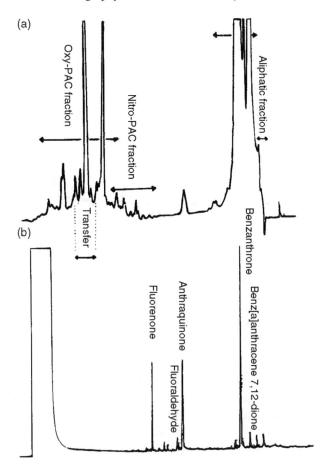

Figure 13.16 LC separation of urban air particulate extract (a), along with the GC/FID chromatogram (b) of an oxy-PAC fraction (transferred via a loop-type interface). Reprinted from *Environmental Science and Technology*, **29**, A. C. Lewis *et al.*, 'On-line coupled LC–GC–ITD/MS for the identification of alkylated, oxygenated and nitrated polycyclic aromatic compounds in urban air particulate extracts', pp. 1977–1981, copyright 1995, with permission from the American Chemical Society.

in water by use of the vaporizer/precolumn solvent split/gas discharge interface (96). The fraction from the LC column, eluted by the mobile phase consisting of water/methanol (15 : 85, v/v), is directly inserted into the PTV interface. The eluent is vaporized in a chamber kept at a high temperature which depends on the transfer flow. The transfer line contained packing material, i.e. Carbofrit, for smooth evaporation to occur. This was situated at the top of the properly heated area in order to prevent the transfer line from entering too far into the hot area. The solvent vapours were largely discharged through an early vapour exit, driven by the flow of carrier gas. Solvent/solute separation occurred in the precolumn. An uncoated retention gap was not necessary since water or water/methanol mixtures do not wet precolumn

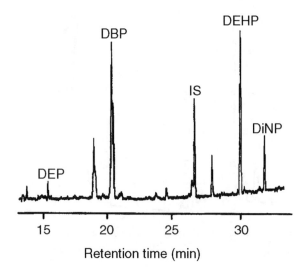

Figure 13.17 LC–GC–MS(EI) chromatogram of a treated drinking water containing 55 and 40 ng l^{-1}, respectively, of DBP and DEHP. Reprinted from *Journal of High Resolution Chromatography*, **20**, T. Hyötyläinen *et al.*, 'Reversed phase HPLC coupled on-line to GC by the vaporizer/precolumn solvent split/gas discharge interface; analysis of phthalates in water', pp. 410–416, 1997, with permission from Wiley-VCH.

surfaces and solvent trapping can not be used to improve the retention of volatile solutes. The experimental conditions for this analysis are shown in Table 13.2. This study has demonstrated that column temperature during transfer is a very important parameter which depends to a large extent on the volatility of the compounds and the composition of the mobile phase. However, problems with volatile solutes can still be significant.

For the LC separation, 10 ml of sample was injected through a loop. The LC flow-rate was 1000 μl min^{-1} and at the end of the enrichment process this was reduced to 100 μl min^{-1}. The mobile phase composition was optimized to eliminate matrix polar compounds and elute the phthalates in a small fraction.

Figure 13.17 shows the LC–GC–MS chromatogram of treated drinking water containing 55 and 40 ng l^{-1} of di-n-butylpthalate (DBP) and di(2-ethylhexyl)phthalate (DEHP), respectively. The PTV system therefore allows the LC eluent to be injected directly and no change in the composition is needed.

However, the most frequently used system is trace enrichment in a short LC column or SPE precolumn. The column is usually filled with C$_{18}$ or PLRP-S and dried with nitrogen prior to elution. The analytes are eluted with an organic solvent, usually ethyl acetate, which is injected into the gas chromatograph through an on-column interface by a retention gap. In the chromatograph there is also a retaining precolumn, to minimize losses of the most volatile compounds, an analytical column and, between the precolumn and the analytical column, a solvent vapour exit (SVE) to eliminate the vapour (Figure 13.18).

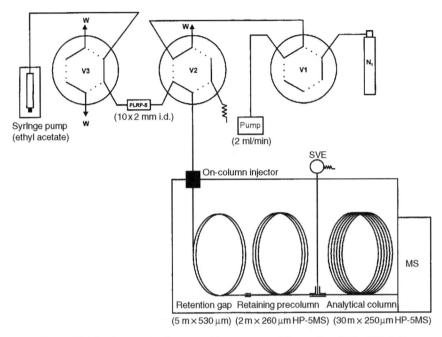

Figure 13.18 Schematic diagram of the set-up used for on-line SPE–GC–MS.

This set-up, or a very similar one, has been used to determine different group of pollutants in environmental waters (45, 83, 93). For example, with 10 ml of sample the limits of detection of a group of pesticides were between 2 and 20 ng l^{-1} (92) in tap and river water, with this system being fully automated. Figure 13.19 shows the chromatograms obtained by on-line SPE–GC–MS under selected ion-monitoring conditions of 10 ml of tap water spiked with pesticides at levels of 0.1 μg l^{-1} (92).

One disadvantage of this system in routine analysis is the long time required for the drying step. An alternative system has therefore been described (90, 102) and used to analyse water samples. This includes a drying precolumn after the SPE precolumn and before the entrance to the GC unit. Silica, copper sulfate and sodium sulfate cartridges were tested, with the results depending on the types of pesticides studied. One disadvantage of this system is the need to dry the precolumn after each sample or after a few samples. However, with an additional switching valve the precolumn can be regenerated by simultaneous heating and purging with a moisture-free gas during the GC run and this increases the sample throughput considerably. This system has been successfully applied to determine different types of pollutants (83).

The low selectivity of the SPE columns currently in use can be increased with more selective sorbents such as the immunosorbents, which have been quite extensively used in SPE–LC (72). Immunoaffinity-based solid-phase extraction (IASPE) sorbents have also been used in coupled gas chromatography for determining

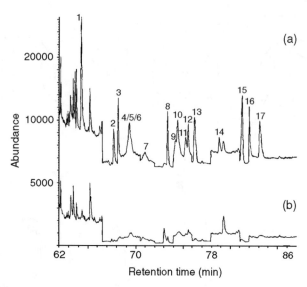

Figure 13.19 Chromatograms obtained by on-line SPE–GC–MS(SIM) of: (a) 10 ml of tap water spiked with pesticides at levels of 0.1 ng l^{-1}; (b) 10 ml of a sample of unspiked tap water. Peak identification for (a) is as follows: 1, molinate; 2, α-HCH; 3, dimethoate; 4, simazine; 5, atrazine; 6, γ-HCH; 7, δ-HCH; 8, heptachlor; 9, ametryn; 10, prometryn; 11, fenitrothion; 12, aldrin; 13, malathion; 14, endo-heptachlor; 15, α-endosulfan; 16, tetrachlorvinphos; 17, dieldrin. Reprinted from *Journal of Chromatography, A* **818**, E. Pocurull *et al.*, 'On-line coupling of solid-phase extraction to gas chromatography with mass spectrometric detection to determine pesticides in water', pp. 85–93, copyright 1998, with permission from Elsevier Science.

micropollutants (94, 103) although such coupling is rather difficult to achieve. In order to remove these difficulties, the cartridge that contained the immobilized antibodies was coupled to the GC unit via a reverse-phase cartridge (PLRP-S) in the determination of *s*-triazines (94). After enrichment of the analytes on the immunoaffinity cartridge, they were desorbed and recollected on the PLRP-S cartridge by using an acidic buffer. The PLRP-S cartridge was then cleaned and dried with nitrogen. Finally, the desorption and transfer steps were carried out with ethyl acetate via an on-column interface.

The high selectivity of this system is demonstrated in Figure 13.20, which shows that a non-selective FID could be used to detect triazines in complex matrices, and that with 10 ml of sample the detection limits were 15–25 ng l^{-1}. The experimental conditions are shown in Table 13.2.

The on-column interface is the one which is most often used in LC–GC of aqueous samples because it can be applied to a wider range of compounds. The loop-type interface is limited for determining volatile compounds that are volatilized together with the solvent and not retained in the retention gap. Several attempts at solving this problem have been made. One option is to add a co-solvent which enters the retention gap before the analytes and thus forms a co-solvent film in front of the eluate.

This is a highly efficient barrier against evaporative losses of volatile compounds, which also improves the peak width of the early eluting compounds. This system has been successfully applied to a group of pesticides, using *n*-decane as the co-solvent and has enabled a group of volatile phosphorus pesticides to be determined (95). The experimental conditions used in this work are shown in Table 13.2.

Another way of introducing water directly into a gas chromatography system, and avoiding the need for a retention gap, is to use thermal desorption instead of solvent desorption, namely on-line solid-phase extraction–thermal desorption (SPETD) (104). This system has been used in environmental analysis with good results (105,106). A PTV injector is filled with an SPE sorbent in which the analytes in water are retained. The choice of sorbent is very important and several different ones have been tested (107). For instance, when injecting 500 μl of water sample into a PTV liner filled with Tenax TA, the detection limits are between 0.01 μg l^{-1} (for dieldrin, using an ECD) and 0.5 μg l^{-1} (for aldimorph, using an NPD) (106). As no sample preparation steps are required with this system, the chances of contamination are reduced and substances with a high water solubility can be enriched because such enrichment takes place out of the gas phase. The major drawback with this system is the long injection time because the injection rate must be below the evaporation rate. An improved set-up has been used to analyse surface water and tap water samples (108). The void volume of the injector was reduced and a make-up gas was added in the backflush mode in order to prevent water from reaching the analytical

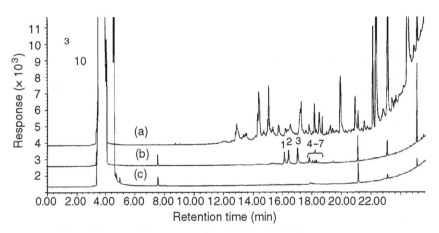

Figure 13.20   GC-FID chromatograms of an extract obtained by (a) SPE and, (b) IASPE of 10 ml of municipal waste water, spiked with 1 μg l^{-1} of seven s-triazines; (c) represents a 'blank' run from IASPE–GC-NPD of 10 ml of HPLC water. Peak identification is as follows: 1, atrazine; 2, terbuthylazine; 3, sebuthylazine; 4, simetryn; 5, prometryn; 6, terbutryn; 7, dipropetryn. Reprinted from *Journal of Chromatography, A* **830**, J. Dallüge *et al.*, 'On-line coupling of immunoaffinity-based solid-phase extraction and gas chromatography for the determination of *s*-triazines in aqueous samples', pp. 377–386, copyright 1999, with permission from Elsevier Science.

GC column. The system enables volumes of up to 1.0 ml to be enriched in the packed liner in the PTV injector without breakthrough. By injecting 100 μl, the detection limits for alachlor and metolachlor in SPETD–GC–MS/MS were $0.1-0.2 \, \mu g \, l^{-1}$.

13.5 CONCLUSIONS AND TRENDS

Multidimensional chromatography is a very powerful technique which can help solve complex problems in environmental analysis. Since it requires more complex instrumentation, it has not been widely used in routine analysis, although some of the coupled techniques may become important in control laboratories in the future.

GC–GC in environmental analysis will be limited to solving specific problems that can not easily be solved by other techniques. Coupling LC to LC or GC will therefore become very common, not only because of the significantly greater sensitivity, but also because of the greater selectivity and high powers of automation which implies fewer sources for possible errors and a higher throughput, both of which are very important in routine analysis.

When these techniques are coupled to selective detection techniques, such as the increasingly used mass spectrometry, very powerful techniques for determining pollutants in environmental samples are achieved.

When such systems are commercialized, their use will become more widespread. Most laboratories have so far only used systems which they have designed themselves.

ACKNOWLEDGEMENTS

The author would like to acknowledge the contributions made to this work by Dr F. Borrull, Dr E. Pocurull and S. Peñalver.

REFERENCES

1. P. Van Zoonen, E. A. Hogendoorn, G. R. Van der Hoof and R. A. Baumann, 'Selectivity and sensitivity in coupled chromatographic techniques as applied in pesticide residue analysis', *Trends. Anal. Chem.* **11**: 11–17 (1992).
2. P. Van Zoonen, 'Coupled chromatography in pesticide residue analysis', *Sci. Total Environ.* **132**: 105–114 (1993).
3. P. De Voogt, 'Chromatographic clean-up methods for the determination of persistent organic compounds in aqueous environmental samples', *Trends. Anal. Chem.* **13**: 389–397 (1994).
4. S. G. Dai and C. R. Jia, 'Application of hyphenated techniques in environmental analysis', *Anal. Sci.* **12**: 355–361 (1996).
5. U. A. Th Brinkman, 'Multidimensional approaches in environmental analysis', *Anal. Commun.* **34**: 9H–12H (1997).

6. U. A. Th Brinkman, 'Multidimensional gas and liquid chromatographic approaches to trace-level environmental analysis', *Chromatographia* **45**: 445–449 (1997).

7. Z. Liu, I. Ostrovsky, P. B. Farnsworth and M. L. Lee, 'Instrumentation for comprehensive two-dimensional capillary supercritical fluid-gas chromatography', *Chromatographia* **35**: 567–573 (1993).

8. E. Pocurull, R. M. Marcé, F. Borrull, J. L. Bernal, L. Toribio and M. L. Serna, 'On-line solid-phase extraction coupled to supercritical fluid chromatography to determine phenol and nitrophenols in water', *J. Chromatogr.* **755**: 67–74 (1996).

9. J. L. Bernal, M. J. Nozal, L. Toribio, M. L. Serna, F. Borrull, R. M. Marcé and E. Pocurull, 'Determination of polycyclic aromatic hydrocarbons in waters by use of supercritical fluid chromatography coupled on-line to solid-phase extraction with disks', *J. Chromatogr.* **778**: 321–328 (1997).

10. G. Schomburg, 'Two-dimensional gas chromatography: principles, instrumentation, methods', *J. Chromatogr.* **703**: 309–325 (1995).

11. J. de Boer, Q. T. Dao, P. G. Wester, S. Bowadt and U. A. Th Brinkman, 'Determination of mono-ortho substituted chlorobiphenyls by multidimensional gas chromatography and their contribution to TCDD equivalents', *Anal. Chim. Acta* **300**: 155–165 (1995).

12. J. C. Duinker, D. E. Schult and G. Petrick, 'Multidimensional gas chromatography with electron capture detection for the determination of toxic congeners in polychlorinated biphenyl mixture', *Anal. Chem.* **60**: 478–482 (1998).

13. Y. V. Gankin, A. E. Gorshteyn and A. Robbat-Jr, 'Identification of PCB congeners by gas chromatography electron capture detection employing a quantitative structure-retention model', *Anal. Chem.* **67**: 2548–2555 (1995).

14. A. Glausch, G. P. Blanch and V. Schurig, 'Enantioselective analysis of chiral polychlorinated biphenyls in sediment samples by multidimensional gas chromatography-electron-capture detection after steam distillation-solvent extraction and sulfur removal', *J. Chromatogr.* **723**: 399–404 (1996).

15. D. E. Wells, '*Environmental analysis. Techniques*, Applications and Quality Assurance. Barceló D, (Eds.), Elsevier Amsterdam, pp. 80–109 (1993).

16. A. C. Lewis, K. D. Bartle and L. Rattner, 'High-speed isothermal analysis of atmospheric isoprene and DMS using online two-dimensional gas chromatography. *Env. Sci. Technol.* **31**: 3209–3217 (1997).

17. W. Schrimpf, K. P. Mueller, F. J. Johnen, K. Lienaerts and J. Rudolph, 'An optimized method for airborne peroxyacetylnitrate (PAN) measurements. *J. Atmos. Chem.* **22**: 303–317 (1995).

18. D. Helmig, 'Air analysis by gas chromatography', *J. Chromatogr.* **843**: 129–146 (1999).

19. Z. Liu, S. R. Sirimanne, D. G. Patterson, L. L. Needham and J. B. Phillips, 'Comprehensive two-dimensional gas chromatography for the fast separation and determination of pesticides from human serum', *Anal. Chem.* **66**: 3086–3092. (1994)

20. K. A. Krock and C. L. Wilkins, 'Qualitative analysis of contaminated environmental extracts by multidimensional gas chromatography with infrared and mass spectral detection (MDGC-IR-MS)', *J. Chromatogr.* **726**: 167–178 (1996).

21. J. B. Phillips and J. Xu, 'Comprehensive multi-dimensional gas chromatography', *J. Chromatogr.* **703**: 327–334 (1995).

22. R. M. Kinghorn, P. J. Marriott and M. Cumbers, 'Multidimensional capillary gas chromatography of polychlorinated biphenyl marker compounds', *J. High Resolut. Chromatogr.* **19**: 622–626 (1996).

23. G. Schomburg, H. Husmann and E. Hübinger, 'Multidimensional separation of isomeric species of chlorinated hydrocarbons such as PCB, PCDD and PCDF', *J. High Resolut. Chromatogr.* **8**: 395–400. (1985)

24. N. Kannan, D. E. Schulz-Bull, G. Petrick and J. C. Duinker, 'High-resolution PCB analysis of Kanechlor, Phenoclor and Sovol mixtures using multidimensional gas chromatography', *Int. J. Environ. Anal. Chem.* **47**: 201–215 (1992).

25. A. Glausch, G. J. Nicholson, M. Fluck and V. Schurig, 'Separation of enantiomers of stable atropisomeric polychlorinated biphenyls (PCBs), by multidimensional gas chromatography on Chirasil-Dex', *J. High Resolut. Chromatogr.* **17**: 347–349 (1994).

26. D. E. Schulz, G. Petrick and J. C. Duinker, 'Complete characterization of polychlorinated biphenyl congeners in commercial Aroclor and Clophen mixtures by multidimensional gas chromatography-electron capture detection. *Environ. Sci. Technol.* **23**: 852–859 (1989).

27. D. Duebelbeis, S. Kapila, T. E. Clevenger and A. F. Yanders, 'Application of a two-dimensional reaction chromatography system for confirmatory analysis of Chlordane constituents in environmental samples. *Chemosphere* **20**: 1401–1408 (1990).

28. J. V. Sancho, F. Hernández, F. J. López, E. A. Hogendoorn, E. Dijkman and P. van Zoonen, 'Rapid determination of glufosinate, glyphosate and aminomethylphosphonic acid in environmental water samples using precolumn fluorogenic labelling and coupled-column liquid chromatography', *J. Chromatogr.* **737**: 75–83 (1996).

29. Y. Zebuehr, C. Naef, D. Broman, K. Lexen, A. Colmsjo and C. Oestman, 'Sampling techniques and cleanup procedures for some complex environmental samples with respect to PCDDs and PCDFs and other organic contaminants', *Chemosphere* **19**: 39–44 (1989).

30. E. A. Hogendoorn and P. van Zoonen, *'Environmental Analysis. Techniques, Applications and Quality Assurance,* Barceló D. (Ed.) Elsevier, Amsterdam, pp. 181–224 (1993).

31. E. A. Hogendoorn and P. van Zoonen, 'Coupled-column reversed-phase liquid chromatography in environmental analysis', *J. Chromatogr.* **703**: 149–166 (1995).

32. J. V. Sancho, C. Hidalgo and F. Hernández, 'Direct determination of bromacil and diuron residues in environmental water samples by coupled-column liquid chromatography and large-volume injection', *J. Chromatogr.* **761**: 322–326 (1997).

33. C. Hidalgo, J. V. Sancho and F. Hernández, 'Trace determination of triazine herbicides by means of coupled liquid chromatography and large volume injection', *Anal. Chim. Acta* **338**: 223–29 (1993).

34. F. Hernández, C. Hidalgo, J. V. Sancho and F. J. López, 'New method for the rapid determination of triazine herbicides and some of their main metabolites in water by using coupled-column liquid chromatography and large volume injection', *J. Chromatogr.* **778**: 171–181 (1997).

35. N. Masqué R. M. Marcé and F. Borrull, 'New polymeric and other types of sorbents for solid-phase extraction of polar organic micropollutants from environmental water', *Trends. Anal. Chem.* **17**: 384–394 (1998).

36. I. Ferrer and D. Barceló, 'Validation of new solid-phase extraction materials for the selective enrichment of organic contaminants from environmental samples', *Trends. Anal. Chem.* **18**: 180–192 (1999).

37. N. Masqué, M. Galià, R. M. Marcé and F. Borrull, 'New chemically modified polymeric resin for solid-phase extraction of pesticides and phenolic compounds from water', *J. Chromatogr.* **803**: 147–155 (1998).

38. E. R. Brouwer, H. Lingeman and U. A. Th Brinkman, 'Use of membrane extraction disks for on-line trace enrichment of organic compounds from aqueous samples', *Chromatographia* **29**: 415–418 (1990).

39. C. Aguilar, F. Borrull and R. M. Marcé, 'On-line and off-line solid-phase extraction with styrene-divinylbenzene-membrane extraction disks for determining pesticides in

water by reversed-phase liquid chromatography-diode-array detection', *J. Chromatogr.* **754**: 77–84 (1996).

40. E. Pocurull, R. M. Marcé and F. Borrull, 'Improvement of on-line solid-phase extraction for determining phenolic compounds in water', *Chromatographia* 41: 521–526 (1995).

41. N. Masqué, R. M. Marcé and F. Borrull, 'Comparison of different sorbents for on-line solid-phase extraction of pesticides and phenolic compounds from natural water followed by liquid chromatography', *J. Chromatogr.* **793**: 257–263 (1998).

42. E. M. Thurman and M. S. Mills, '*Solid-Phase Extraction: Principles and Practice*, John Wiley & Sons, New York (1998).

43. J. R. Dean, '*Extraction Methods for Environmental Analysis*, John Wiley & Sons Chichester (1998).

44. D. Barceló and M. C. Hennion (Eds), '*Trace Determination of Pesticides and their Degradation Products in Water* (Techniques and Instrumentation in Analytical Chemistry, Vol 19), Elsevier Oxford UK (1997).

45. J. Slobodnik, A. J. H. Louter, J. J. Vreuls, I. Liska and U. A. Th Brinkman, 'Monitoring of organic micropollutants in surface water by automated on-line trace-enrichment liquid and gas chromatographic systems with ultraviolet diode-array and mass spectrometric detection', *J. Chromatogr.* **768**: 239–258 (1997).

46. E. A. Hogendoorn, U. A. Brinkman, Th and P. van Zoonen, 'Coupled-column reversed-phase liquid-chromatography-UV analyser for the determination of polar pesticides in water', *J. Chromatogr.* **644**: 307–314 (1993).

47. E. A. Hogendoorn, R. Hoogerbrugge, R. A. Baumann, H. D. Meiring, A. P. J. M. de Jong and P. van Zoonen, 'Screening and analysis of polar pesticides in environmental monitoring programmes by coupled-column liquid chromatography and gas chromatography-mass spectrometry', *J. Chromatogr.* **754**: 49–60 (1996).

48. E. A. Hogendoorn, P. van Zoonen and U. A. Th Brinkman, 'Column-switching RPLC for the trace-level determination of ethylenethiourea in aqueous samples', *Chromatographia* **31**: 285–292 (1991).

49. E. A. Hogendoorn, C. E. Goewie and P. van Zoonen, 'Application of HPLC column switching in pesticide residue analysis, Fresenius', *J. Anal. Chem.* **339**: 348–356 (1991).

50. J. V. Sancho, C. Hidalgo, F. Hernández, F. J. López, E. A. Hogendoorn and E. Dijkman, 'Rapid determination of glyphosate residues and its main metabolite aminomethylphosphonic acid (AMPA)', in soil samples by liquid chromatography, *Int. J. Environ. Anal. Chem.* **62**: 53–63 (1996).

51. E. A. Hogendoorn, C. Verschraagen, U. A. Th Brinkman and van P. Zoonen 'Coupled column liquid chromatography for the trace determination of polar pesticides in water using direct large-volume injection: method development strategy applied to methyl isothiocyanate', *Anal. Chim. Acta* **268**: 205–215 (1992).

52. J. V. Sancho-Llopis, F. Hernández-Hernández, E. A. Hogendoorn and P. van Zoonen, 'Rapid method for the determination of eight chlorophenoxy acid residues in environmental water samples using off-line solid-phase extraction and on-line selective pre-column switching', *Anal. Chim. Acta* **283**: 287–296(1993).

53. P. R. Kootstra and H. A. Herbold, 'Automated solid-phase extraction and coupled-column reversed-phase liquid chromatography for the trace-level determination of low-molecular-mass carbonyl compounds in air', *J. Chromatogr.* **697**: 203–211 (1995).

54. E. A. Hogendoorn, E. Dijkman, B. Baumann, C. Hidalgo, J. V. Sancho and F. Hernández, 'Strategies in using analytical restricted access media columns for the removal of humic acid interferences in the trace analysis of acidic herbicides in water

samples by coupled column liquid chromatography with UV detection', *Anal. Chem.* **71**: 1111–1118 (1999).

55. C. Aguilar, I. Ferrer, F. Borrull, R. M. Marcé and D. Barceló, 'Monitoring of pesticides in river water based on samples previously stored in polymeric cartridges followed by on-line solid-phase extraction–liquid chromatography–diode array detection and confirmation by atmospheric pressure chemical ionization mass spectrometry', *Anal. Chim. Acta* **386**: 237–248 (1999).

56. J. Slobodnik, Ö. Östezkizan, H. Lingeman and U. A. Th Brinkman, 'Solid-phase extraction of polar pesticides from environmental water samples on graphitised carbon and Empore-activated carbon disks and on-line coupling to octadecyl-bonded silica analytical columns', *J. Chromatogr.* **750**: 227–238 (1996).

57. H. Bagheri, J. Slobodnik, R. M. Marcé Recasens, R. T. Ghijsen and U. A. Th Brinkman, 'Liquid chromatography–particle beam mass spectrometry for identification of unknown pollutants in water', *Chromatographia* **37**: 159–167 (1993).

58. R. M. Marcé, H. Prosen, C. Crespo, M. Calull, F. Borrull and U. A. Th Brinkman, 'On-line trace enrichment of polar pesticides in environmental waters by reversed-phase liquid chromatography–diode array detection–particle beam mass spectrometry', *J. Chromatogr.* **696**: 63–74 (1995).

59. S. Lacorte, J. J. Vreuls, J. S. Salau, F. Ventura and D. Barceló, 'Monitoring of pesticides in river water using fully automated on-line solid-phase extraction and liquid chromatography with diode array detection with a novel filtration device', *J. Chromatogr.* **795**: 71–82 (1998).

60. H. Bagheri, E. R. Brouwer, R. T. Ghijsen and U. A. Th Brinkman, 'Low-level multi-residue determination of polar pesticides in aqueous samples by column liquid chromatography–thermospray mass spectrometry', *J. Chromatogr.* **657**: 121–129 (1993).

61. C. Aguilar, F. Borrull and R. M. Marcé, 'Determination of pesticides by on-line trace enrichment–reversed-phase liquid chromatography–diode-array detection and confirmation by particle-beam mass spectrometry', *Chromatographia* **43**: 592–598 (1996).

62. S. Lacorte and D. Barceló, 'Determination of parts per trillion levels of organophosphorus pesticides in groundwater by automated on-line liquid–solid extraction followed by liquid chromatography/atmospheric pressure chemical ionization mass spectrometry using positive and negative ion modes of operation', *Anal. Chem.* **68**: 2464–2470 (1996).

63. E. Pocurull, R. M. Marcé and F. Borrull, 'Determination of phenolic compounds in natural waters by liquid chromatography with ultraviolet and electrochemical detection after on-line trace enrichment', *J. Chromatogr.* **738**: 1–9 (1996).

64. E. R. Brouwer and U. A. Th Brinkman, 'Determination of phenolic compounds in surface water using on-line liquid chromatographic precolumn-based column-switching tecniques', *J. Chromatogr.* **678**: 223–231 (1994).

65. D. Puig, L. Silgoner, M. Grasserbauer and D. Barceló, 'Part-per-trillion level determination of priority methyl-, nitro-, and chlorophenols in river water samples by automated online liquid/solid extraction followed by liquid chromatography/mass spectrometry using atmospheric pressure chemical ionization and ion spray interfaces', *Anal. Chem.* **69**: 2756–2761 (1997).

66. E. R. Brouwer, A. N. J. Hermans, H. Lingeman and U. A. Th Briknman, 'Determination of polycyclic aromatic hydrocarbons in surface water by column liquid chromatography with fluorescence detection, using on-line micelle-mediated sample preparation', *J. Chromatogr.* **669**: 45–57 (1994).

67. R. El Harrak, M. Calull, R. M. Marcé and F. Borrull, 'Influence of the organic solvent in on-line solid phase extraction for the determination of PAHs by liquid chromatography and fluorescence detection', *J. High Resolut. Chromatogr.* **21**: 667–670 (1998).

68. R. El Harrak, M. Calull, R. M. Marcé and F. Borrull, 'Determination of naphthalene-sulphonates in water by on-line ion-pair solid-phase extraction and ion-pair liquid chromatography with diode-array UV detection', *Int. J. Environ Anal. Chem.* **69**: 295–305 (1998).

69. E. Pocurull, C. Aguilar, M. C. Alonso, D. Barceló, F. Borrull and R. M. Marcé, 'On-line solid-phase extraction–ion-pair liquid chromatography–electrospray mass spectrometry for the trace determination of naphthalene monosulphonates in water', *J. Chromatogr.* **854**: 187–195 (1999).

70. N. Masqué, M. Galià, R. M. Marcé and F. Borrull, 'Chemically modified polymeric resin used as sorbent in a solid-phase extraction process to determine phenolic compounds in water', *J. Chromatogr.* **771**: 55–61 (1997).

71. N. Masqué R. M. Marcé and F. Borrull, 'Chemical removal of humic substances interfering with the on-line solid-phase extraction-liquid chromatographic determination of polar water pollutants', *Chromatographia* **48**: 231–236 (1998).

72. V. Pichon, M. Bouzige, C. Miège and M. C. Hennion, 'Immunosorbents: natural molecular recognition materials for sample preparation of complex environmental matrices', *Trends. Anal. Chem.* **18**: 219–235 (1999).

73. I. Ferrer, V. Pichon, M. C. Hennion and D. Barceló, 'Automated sample preparation with extraction columns by means of anti-isoproturon immunosorbents for the determination of phenylurea herbicides in water followed by liquid chromatography diode array detection and liquid chromatography–atmospheric pressure chemical ionization mass spectrometry', *J. Chromatogr.* **777**: 91–98 (1997).

74. I. Ferrer, M. C. Hennion and D. Barceló, 'Immunosorbents coupled on-line with liquid chromatography/atmospheric pressure chemical ionization/mass spectrometry for the part per trillion level determination of pesticides in sediments and natural waters using low preconcentration volumes', *Anal. Chem.* **69**: 4508–4514 (1997).

75. E. R. Brouwer, I. Liska, R. B. Geerdink, P. C. M. Frintrop, W. H. Mulder, H. Lingeman and U. A. Th Brinkman, 'Determination of polar pollutants in water using an on-line liquid chromatographic preconcentration system', *Cromatographia* **32**: 445–452 (1991).

76. F. Hernández, C. Hidalgo, J. V. Sancho and F. J. López, 'Coupled-column liquid chromatography applied to the trace-level determination of triazine herbicides and some of their metabolites in water samples', *Anal. Chem.* **70**: 3322–3328 (1998).

77. L. Mondello, G. Dugo and K. D. Bartle, 'On-line microbore high performance liquid chromatography–capillary gas chromatography for food and water analyses: a review', *J. Microcolumn Sep.* **8**: 275–310 (1996).

78. L. Mondello, P. Dugo, G. Dugo, A. C. Lewis and K. D. Bartle, 'High-performance liquid chromatography coupled on-line with high resolution gas chromatography. State of the art', *J. Chromatogr.* **842**: 373–390 (1999).

79. K. Grob (Ed.), *On-Line Coupled LC–GC,* Hüthig, Heidelberg, Germany (1991).

80. K. Grob and M. Biedermann, 'Vaporising systems for large volume injection or on-line transfer into gas chromatography: classification, critical remarks and suggestions', *J. Chromatogr.* **750**: 11–23 (1996).

81. K. Grob, 'Development of the transfer techniques for on-line high-performance liquid chromatography–capillary gas chromatography', *J. Chromatogr.* **703**: 265–276 (1995).

82. J. Slobodnik, A. C. Hogenboom, A. J. H. Louter and U. A. Th Brinkman, 'Integrated system for on-line gas and liquid chromatography with a single mass spectrometric detector for the automated analysis of environmental samples', *J. Chromatogr.* **730**: 353–371 (1996).

83. T. Hankemeier, A. J. H. Louter, J. Dallüge, R. J. J. Vreuls and U. A. Th Brinkman, 'Use of a drying cartridge in on-line solid-phase extraction gas chromatography – mass spectrometry', *J. High Resolut. Chromatogr.* **21**: 450 – 456 (1998).

84. H. G. J. Mol, H.-G. M. Janssen, C. A. Cramers, J. J. Vreuls and U. A. Th Brinkman, 'Trace level analysis of micropollutants in aqueous samples using gas chromatography with on-line sample enrichment and large volume injection', *J. Chromatogr.* **703**: 277 – 307 (1995).

85. J. J. Vreuls, A. J. H. Louter and U. A. Th Brinkman, 'On-line combination of aqueous-sample preparation and capillary gas chromatography', *J. Chromatogr.* **856**: 279 – 314 (1999).

86. A. J. H. Louter, S. Ramalho, R. J. J. Vreuls, D. Jahr and U. A. Th Brinkman, 'An improved approach for on-line solid-phase extraction-gas chromatography', *J. Microcolumn Sep.* **8**: 469 – 477 (1996).

87. E. C. Goosens, D de Jong, G. J. de Jong and U. A. Th Brinkman, 'On-line sample treatment – capillary gas chromatography', *Chromatographia* **47**: 313 – 345 (1998).

88. J. J. Vreuls, W. J. G. M. Cuppen, G. J. de Jong and U. A. Th Brinkman, 'Ethyl acetate for the desorption of a liquid chromatographic precolumn on-line into a gas chromatograph', *J. High Resolut. Chromatogr.* **13**: 157 – 161 (1990).

89. A. J. H. Louter, J. J. Vreuls and U. A. Th Brinkman, 'On-line combination of aqueous-sample preparation and capillary gas chromatography', *J. Chromatogr.* **842**: 391 – 426 (1999).

90. Y. Picó, A. J. H. Louter, J. J. Vreuls and U. A. Th Brinkman, 'On-line trace-level enrichment gas chromatography of triazine herbicides, organophosphorus pesticides and organosulfur compounds from drinking and surface waters', *Analyst* **119**: 2025 – 2031 (1994).

91. P. J. M. Kwakman, J. J. Vreuls, U. A. Th Brinkman and R. T. Ghijsen, 'Determination of organophosphorus pesticides in aqueous samples by on-line membrane disk extraction and capillary gas chromatography', *Chromatographia* **34**: 41 – 47 (1992).

92. E. Pocurull, C. Aguilar, F. Borrull and R. M. Marcé, 'On-line coupling of solid-phase extraction to gas chromatography with mass spectrometric detection to determine pesticides in water', *J. Chromatogr.* **818**: 85 – 93 (1998).

93. A. J. H. Louter, C. A. Beekelt, P. Cid Montanes, J. Slobodnik, J. J. Vreuls and U. A. Th Brinkman, 'Analysis of microcontaminants in aqueous samples by fully automated on-line solid-phase extraction – gas chromatography – mass selective detection', *J. Chromatogr.* **725**: 67 – 83 (1996).

94. J. Dallüge, T. Hankemeier, J. J. Vreuls and U. A. Th Brinkman, 'On-line coupling of immunoaffinity-based solid-phase extraction and gas chromatography for the determination of *s*-triazines in aqueous samples', *J. Chromatogr.* **830**: 377 – 386 (1999).

95. T. H. M. Noij, M. E. Margo and M. E. van der Kooi, 'Automated analysis of polar pesticides in water by on-line solid phase extraction and gas chromatography using the co-solvent effect', *J. High Resolut. Chromatogr.* **18**: 535 – 539 (1995).

96. T. Hyötyläinen, K. Grob, M. Biedermann and M. L. Riekkola, 'Reversed phase HPLC coupled on-line to GC by the vaporizer/precolumn solvent split/gas discharge interface; analysis of phthalates in water', *J. High Resolut. Chromatogr.* **20**: 410 – 416 (1997).

97. A. C. Lewis, R. E. Robinson, K. D. Bartle and M. J. Pilling, 'On-line coupled LC – GC – ITD/MS for the identification of alkylated, oxygenated and nitrated polycyclic aromatic compounds in urban air particulate extracts', *Environ. Sci. Technol.* **29**: 1977 – 1981 (1995).

98. K. Grob-Jr and Z. Li, 'Coupled reversed-phase liquid chromatography – capillary gas chromatography for the determination of atrazine in water', *J. Chromatogr.* **473**: 423 – 430 (1989).

99. J. J. Vreuls, V. P. Goudriaan, U. A. Th Brinkman and G. J. de Jong, 'A trapping column for the coupling of reversed-phase liquid chromatography and capillary gas chromatography', *J. High Resolut. Chromatogr.* **14**: 475–480 (1991).

100. K. K. Verma, A. J. H. Louter, A. Jain, E. Pocurull, J. J. Vreuls and U. A. Th Brinkman, 'On-line solid-phase extraction–gas chromatography–ion trap tandem mass spectrometric detection for the nanogram per litre analysis of trace pollutants in aqueous samples', *Chromatographia* **44**: 372–380 (1997).

101. D. Jahr, 'Determination of alkyl, chloro and mononitrophenols in water by sample-acetylation and automatic on-line solid phase extraction-gas chromatography-mass spectrometry', *Chromatographia* **47**: 49–56 (1998).

102. Y. Picó, J. J. Vreuls, R. T. Ghijsen and U. A. Th Brinkman, 'Drying agents for water-free introduction of desorption solvent into a GC after on-line SPE of aqueous samples', *Chromatographia* **38**: 461–469 (1994).

103. A. Farjam, J. J. Vreuls, W. J. G. M. Cuppen, U. A. Th Brinkman and G. J. De Jong, 'Direct introduction of large-volume urine samples into an on-line immunoaffinity sample pretreatment–capillary gas chromatography system', *Anal. Chem.* **63**: 2481–2487 (1991).

104. J. J. Vreuls, G. J. de Jong, U. A. Th Brinkman, K. Grob and A. Artho, 'On-line solid phase extraction-thermal desorption for introduction of large volumes of aqueous samples into a gas chromatograph', *J. High Resolut. Chromatogr.* **14**: 455–459 (1991).

105. J. J. Vreuls, G. J. de Jong, R. T. Ghijsen and U. A. Th Brinkman, 'On-line solid-phase extraction–thermal desorption for introduction of large volumes of aqueous samples into a capillary gas chromatograph', *J. Microcolumn Sep.* **5**: 317–322 (1993).

106. S. Müller, J. Efer and W. Engewald, 'Gas chromatographic water analysis by direct injection of large sample volumes in an adsorbent-packed PTV injector', *Chromatographia* **38**: 694–700 (1994).

107. H. G. J. Mol, P. J. M. Hendrinks, H. G. Janssen, C. A. Cramers and U. A. Th Brinkman, 'Large volume injection in capillary GC using PTV injectors: comparison of inertness of packing materials', *J. High Resolut. Chromatogr.* **18**: 124–128 (1995).

108. A. J. H. Louter, J. van Doornmalen, J. J. Vreuls and U. A. Th Brinkman, 'On-line solid-phase extraction–thermal desorption–gas chromatography with ion trap detection tandem mass spectrometry for the analysis of microcontaminants in water', *J. High Resolut. Chromatogr.* **19**: 679–685(1996).

14 Multidimensional Chromatographic Applications in the Oil Industry

J. BEENS

Free University de Boelelaan 1083, Amsterdam, The Netherlands

14.1 INTRODUCTION

From its very beginnings, chromatography has played an important role in the oil industry, with workers in this field having developed many important fundamentals of the technique. The names of van Deemter, Keulemans, Rijnders and Sie are still remembered from those early days of chromatography, where some quite fundamental work had been performed in an industrial environment. The main reason for this is the fact that chromatography is an outstanding technique for analysing the complex samples that are present in the oil industry. Csaba Horváth stressed this in his Golay Award Lecture at the *21st International Symposium on Capillary Chromatography and Electrophoresis* in Salt Lake City in June 1999: 'The rapid growth of gas chromatography was fuelled by the exploding need of the petroleum based industries for suitable tools after the war'.

The complexity of oil fractions is not so much the number of different classes of compounds, but the total number of components that can be present. Even more challenging is the fact that, unlike the situation with other complex samples, in which only a few specific compounds have to be separated from the matrix, in oil fractions the components of the matrix itself are the analytes. Figure 14.1 presents an estimation (by extrapolation) of the total number of possible hydrocarbon isomers with up to twenty carbon atoms present in oil fractions. Although probably not all of these isomers are always present, these numbers are nevertheless somewhat overwhelming. This makes a complete compositional analysis using a single column separation of unsaturated fractions with boiling points above 100 °C utterly impossible.

For this reason, multidimensional gas chromatography (GC) has been introduced as a means of increasing the separating efficiency. This was already explored in the late 1950s on chlorinated hydrocarbons with two packed columns and designated as two-stage chromatography (1, 2). A fully integrated system for the complete analysis of all of the constituents of a refinery off-gas by using four different columns was presented in 1961 by Bloch (3). Since then the number and type of multidimensional systems for the analysis of petroleum fractions has steadily increased, but most of

Multidimensional Chromatography, edited by L. Mondello, A. C. Lewis and K. D. Bartle
©2002 John Wiley & Sons Ltd.

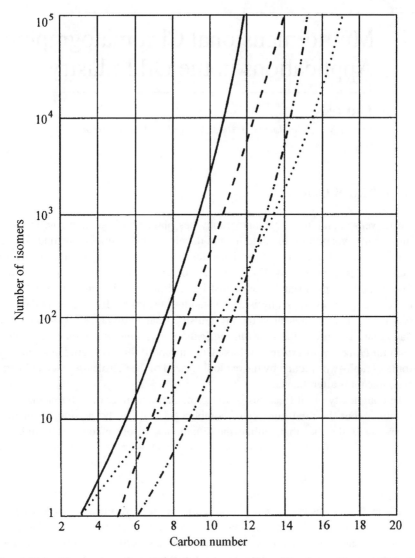

Figure 14.1 The number of possible hydrocarbon isomers of fractions with up to 20 carbon atoms in oil fractions: ————, paraffins; – – – – –, naphthenes; . . . . . ., mono-olefins; —··—··, aromatics.

the details of these developments has been kept inside the walls of the research institutes and have never been reported in the open literature.

Since the boiling range of a petroleum fraction is its main characteristic and the fractions are generally designated according to these boiling ranges, this present chapter will be divided into sections according to boiling range.

14.2   GASES

The analyses of gases in the oil industry comprises the determination of the inert gases (He, H_2, O_2, Ar and N_2), low-boiling compounds (CO, CO_2, H_2S, COS) and the lower hydrocarbons, saturated and unsaturated, up to hexane. Some special samples, such as natural gas, have to be analysed for low concentrations of higher-boiling compounds (up to $C_{10}s$) since such compounds have an important influence on the calorific value and dew point.

For measuring the inert species, some of which are present in the majority of gases, the thermal-conductivity detector (TCD) is often the detector of choice for gas analyses. Since the TCD is a concentration detector and its sensitivity is lower than that of mass-flow detectors such as the flame-ionization detector (FID), relatively high concentrations of compounds in the carrier gas are needed. This means that packed columns, with their high loadability, are still quite popular for such analyses.

Some of the analysis configurations in use enable the analysis of specific compounds in gas samples, such as sulfur compounds in hydrocarbon gases and various impurities in main compounds. Other configurations aim at the determination of all of the different constituents of refinery gases. Various standardization organisations, such as the American Society for Testing and Materials (ASTM), Institute of Petroleum (IP), Universal Oil Products (UOP), Deutsches Institut für Normung (DIN) and Gas Processors Association (GPA) (4–8) have published a number of these configurations as standardized methods.

14.2.1   THE ANALYSIS OF TRACES OF SULFUR COMPOUNDS IN ETHENE AND PROPENE

Ethene and propene are produced as bulk feedstocks for the chemical (polymer) industry and therefore their purities are important parameters. In particular, H_2S and COS are compounds which may not only cause corrosion problems in processing equipment, but also may have detrimental effects on the catalysts in use. Furthermore, air pollution regulations issued by, among others, the US Environmental Protection Agency (EPA) require that most of the sulfur gases should be removed in order to minimize sulfur emissions into the atmosphere. Therefore, these compounds have to be determined to the ppb level.

Several methods are available for the determination of sulfur compounds in refinery gas streams. Figure 14.2 depicts a one-column system with column preflush and a sulfur specific detector. In order to prevent adsorption of the acidic sulfur compounds, a H_3PO_4-treated Carbopack phase is used in a Teflon column. The compounds eluting in front of the H_2S and COS are flushed to vent, after which the second valve V2 is switched in order to direct the sulfur compounds towards the sulfur chemiluminescence detector (SCD). The remaining hydrocarbons, possibly coeluting with the sulfur compounds do not interfere, since the selectivity of the SCD for sulfur to carbon is about 10^7 (9). An example of a typically resulting chromatogram is presented in Figure 14.3, where the lower limit of detection is 50 ppb for both compounds.

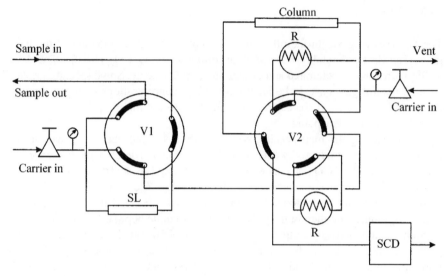

Figure 14.2 Schematic diagram of the chromatographic system used for the analysis of low concentrations of sulfur compounds in ethene and propene: V1, injection valve; V2, column switching valve; SL, sample loop; R, restriction to replace the column; SCD, sulfur chemiluminescence detector.

A more sensitive separation scheme is presented in Figure 14.4. This valveless switching system was originally developed by Deans (10, 11) and utilizes pressures to direct the flows through the different columns. Two capillary columns are used, of which the first separates the hydrocarbons from each other and from the sulfur gases.

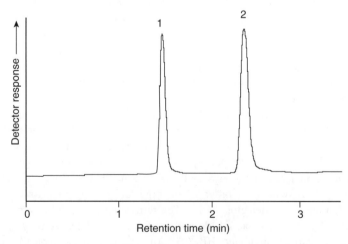

Figure 14.3 Chromatographic separation of sulfur compounds in propene, obtained by using the system illustrated in Figure 14.2: 1, hydrogen sulfide; 2, carbonyl sulfide.

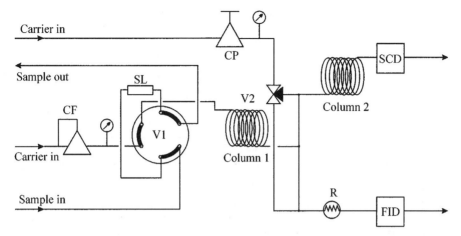

Figure 14.4 Schematic diagram of the chromatographic system used for the analysis of very low concentrations of sulfur compounds in ethene and propene: CP, pressure regulator; CF, flow regulator; SL, sample loop; R, restriction to replace column 2; V1, injection valve; V2, three-way valve to direct the effluent of column 1 to either column 2 or the restriction; column 1, non-polar capillary column; column 2, thick-film capillary column; SCD, sulfur chemiluminescence detector; FID, flame-ionization detector.

The hydrocarbons are directed to the FID and identified. After this separation, the second value (V2) is switched such that the effluent of column 1 is directed to column 2, a thick-film capillary column, which separates the sulfur compounds from the remaining hydrocarbons and from each other. This effluent is directed to the SCD, where the sulfur compounds are identified. The advantage of this approach above that of the one shown in Figure 14.2 is the fact that the analytes do not pass any valves in the separation process, and thus adsorption of acidic compounds is avoided. Moreover, the hydrocarbons are not disregarded, but can be determined by using the FID, as is shown in Figure 14.5. The lower limit of detection of sulfur compounds when using this configuration is around 1 ppb.

14.2.2 THE ANALYSIS OF REFINERY GAS

Refinery gas is the collective noun used for a range of off-gases originating from the various oil processes. A detailed knowledge of the composition of these gases is needed for three reasons, as follows:

- to monitor and adjust individual processes or products;
- to comply with legislative measures concerning environmentally hazardous compounds;
- to establish the economic value of the gas.

For most gas process environments – refineries, catalyst development sites, research and development and plant laboratories – a knowledge of the exact composition of the

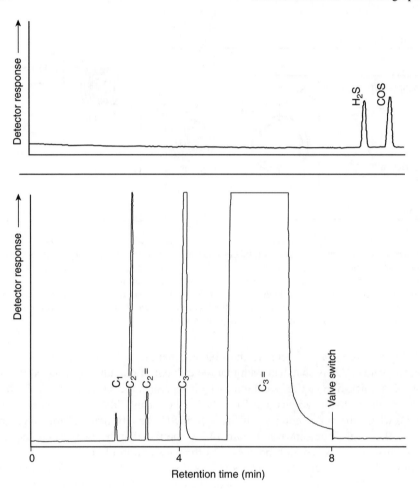

Figure 14.5 Chromatographic separation of (a) sulfur compounds, and (b) hydrocarbons, obtained by using the system illustrated in Figure 14.4.

gas stream components is a prerequisite for achieving optimum control and quality assurance. The refinery analyser configurations therefore have to cope with a wide variety of gas compositions and concentrations.

A five-column configuration of such an analyser system is depicted in Figure 14.6. The first event in the process is the analysis of H_2 by injection of the contents of sample loop 2 (SL2) onto column 5 (a packed molecular sieve column). Hydrogen is separated from the other compounds and detected by TCD 2, where nitrogen is used as a carrier gas. The next event is the injection of the contents of sample loop 1 (SL1), which is in series with SL2, onto column 1. After the separation of compounds up to and including C_5, and backflushing the contents of column 1, all compounds above C_5 (C_{6+}) are detected by TCD1. The fraction up to and including C_5 is directed to column 2, where air, CO, CO_2, C_2, and $C_2=$ (ethene) are separated from

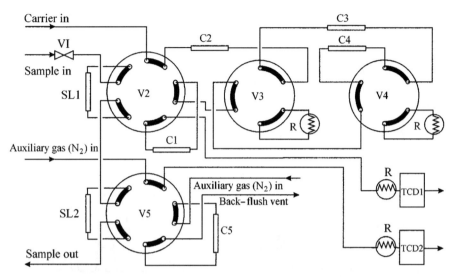

Figure 14.6   Schematic diagram of a five-column chromatographic refinery analyser system: SL, sample loop; V1, two-way valve to block the sample line; V2 and V5, ten-port valves; V3 and V4, six-way valves; C1 C5, packed columns; R, restriction; TCD, thermal conductivity detector.

the rest. This fraction is directed to column 3. The remainder of the sample C_3s, C_4s and C_5s on column 2 is separated and detected by TCD1. Meanwhile, the contents of column 3 are separated, while the air is directed to column 4. The remainder, CO_2, $C_2=$ and C_2, is detected by TCD1. Finally, the contents of column 5 (Ar + O_2, N_2, C_1 and CO) are separated and detected by TCD1. The resulting chromatogram is presented in Figure 14.7.

A complicated analyser system such as that described above can only be maintained if all of the valve-switching events are scheduled in the correct positions in the chromatogram. Mismatch of one of the events will cause (parts of) components to be directed to the wrong columns and thus possible misidentifications. Therefore, accurate determination and maintenance of the cutting windows are essential. This can only be accomplished in a fully automated system with accurate flow and temperature controls. Once these prerequisites are fulfilled, the system will operate unattended and produce results of high quality. The repeatabilities generally achieved are of the order of $\pm 1\%$ rel.

14.2.3 THE ANALYSIS OF NATURAL GAS

Natural gas, found in geological accumulations, normally refers to the gaseous fossil-based equivalent of oil. Its composition varies widely, from high concentrations of nitrogen and carbon dioxide to (almost) pure methane. In general, it contains low concentrations of the higher (saturated) hydrocarbons, which influence the physical properties and may present condensation problems in high-pressure transport lines.

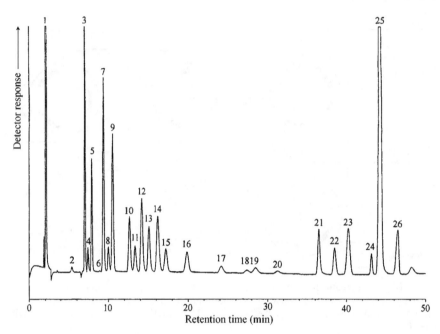

Figure 14.7   Typical chromatogram obtained by using the refinery analyser system shown in Figure 14.6. Peak identification is as follows: 1, hydrogen; 2, C_6+, 3, propane; 4, acetylene; 5, propene; 6, hydrogen sulfide; 6, iso-butane; 8, propadiene; 9, n-butane, 10. iso-butene; 11, 1-butene; 12, *trans*-2-butene; 13, *cis*-2-butene; 14, 1,3-butadiene; 15, iso-pentane; 16, *n*-pentane; 17, 1-pentene; 18, *trans*-2-pentene; 19, *cis*-2-pentene; 20, 2-methyl-2-butene; 21, carbon dioxide; 22, ethene; 23, ethane; 24, oxygen + argon, 25, nitrogen, 26, carbon monoxide.

The economic value of natural gas is primarily determined by the thermal energy it contains, which is expressed in British thermal units (Btu) or calorific value (CV). Other important physical properties comprise the liquid content, the burning characteristics, the dew point and the compressibility. In order to enable the calculation of these properties from its composition, a natural gas analysis should contain a detailed determination of all of the individual components, even in the low-concentration range.

Figure 14.8 shows a detailed schematic representation of a natural gas analysis system, which fully complies with GPA standardization (8). This set-up utilizes four packed columns in connection with a TCD and one capillary column in connection with an FID. The contents of both sample loops, which are connected in series, are used to perform two separate analyses, one on the capillary column and one on the packed columns. The resulting chromatograms are depicted in Figure 14.9.

The packed column section contains a stripper pre-column (column 1), which separates the C_{6+} fraction by back-flushing all compounds above *n*-pentane in one peak. H_2S, CO_2, C_2, O_2, N_2 and C_1 are trapped in columns 3 and 4, while C_3-C_5 hydrocarbons elute from column 2 to the TCD. The remaining components are

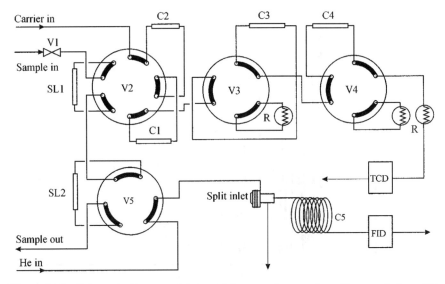

Figure 14.8 Schematic diagram of the natural gas analyser system: SL, sample loop; V1, two-way valve to block the sample lines; V2, ten-port valve; V3, V4 and V5, six-port valves; R, restriction; TCD, thermal-conductivity detector; FID, flame-ionization detector.

separated by columns 3 and 4 and detected by the TCD. The second channel (through valve 5 (V5)) uses a split/splitless injector in order to decrease the bandwidth of the injected sample from the sample loop. The capillary column separates the individual components up to C_{10}. The lower limit of detection of this second channel is 0.001 vol%.

The previous comments concerning maintenance of the refinery analyser described in Section 14.2.2, also apply to this analysis configuration, i.e. proper automation and the use of accurate flow and temperature controls are a necessity.

14.3 GASOLINES AND NAPHTHAS

The US Clean Air Act of 1990 mandates the reformulation of gasoline, being aimed at a reduction of the emissions of toxic compounds from combustion engines. Benzene and toluene are classified as toxic air pollutants under this act. In view of the requirements formulated by the American Environmental Protection Agency (EPA), a quantitative determination of benzene concentrations at all process stages is essential to meet the target values and to achieve optimum control of the refining and blending processes. Similar legislation is pending in other countries. The pollutant limits and corresponding test methods from the (US) Clean Air Act may be expected to serve as guidelines in other jurisdictions. In the US, the EPA regulations specify the level of benzene as not to exceed 1% vol. Similar regulations also exist in Europe.

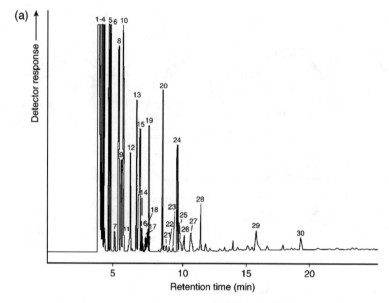

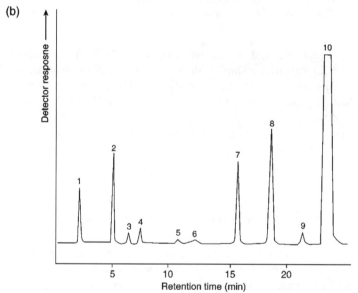

Figure 14.9 (a) Typical chromatogram obtained from the natural gas analyser, capillary column channel. Peak identification is as follows: 1, methane; 2, ethane; 3, propane; 4, butane; 5, 2-methyl-butane; 6, n-pentane; 7, 2,2-dimethyl-butane; 8, 2-methyl-pentane; 9, 3-methyl-pentane; 10, n-hexane; 11, methylcyclopentane; 12, 2,2-dimethyl-pentane; 13, benzene; 14–19, C_7; 20–23, C_8s; 24, toluene; 25–28, C_8s; 29, C_9, 30, C_{10}. (b) Corresponding (typical) chromatogram obtained from the packed column channel. Peak identification is as follows: 1, C_{6+}; 2, propane; 3, isobutane; 4, n-butane; 5, 2-methyl-butane; 6, n-pentane; 7, carbon dioxide; 8, ethane; 9, nitrogen; 10, methane.

14.3.1 THE ANALYSIS OF BENZENE, TOLUENE AND HIGHER AROMATICS IN LOW-BOILING FRACTIONS

The analysis system depicted in Figure 14.10 is able to analyse the aromatics in oxygenated gasolines by using the Deans switching system. This system pre-separates the non-aromatics from the sample on the polar first capillary column, and FID1 detects these (group of) compounds. After switching the valve, the eluting aromatics from the first column are further separated on the second capillary column and detected by FID2. The remaining high-boiling components on the first column are transferred directly to FID1. Figure 14.11 presents the resulting chromatogram. In order to enable an accurate quantitation, methyl ethyl ketone (MEK) is used as an internal standard. The lower limit of detection of this system is 0.001 mass%.

14.3.2 PIONA-ANALYSER AND REFORMULYSER

In order to find a balance between environmental legislation and economical refinery requirements, many refiners blend modern stocks into new gasolines for use in high-performance spark-ignition engines. These blended stocks comprise platformates, isomerates, alkylates, light fluid cat(alyst) cracked (FCC) naphtha, FCC gas and oxygenates, such as methyl *tert*-butyl ether (MTBE), *tert*-amyl methyl ether (TAME), and alcohols. A detailed analytical characterization of gasoline and naphtha is therefore of the utmost importance for process engineers and blenders.

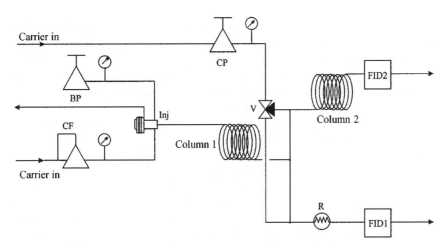

Figure 14.10 Schematic diagram of the aromatics analyser system: BP, back-pressure regulator; CF, flow controller; CP, pressure controller; Inj, splitless injector with septum purge; V, three-way valve; column 1, polar capillary column; column 2, non-polar capillary column; R, restrictor; FID1, and FID2, flame-ionization detectors.

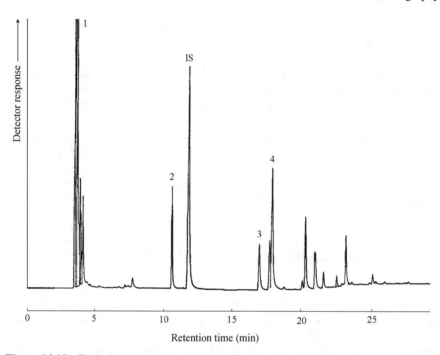

Figure 14.11 Typical chromatogram obtained by using the aromatics analyser system. Peak identification is as follows: 1, non-aromatics; 2, benzene; IS, internal standard (MEK); 3, ethylbenzene; 4, *p*-and *m*-xylenes.

A variety of multidimensional GC systems have been developed for the complete characterization of gasoline and naphtha-type samples. The limit of these multidimensional systems has been the introduction of the with PIONA-analyser in 1971 (12), with PIONA standing for Paraffins – Iso-paraffins – Olefins – Naphthenes – Aromatics. This system has exploited the unique separation of naphthenes and paraffins as a function of carbon number on a column packed with zeolites of a very specific pore size (13X molecular sieves) (13). In later years, the technique has been expanded to samples having boiling points up to 270 °C (14) and implemented in a commercial instrument (15), which is still in use in the majority of the refinery laboratories for the compositional analyses of gasolines and naphthas. Other investigators have developed comparable systems with capillary columns (16–22), some of which incorporated a mass spectrometer, but these were never commercialized. More recently, with the introduction of oxygenates in gasolines, all of these analyser systems have experienced the shortcoming that they are not able to separate the oxygenates from the hydrocarbon matrix. A new multi-column system has therefore been developed, i.e. the Reformulyser, which overcomes this shortcoming. Figure 14.12 depicts a schematic diagram of the Reformulyser system, with a typical resulting chromatogram obtained from this set-up being shown in Figure 14.13.

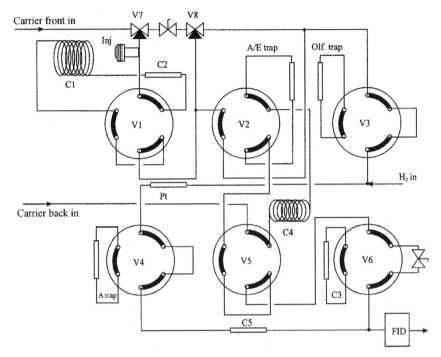

Figure 14.12 Schematic diagram of the Reformulyser system: Inj, split injector; C1, polar capillary column; C2, packed column to retain the alcohols; C3, packed Porapak column for the separation of the oxygenates; C4, non-polar capillary column; C5, packed 13X column; A/E trap, Tenax trap to retain the aromatics; Olf. trap, trap to retain the olefins; Pt, olefins hydrogenator; A trap, trap to retain the *n*-alkanes; FID, flame-ionization detector.

14.3.3 A DUAL SFC SYSTEM FOR THE ANALYSIS OF TOTAL OLEFINS IN GASOLINES

A different approach for the analysis of the total olefin content of a gasoline, using supercritical carbon dioxide as the carrier, is given in a proposed ASTM method. This utilizes the switching of two packed supercritical fluid chromatography (SFC) columns. The first column is a high-surface-area silica column, capable of separating alkanes and olefins from aromatics, while the second column is a silver-loaded silica column or a cation-exchange column in the silver form. A schematic diagram of such a system is presented in Figure 14.14. In this technique, the sample is injected onto the silica column, which is in series with the silver-loaded column. After elution of the saturates, the aromatics remain on the silica column and the olefins have moved to the silver-loaded column, where they are trapped. The silica column is then back-flushed to elute the aromatics to the detector. After switching the silica column out of the flowpath, the silver-loaded column is back-flushed to elute the olefins to the detector. A typical chromatogram obtained from such a system is presented in Figure 14.15.

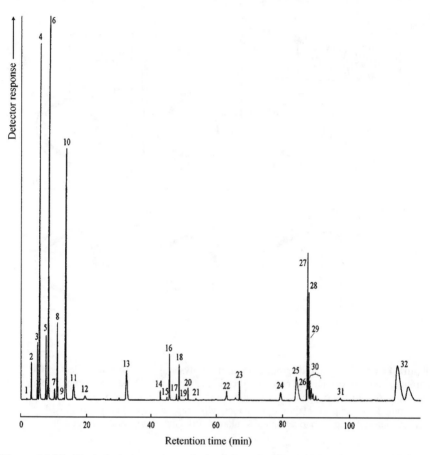

Figure 14.13 Typical chromatogram obtained by using the Reformulyser system. Peak identification is as follows: 1, P3; 2, P4; 3, N5; 4, P5; 5, N6; 6, P6; 7, N7; 8, P7; 9, N8; 10, P8; 11, P9; 12, P10; 13, MTBE; 14, OP4; 15, ON5; 16, OP5; 17, ON6; 18, OP6; 19, ON7; 20, OP7; 21, OP8; 22, P8; 23, A8; 24, A6; 25, A7; 30, A9/10; 31, A10/11; 32, A8 (where P is alkane, N is cycloalkane, MTBE is methyl *tert*-butyl ether, OP is alkene, ON is cycloalkene and A is aromatic).

14.4 MIDDLE DISTILLATES

Middle distillates are defined as those fractions having a boiling range between 150 °C and 370 °C. Various products fall within this range, e.g. kerosine, jet fuel, diesel fuel, light gasoil and heavy gasoil. Diesel fuels in particular are subjected to several environmental regulations. Because the aromatics present in diesel fuel promote the emission of particulates, and legislative measures restrict this emission, refiners thus require a rapid and sensitive method for monitoring these aromatics. In light of the year-2000 and (projected) year-2005 legislations, the European fuel market is very interested in so-called city-diesel. The latter is a petroleum-based, lower-emission diesel fuel developed in Sweden. This fuel is available in many

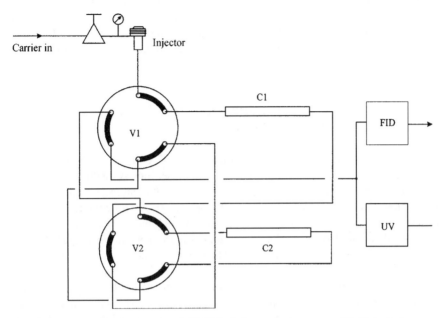

Figure 14.14 Schematic diagram of the SFC olefins analyser system: C1, high-surface-area silica column; C2, silver-loaded silica column; V1 and V2, six-port valves; FID, flame-ionisation detector; UV, ultraviolet monitor detector.

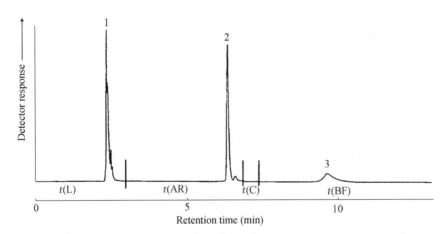

Figure 14.15 Typical SFC chromatogram of total olefins in gasoline: 1, saturates; 2, aromatics; 3, olefins; $t(L)$, time of loading sample on to columns and eluting saturates; $t(AR)$, time of eluting aromatics; $t(C)$, time of eluting remaining saturates from olefin trap; $t(BF)$, time of eluting olefins by back-flush.

European countries, including the UK, and covers about 5% of the total European diesel market. It contains less than 5 vol%. of mono-ring aromatics and less than 0.1 vol%. of di- and higher-ring aromatics. Furthermore it is low in sulfur. Both LC and SFC methods have been developed for the rapid analysis of aromatics in diesel fuel.

14.4.1   AN LC–SFC SYSTEM FOR THE ANALYSIS OF TOTAL AROMATICS IN DIESEL FUEL

Different approaches utilizing multidimensional LC or SFC systems have been reported for the analysis of middle distillates in diesel fuel. A method, based on the LC separation of paraffins and naphthenes by means of a micro-particulate, organic gel column has been described (23, 24). The complete system contained up to four different LC columns, a number of column-switching valves and a dielectric constant detector. However, the LC column for the separation of paraffins and naphthenes, which is an essential part of the system, is no longer commercially available.

In Figure 14.16, the chromatograms of a normal-phase LC analysis, using a straight-phase silica column with back-flush are shown. The same method, using carbon dioxide as the mobile phase and a cyanopropyl-modified silica column for the SFC stage, is employed for the second separation. When the column is not back-flushed, a clear distinction can be made between the mono-ring and higher-ring aromatics.

14.4.2   ON-LINE COUPLING OF LC AND GC FOR HYDROCARBON CHARACTERIZATION

A more sophisticated method, giving a much more detailed characterization, involves the on-line coupling of LC and GC (LC–GC). Analysis schemes for middle distillates (kerosine, diesel and jet fuels) combining LC and GC have been reported by various authors (25–31). However, only Davies et al. (25) and Munari et al. (27) have reported on the required automatic transfer of all of the individual separated fractions from the LC unit the GC system. Davies used the loop-type interface and Munari the on-column interface. Only Beens and Tijssen report a full quantitative characterzation by means of LC-GC (31).

The system utilized is depicted schematically in Figure 14.17. In this set-up the sample is injected onto the cyanopropyl-derivatized LC column. This column separates hydrocarbons according to the number of π-electrons in the molecules, which is more or less according to the number of fused aromatic rings. The outlet of the column is monitored by the UV detector for the start of eluting aromatics, and is connected to the GC system by means of valve 1 (V1) through a narrow capillary into the on-column injector. The effluent of the column–the mobile phase, n-heptane, together with the first eluting fraction, the saturates–flows directly into the GC

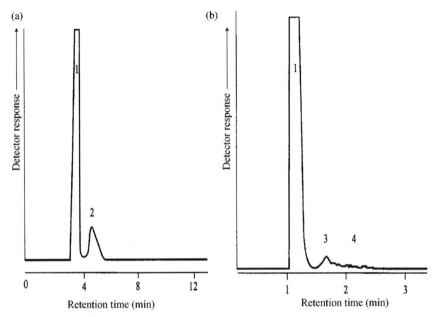

Figure 14.16   Typical chromatograms of LC (a) and SFC (b) analysis of aromatics in diesel fuel. Peak identification is as follows: 1, total saturates; 2, total aromatics; 3, mono-aromatics; 4, higher-ring aromatics.

retention gap. Here, the heptane will start the 'solvent-effect', thus focusing the saturates into a narrow band. The heptane on the front of the plug will evaporate and leave the system through the SVE valve. As soon as the saturates have completely left the LC column, i.e. at the start of the mono-aromatics fraction, the LC pump is stopped. The remaining fractions in the column will not broaden, since the diffusion coefficients in the liquid phase are very low. Programming the GC oven will then separate the focused fraction in the GC system. After cooling down the GC oven, the whole process subsequently starts again for the mono-aromatics, di-aromatics, tri-aromatics, etc. Clear separated chromatograms of all the fractions can be obtained, as can be seen in Figure 14.18, for the mono-aromatics, and in Figure 14.19, for the di-aromatics. No overlap from one fraction into the other is detectable. This means that not only a quantitative group-type separation can be provided, but also that within the various groups a clear carbon distribution can be obtained.

14.4.3   ON-LINE COUPLED LC–GC-FID-SCD FOR SULFUR COMPOUND CHARACTERIZATION

Since the majority of middle distillates are used as a fuel, combustion of these products will contribute to SO_2/SO_3 air pollution and acid rain. However, in catalytic processes of petroleum fractions sulfur levels are also important. For instance, quantities

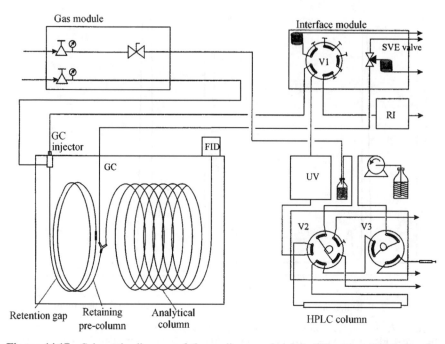

Figure 14.17 Schematic diagram of the on-line coupled LC–GC system: V1, valve for switching the LC column outlet to the GC injector; V2, valve for switching the LC column to back-flush mode; V3, LC injection valve; RI, refractive index monitor detector; UV, ultraviolet monitor detector; FID, flame-ionization detector.

of as little as 1 ppm sulfur in feedstocks may have a detrimental effect on modern bimetallic catalysts. For these reasons, petroleum fractions are often desulfurized.

Although desulfurization is a process, which has been in use in the oil industry for many years, renewed research has recently been started, aimed at improving the efficiency of the process. Environmental pressure and legislation to further reduce sulfur levels in the various fuels has forced process development to place an increased emphasis on hydrodesulfurization (HDS). For a clear comprehension of the process kinetics involved in HDS, a detailed analyses of all the organosulfur compounds clarifying the desulfurization chemistry is a prerequisite. The reactivities of the sulfur-containing structures present in middle distillates decrease sharply in the sequence thiols $\gg$ sulfides $\gg$ thiophenes $\gg$ benzothiophenes $\gg$ dibenzothiophenes (32). However, in addition, within the various families the reactivities of the substituted species are different.

It is for this reason that not only the various sulfur-containing groups present, but also the mono- and dimethyl-substituted species of benzothiophenes and dibenzothiophenes have to be separated and quantified individually. As the number of sulfur compounds present in (heavy) middle distillate fractions may easily exceed 10 000 species, a single high resolution GC capillary column is unable to perform such a separation.

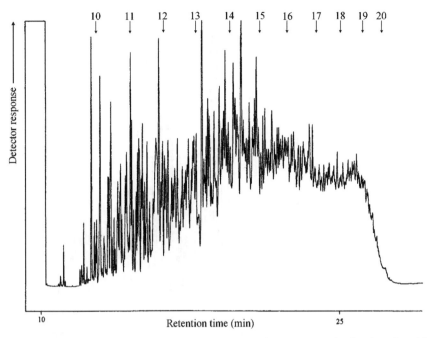

Figure 14.18 Typical GC chromatogram of the separated mono-aromatics fraction of a middle distillate sample; the numbers indicate the retention time of the various *n*-alkanes.

The system described in the previous section has been extended with a sulfur chemiluminescence detector (SCD) for the detection of sulfur compounds (32). The separated fractions were thiols + sulfides + thiophenes (as one group), benzothiophenes, dibenzothiophenes and benzonaphtho-thiophenes. These four groups have been subsequently injected on-line into and separated by the GC unit. Again, no overlap between these groups has been detected, as can be seen from Figure 14.20, in which the total sulfur compounds are shown and from Figure 14.21 in which the separated dibenzothiophenes fraction is presented. The lower limit of detection of this method proved to be 1 ppm (mg kg^{-1}) sulfur per compound.

14.4.4 COMPREHENSIVE TWO-DIMENSIONAL GAS CHROMATOGRAPHY (GC × GC)

When John Phillips, in 1991, presented the practical possibility of acquiring a real comprehensive two-dimensional gas chromatographic separation (33), the analytical chemists in the oil industry were quick to pounce upon this technique. Venkatramani and Phillips (34) subsequently indicated that GC × GC is a very powerful technique, which offers a very high peak capacity, and is therefore eminently suitable for analysing complex oil samples. These authors were able to count over 10 000 peaks in a GC × GC chromatogram of a kerosine. Blomberg, Beens and co-workers

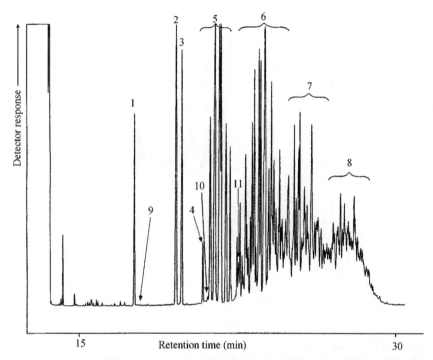

Figure 14.19 Typical GC chromatogram of the separated di-aromatics fraction of a middle distillate sample: Peak identification is as follows: 1, naphthalene; 2, 2-methylnaphthalene; 3, 1-methylnaphthalene; 4, biphenyl; 5, C_2-naphthalenes; 6, C_3-naphthalenes; 7, C_4-naphthalenes; 8, C_{5+}-naphthalenes; 9, benzothiophene; 10, methylbenzothiophenes; 11, C_2-benzothiophenes. Note the clean baseline between naphthalene and the methylnaphthalenes, which means that no overlap with the previous (mono-aromatics) fraction has occurred.

(35, 36), who presented a number of separations of different oil fractions by using GC × GC, later confirmed this. The latter authors also demonstrated (36) that the quantitative results produced by GC × GC equipped with a flame-ionization detector are at least of the same quality, if not better, than those obtained with 'conventional' one-dimensional capillary GC.

In GC × GC, a sample is separated into a large number of small fractions and each of these is subsequently quantitatively transferred to a secondary column to be further separated. The second separation is very much faster than the first separation, so that the fractions can be narrow and the separation obtained on the first column can be maintained. The collection of the fractions from the first column is achieved by focusing, rather than by valve switching, and the entire sample reaches the detector. The consequence is a chromatogram, with a two-dimensional plane, rather than a one-dimensional axis, as the time domain. One dimension of this plane represents the retention time on the first column, while the second dimension represents the retention time on the second column. Every separated peak can be presented as a

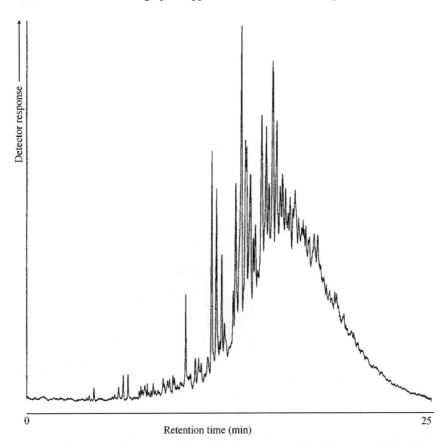

Figure 14.20   GC chromatogram of the total sulfur compounds in a heavy gas oil sample.

spot in a contour plot. The very many components present in oil fractions give rise to bands spread across the plane. As a result of the separation mechanism of GC × GC, the resulting chromatograms show a great deal of structure, which can be used to assign the (components in) different bands with high analytical certainty.

An example of such a separation, showing the different groups in separate bands along the plane, is presented in Figure 14.22. The saturates are present in a band which crosses over from the right of the contour plot to the left. This band, upon closer inspection at the first retention time (above 30 min), appears to consists of two bands; i.e. the alkanes and the cycloalkanes. Each individual spot represents a component of the original mixture. When integrated, the data provide a true quantitative result. Although only a few spots can be designated to individual compounds, the rest of the spots can be identified as being members of a group. By including the retention of the first dimension (non-polar column), this also provides an indication of the boiling points and/or carbon numbers.

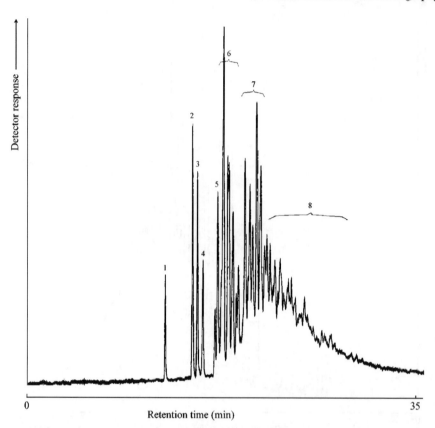

Figure 14.21   GC chromatogram of the dibenzothiophenes fraction of a heavy gas-oil sample. Peak identification is as follows: 1, dibenzothiophene; 2, 4-methyldibenzothiophene; 3, 2-methyldibenzothiophene; 4, 3-methyldibenzothiophene; 5, 4,6-dibenzothiophene; 6, other C_2-dibenzothiophenes, 7, C_3-dibenzothiophenes; 8, C_4- and higher dibenzothiophenes.

Phillips and Xu have presented two-dimensional (2D) chromatograms of kerosines, separated with different stationary phase combinations, in many thousands of components (37). Frysinger *et al.* have separate benzene–toluene–ethyl benzene–xylenes (BTEX) and total aromatics in gasolines by using GC × GC (38). These authors also analysed marine diesel fuel with GC × GC, connected to a quadrupole mass spectrometer for identification purposes, although the scan speed of the spectrometer was not quite suited for the fast second-dimension peaks (39). GC × GC was used as an excellent tool for identifying oil spill sources by Gaines *et al.* (40). Synovec and co-workers used the GC × GC separation of mixtures of toluene, ethylbenzene, *m*- and *o*-xylenes and propylbenzene in white gas to investigate the use of generalized rank annihilation methods (GRAMs) for quantitation purposes (41), while Kinghorn and Marriott used the separation of kerosine to demonstrate the cryogenic modulator (42). Although quantitation of

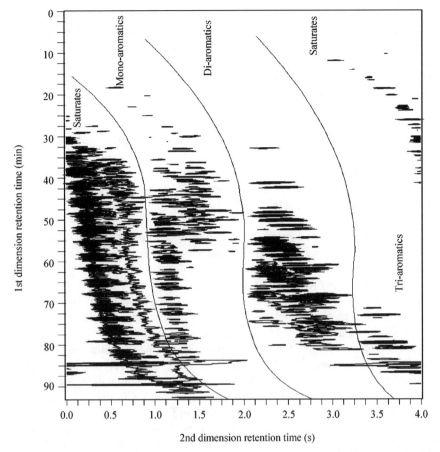

Figure 14.22   Contour plot of the GC × GC separation of a heavy gas oil sample; the thin lines indicate the borders between the groups.

the massive data matrix is possible and provides excellent results (36), no software, which is compatible with all of the various GC × GC analyses, is yet available.

An overview of the state-of-the-art of GC × GC is given in (43).

14.5   RESIDUE-CONTAINING PRODUCTS

These products comprise whole crude oils, as well as bottom fractions of distilling units and (partly) converted materials. The common property of these products is the fact that they contain high-boiling material which is not amenable to gas chromatographic analysis.

Nevertheless, a number of gas chromatographic applications exist, epecially those for the determination of crude oil indicators. Such indicators are used as geochemical parameters for the thermal history of the crude as well as to indicate the possible relationship between crudes from different wells. These indicators comprise a number of isomeric aromatic species, such as the individual alkylnaphthalenes (44, 45), the individual C_{10}-mono-aromatics or the individual C_9-mono-aromatics. The ratio between these isomers gives a definite indication of the crude oil. In general, these systems use a Deans switching unit to make a heart-cut, which then is focused, re-injected and separated on a second column with a different polarity.

14.5.1 A DETAILED HYDROCARBON ANALYTICAL SYSTEM COUPLED TO A SIMULATED DISTILLATION PROCESS

The main characteristic of a crude oil for processing is its boiling range, which is generally determined by (single-column) gas chromatography, and is designated as 'simulated distillation'. Because this analysis is performed with a 'high-temperature' (short, thin film, highly temperature stable, etc.) column, the lower part of the boiling range is not too well separated and defined. In order to determine the economic value, as well as to predict and control the optimum (crude) distillation cut points, this low-boiling part can be determined with a valve switched pre-column. The 'high temperature' column is fitted in the GC oven, which is temperature programmed up to temperatures of around 400 °C. In a separate valve oven, a second, high resolution capillary column with a valve is accommodated. This column will separate isothermally the light end (in which an internal standard, 3,3-dimethyl-1-butene, is included) of the sample, e.g. up to C_9, as shown in Figure 14.23. The remaining part is back-flushed to vent. A separate injection is performed on to the 'high-temperature' column for the simulated distillation. A software program then enables the incorporation of the two data sets into a final report.

14.5.2 AN SEC–LC–GC SYSTEM FOR THE ANALYSIS OF 'LOW BOILING' MATERIALS IN RESIDUAL PRODUCTS

A more complicated, but flexible, system has been reported by Blomberg *et al.* (46). Here, size exclusion chromatography (SEC), normal phase LC (NPLC) and GC were coupled for the characterization of restricted (according to size) and selected (according to polarity) fractions of long residues. The seemingly incompatible separation modes, i.e. SEC and NPLC, are coupled by using an on-line solvent-evaporation step.

Another interesting, but rather complex system, which couples flow injection analysis, LC and GC has been recently reported (47). This system allows the determination of the total amount of potentially carcinogenic polycyclic aromatic compounds (PACs) in bitumen and bitumen fumes. This system could also be used for the analysis of specific PACs in other residual products.

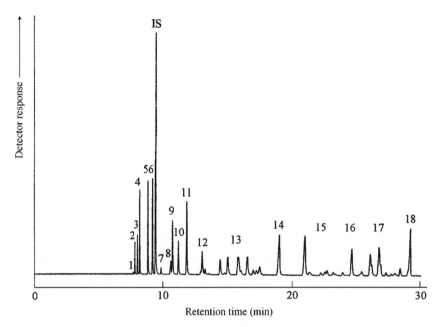

Figure 14.23   GC Light-end separation of a sample of crude oil. Peak identification is as follows: 1, C_3; 2, i-C_4; 3, n-C_4; 5, 2-methyl-C_4; 6, n-C_5; 7, 2,2-dimethyl-C_4; 8, cyclo-C_5 + 2,3-dimethyl-C_4; 9, 2-methyl-C_5; 10, 3-methyl-C_5; 11, n-C_6; 12, methyl-cyclo-C_5; 13, C_7s; 14, n-C_7; 15, C_8s; 16, n-C_8; 17, C_9s; 18, n-C_9; IS, internal standard (3,3-dimethyl-1-C_4–).

ACKNOWLEDGEMENTS

The author would like to thank Jan Blomberg of the Shell Research and Technology Center, Amsterdam and Dolf Grutterink of Analytical Controls, Rotterdam for kindly supplying some of the chromatograms and schematic diagrams.

REFERENCES

1. S. W. Green, 'The quantitative analysis of mixtures of chlorofluoromethanes in vapour phase chromatography', in *Vapour Phase Chromatography* D. H. Desty (Ed.), Butterworths, London, pp. 388 (1957).

2. F. Harrison, P. Knight, R. P. Kelly and M. T. Heath, 'The use of multiple columns and programmed column heating in the analysis of wide-boiling range halogenated hydrocarbon samples in gas chromatography', in *Gas Chromatography* D. H. Desty (Ed.), Butterworths, London, pp. 216–247 (1958).

3. M. G. Bloch, 'Determination of C_1-through C_7-hydrocarbons in a single run by a four-stage gas chromatograph', in *Proceeding of the 2nd International Symposium on Gas Chromatography*, East Lansing, MI, USA H. J. Noebels (Ed.) Academic Press, New York pp. 133–162 (1961).

4. American Society for Testing and Materials, *Annual Book of ASTM Standards*, Section 5, American Society for Testing and Materials, Philadelphia, PA, USA (1999).

5. The Institute of Petroleum, *IP Standard Methods for Analysis and Testing of Petroleum and Related Products and British Standards,* John Wiley & Sons, Chichester, UK (1999).

6. Universal Oil Products Process Division, Des Plaines, IL, USA.

7. Deutsches Institut für Normung Fachausschuss Mineralöl- und Brennstoffnormung, Hamburg, Germany.

8. Gas Processors Association (GPA) Standards, Tulsa, OK, USA.

9. R. S. Hutte, 'The sulfur chemiluminescence detector', *J. Chromatogr. Libr.* **56**: 201–229 (1995).

10. D. R. Deans, 'An improved technique for back-flushing gas chromatographic columns', *J. Chromatogr.* **18**: 477–481 (1965).

11. D. R. Deans, 'A new technique for heart cutting in gas chromatography', *J. Chromatogr.* **1**: 18–22 (1968).

12. H. Boer and P. van Arkel, 'Automatic PNA (paraffins–naphthenes–aromatics), analyser for (heavy), "naphtha"', *Chromatographia* **4**: 300–308 (1971).

13. J. V. Brunnock, 'Separation and distribution of normal paraffins from petroleum heavy distillates by molecular sieve adsorption and gas chromatography', *Anal. Chem.* **38**: 1648–1652 (1966).

14. H. Boer, P. van Arkel and W. J. Boersma, 'An automatic paraffins–naphthenes–aromatics (PNA), analyzer for the under 200 °C fraction contained in a higher boiling product', *Chromatographia* **13**: 500–512 (1980).

15. P. van Arkel, J. Beens, H. Spaans, D. Grutterink and R. Verbeek, 'Automated PNA analysis of naphthas and other hydrocarbon samples', *J. Chromatogr. Sci.* **25**: 141–148 (1988).

16. J. J. Szakasits and R. E. Robinson, 'Hydrocarbon type determination of naphthas and catalytically reformed products by automated multidimensional gas chromatography', *Anal. Chem.* **63**: 114–120 (1991).

17. M. G. Block, R. B. Callen and J. H. Stockinger, 'The analysis of hydrocarbon products obtained from methanol conversion to gasoline using open tubular GC columns and selective olefin absorption', *J. Chromatogr. Sci.* **15**: 504–512 (1977).

18. F. P. DiSanzo, J. L. Lane and R. E. Yoder, 'Application of state-of-the-art multidimensional high resolution gas chromatography for individual component analysis of gasoline range hydrocarbons', *J. Chromatogr. Sci.* **26**: 206–209 (1988).

19. F. P. DiSanzo and V. J. Giarrocco, 'Analysis of pressurized gasoline-range liquid hydrocarbon samples by capillary column and PIONA analyzer gas chromatography', *J. Chromatogr. Sci.* **26**: 258–266 (1988).

20. T. J. Lechner-Fish and S. L. Ryder, 'Analysis of simulated petroleum wellhead fluids using multidimensional gas chromatography', *Am. Lab.* **29**: 33X–33EE (1997).

21. H. J. W. Henderickx and J. J. M. Ramaekers, 'Analysis of a C_9–C_{10} aromatic hydrocarbon pyrolysis distillate by multidimensional capillary GC and multidimensional capillary GC–MS', *J. High Resolut. Chromatogr.* **17**: 407–410 (1994).

22. S. T. Teng, A. D. Williams and K. Urdal, 'Detailed hydrocarbon analysis of gasoline by GC–MS (SI-PIONA)', *J. High Resolut. Chromatogr.* **17**: 469–475 (1994).

23. J. A. Apffel and H. McNair, 'Hydrocarbon group-type analysis by on-line multidimensional chromatography. II. Liquid chromatography–gas chromatography', *J. Chromatogr.* **279**: 139–144 (1983).

24. P. C. Hayes-Jr and S. D. Anderson, 'Paraffins, olefins, naphthenes and aromatics analysis of selected hydrocarbon distillates using on-line column switching high-performance

liquid chromatography with dielectric constant detection', *J. Chromatogr.* **437**: 365–377 (1988).

25. I. L. Davies, K. D. Bartle, P. T. Williams and G. E. Andrews, 'On-line fractionation and identification of diesel fuel polycyclic aromatic compounds by two-dimensional microbore high-performance liquid-chromatography/capillary gas-chromatography', *Anal. Chem.* **60**: 204–209 (1988).

26. A. Trisciani and F. Munari, 'Characterization of fuel samples by on-line LC–GC with automatic group-type separation of hydrocarbons', *J. High Resolut. Chromatogr.* **17**: 452–456 (1994).

27. F. Munari, A. Trisciani, G. Mapelli, S. Trestianu, K. Grob-Jr and J. M. Colin, 'Analysis of petroleum fractions by on-line micro HPLC–HRGC coupling, involving increased efficiency in using retention gaps by partially concurrent solvent evaporation', *J. High Resolut. Chromatogr.* **8**: 601–606 (1985).

28. I. L. Davies, M. Raynor, P. T. Williams, G. E. Andrews and K. D. Bartle, 'Application of an automated on-line microbore high-performance liquid chromatography/capillary gas chromatography to diesel exhaust particulates', *Anal. Chem.* **59**: 2579–2583 (1987).

29. I. L. Davies, K. D. Bartle, G. E. Andrews and G. T. Williams, 'Automated chemical class characterization of kerosene and diesel fuels by on-line coupled microbore HPLC/capillary GC', *J. Chromatogr. Sci.* **26**: 125–130 (1988).

30. G. W. Kelly and K. D. Bartle, 'The use of combined LC–GC for the analysis of fuel products: a review', *J. High Resolut. Chromatogr.* **17**: 390–397 (1994).

31. J. Beens and R. Tijssen, 'An on-line coupled HPLC–HRGC system for the quantitative characterization of oil fractions in the middle distillate range', *J. Microcolumn Sep.* **7**: 345–354 (1995).

32. J. Beens and R. Tijssen, 'The characterization and quantitation of sulfur-containing compounds in (heavy) middle distillates by LC–GC-FID-SCD', *J. High Resolut. Chromatogr.* **20**: 131–137 (1997).

33. Z. Liu and J. B. Phillips, 'Comprehensive two-dimensional gas chromatography using an on-column thermal modulator interface', *J. Chromatogr. Sci.* **29**: 227–231 (1991).

34. C. J. Venkatramani and J. B. Phillips, 'Comprehensive two-dimensional gas chromatography applied to the analysis of complex mixtures', *J. Microcolumn Sep.* **5**: 511–516 (1993).

35. J. Blomberg, P. J. Schoenmakers, J. Beens and R. Tijssen, 'Comprehensive two-dimensional gas chromatography (GC × GC), and its applicability to the characterization of complex (petrochemical) mixtures', *J. High Resolut. Chromatogr.* **20**: 539–544 (1997).

36. J. Beens, H. Boelens, R. Tijssen and J. Blomberg, 'Quantitative aspects of comprehensive two-dimensional gas chromatography (GC × GC)', *J. High Resolut. Chromatogr.* **21**: 47–54 (1998).

37. J. B. Phillips and J. Xu, 'Comprehensive multi-dimensional gas chromatography', *J. Chromatogr.* **703**: 327–334 (1995).

38. G. S. Frysinger, R. B. Gaines and E. B. Ledford-Jr, 'Quantitative determination of BTEX and total aromatic compounds in gasoline by comprehensive two-dimensional gas chromatography (GC × GC)', *J. High Resolut. Chromatogr.* **22**: 195–200 (1999).

39. G. S. Frysinger and R. B. Gaines, 'Comprehensive two-dimensional gas chromatography with mass spectrometric detection (GC × GC/MS), applied to the analysis of petroleum', *J. High Resolut. Chromatogr.* **22**: 251–255 (1999).

40. R. B. Gaines, G. S. Frysinger, M. S. Hendrick-Smith and J. D. Stuart, 'Oil spill source identification by comprehensive two-dimensional gas chromatography', *Environ. Sci. Technol.* **33**: 2106–2112 (1999).

41. C. A. Bruckner, B. J. Prazen and R. E. Synovec, 'Comprehensive two-dimensional high-speed gas chromatography with chemometric analysis', *Anal. Chem.* **70**: 2796–2804 (1998).

42. R. M. Kinghorn and P. J. Marriott, 'Enhancement of signal-to-noise ratios in capillary gas chromatography by using a longitudinally modulated cryogenic system', *J. High Resolut. Chromatogr.* **21**: 620–631 (1998).

43. J. B. Phillips and J. Beens, 'Comprehensive two-dimensional gas chromatography: a hyphenated method with strong coupling between the two dimensions', *J. Chromatogr.* **856**: 331–347 (1999).

44. R. G. Schaefer and J. Höltkemeier, 'Direct analysis of alkylnaphthalenes in crude oils by two-dimensional capillary gas chromatography', *Chromatographia* **26**: 311–315 (1988).

45. R. G. Schäfer and J. Höltkemeier, 'Direkte Analyse von Dimethylnaphtalinen in Erdölen mittels zweidimensionaler kapillar Gas-chromatographie', *Anal. Chim. Acta* **260**: 107–112 (1992).

46. J. Blomberg, E. P. C. Mes, P. J. Schoenmakers and J. J. B. van der Does, 'Characterization of complex hydrocarbon mixtures using on-line coupling of size-exclusion chromatography and normal-phase liquid chromatography to high-resolution gas chromatography', *J. High Resolut. Chromatogr.* **20**: 125–130 (1997).

47. J. Blomberg, P. C. de Groot, H. C. A. Brandt, J. J. B. van der Does and P. J. Schoenmakers, 'Development of an on-line coupling of liquid–liquid extraction, normal-phase liquid chromatography and high-resolution gas chromatography producing an analytical marker for the prediction of mutagenicity and carcinogenicity of bitumen and bitumen fumes', *J. Chromatogr.* **849**: 483–494 (1999).

15 Multidimensional Chromatography: Forensic and Toxicological Applications

NICHOLAS H. SNOW

Seton Hall University, South Orange, NJ, USA

15.1 INTRODUCTION

In forensic science and toxicology, chemical analysis involves the detection and identification of compounds that are indicators of disease, poisons, and many types of illegal activities. Often, the analytes are found in difficult matrices such as tissues, urine, blood, serum, hair, arson debris, and shards of a variety of materials. Since there may also be many possible analytes for a given sample, extraction and chromatographic techniques are widely employed in forensic and toxicological analysis as a means for separating the analytes of interest from the complex matrix, and from each other, prior to identification. Often, identification by an orthogonal technique such as mass spectrometry is required. In classical chromatographic analysis, the complexity of the samples and matrices, combined with the need for positive identification and quantitative detection of the analytes, generally means that sample preparation methods must be extensive and laborious. Several authors have recently reviewed classical methods for forensic and toxicological analysis by using chromatography (1–3).

In order to reduce or eliminate off-line sample preparation, multidimensional chromatographic techniques have been employed in these difficult analyses. LC–GC has been employed in numerous applications that involve the analysis of poisonous compounds or metabolites from biological matrices such as fats and tissues, while GC–GC has been employed for complex samples, such as arson propellants and for samples in which special selectivity, such as chiral recognition, is required. Other techniques include on-line sample preparation methods, such as supercritical fluid extraction (SFE)–GC and LC–GC–GC. In many of these applications, the chromatographic method is coupled to mass spectrometry or another spectrometric detector for final confirmation of the analyte identity, as required by many courts of law.

In this present chapter, the applications of multidimensional chromatography to forensic and toxicological analysis are described in detail, being organized by technique. While multidimensional chromatography has not been as widely applied in

Multidimensional Chromatography, edited by L. Mondello, A. C. Lewis and K. D. Bartle
©2002 John Wiley & Sons Ltd.

forensic science as in environmental and food science, many of the methods and ideas presented in those areas could be readily applied to forensic problems. Little detail in instrumental set-ups is given here, as these are described elsewhere in this volume.

15.2   LIQUID CHROMATOGRAPHY – GAS CHROMATOGRAPHY

Liquid chromatography (LC) has been used primarily as an on-line sample clean-up method prior to high resolution capillary GC analysis. Primarily, this has involved the retention of fats and other high-molecular-weight matrix components in the analysis of drugs or pesticides from a variety of matrices such as foods, tissues and fat. LC – gas chromatography (GC) has not been widely employed in forensic and toxicological analysis, although this technique shows promise for simplifying sample preparation, especially in cases where sample volume is limited.

Van der Hoff, and co-workers (4 – 6) have used on-line LC – GC to separate and determine organochlorine pesticides and polychlorinated biphenyls (PCBs) from a variety of fatty matrices, including cow's and human milk, and liposuction biopts. In the most recent method, following treatment with sodium oxalate and extraction with hexane, these authors used a normal phase HPLC column, (typically, 50 mm × 1 mm × 3μm Hypersil 50, or 30 mm × 2.1 mm × 5 μm Spherisorb) followed by direct transfer to capillary GC with electron-chapture detector (ECD) detection. They found detection limits of about 0.3 – 2 μg/kg of fat for a variety of pesticides and PCB's from milk. This paper provides an excellent example of how LC – GC could be employed for the analysis of many other compounds of toxicological interest from fatty matrices. Barcarolo (7) also provides an LC – GC method for pesticide residues from fat, using a reversed phase clean-up prior to high resolution capillary GC. A description of a home-built LC – GC system is provided, along with extensive experimental detail. An LC – GC – ECD chromatogram of a butter sample, which may be considered representative of other high-fat-content matrices is shown in Figure 15.1. Several pesticides are successfully extracted from the fat matrix and separated with a minimum of prior sample preparation. They also observed a limited lifetime for the C_{18} LC column of about 25 – 30 injections before washing was required.

Chappell et al. (8) used LC coupled to two-dimensional GC in the analysis of illegal growth hormones from corned beef. In order to re-focus the chromatographic bands between stages, cryogenic focusing was employed. Figure 15.2 shows the three stages of separation of stilbene hormones form corned beef. The HPLC separation is shown in stage 1, in which a small sample is heart-cut into the first GC column. During the first GC separation, another sample is heart-cut into the second GC column. The third chromatogram shows the complete separation of several hormones, with a total analysis time of about 40 min. Chappell and co-workers also show examples of the removal of matrix interferences by cryogenically focusing them between the two GC columns.

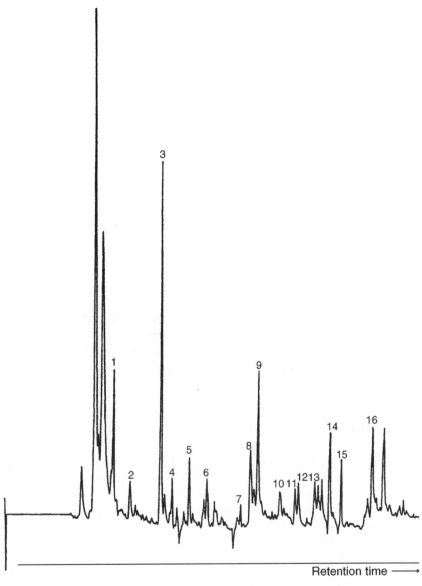

Figure 15.1 Separation of pesticides from butter by using LC–GC-ECD. Peak identification is as follows: 1, HCB; 2, lindane; 5, aldrin; 7, *o,p'*-DDE; 10, endrin; 11, *o,p'*-DDT; 13, *p,p'*-DDT: peaks 3, 4, 6, 8, 9, 12, 14, 15 and 16 were not identified. Adapted from *Journal of High Resolution Chromatography*, **13**, R. Barcarolo, 'Coupled LC–GC: a new method for the on-line analysis of organchlorine pesticide residues in fat', pp. 465–469, 1990, with permission from Wiley-VCH.

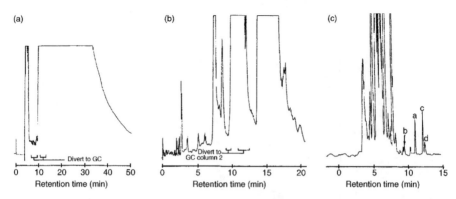

Figure 15.2 Three-stage LC–GC–GC separation of stilbenes from corned beef: (a) initial HPLC stage; (b) first GC stage; (c) second GC stage showing separated analytes. Peak identification in (c) is as follows: a, *trans*-diethylstilbesterol; b, *cis*-diethylstilberterol; c, hexestrol; d, dienestrol. Adapted from *Journal of High Resolution Chromatography*, **16**, G.C. Chappell *et al.*, 'On line high performance liquid chromatography–multidimensional gas chromatography and its application to the determination of stilbene hormones in corned beef, pp. 479–482, 1993, with permission from Wiley-VCH.

LC–GC, therefore, shows promise for forensic science applications, reducing sample handling and preparation steps by essentially using an on-line LC column in place of one or more extraction steps. This is followed by a traditional high resolution GC analysis. The methods described here for pesticides and hormones could be readily adapted to a variety of analyses, especially those involving fatty matrices, such as tissues, food or blood.

15.3 LIQUID–CHROMATOGRAPHY–LIQUID CHROMATOGRAPHY

Multidimensional liquid chromatography (LC–LC) has been employed mainly in efforts to directly inject biological materials such as urine, blood, plasma or serum, without prior clean-up steps. The particular *advantage* of direct injection methods actually lies in the *disadvantages* of classical extraction and preparation methods, which require extensive handling of potentially infectious samples, and losses of material due to these handling steps. Multidimensional chromatography also lends itself well to automation, which may be required in forensic or drug-monitoring laboratories. The analytes of interest in the direct injection of biological fluids include mostly drugs and other compounds of clinical interest. The use of guard columns and other pre-column devices is not described here, as these are generally used only for sample clean-up, rather than for improvement of fundamental chromatographic performance. Most systems for multidimensional LC–LC consist of two or more columns connected by column switching pneumatics, and these are described in detail in Chapter 5 of this text. Several examples of drug analysis from blood and plasma are described here, showing the utility of the first HPLC column to remove matrix components.

LC–LC has been widely applied for applications such as the determination of drugs and metabolites from blood. In this case, it is common to use an alkyl-bonded first column, to allow the proteinaceous material to pass through, while retaining the drugs of interest. Fouling of the first column with proteins and other high-molecular-weight interferences is a major problem in this technique. In order to avoid this problem, size-exclusion chromatography (SEC) is also commonly used as the pre-column procedure. Again, the large interferences elute rapidly, while the smaller drug molecules are retained.

Micellar HPLC, developed in the 1970s and 1980s has provided both alternatives to and interesting methods for multidimensional analysis. Micellar systems have been used by many workers for on-line sample clean-up, with an emphasis on the direct injection of biological fluids such as urine or blood. In a typical application, Posluszny and Weinberger (9) used sodium dodecyl sulfate above the critical micelle concentration as a mobile phase modifier. The micelles form a pseudo-stationary phase, which moves along with the solvent flow. This third phase can effectively partition interfering components, thus removing them from the column and avoiding fouling of the stationary phase. In multidimensional applications, the second column can mitigate the well-known efficiency limitations of single-dimension micellar chromatography. When the process is automated by the use of switching valves, automated, on-line enrichment and separation of drugs from complex urine and blood samples can be performed. Posluszny and Weinberger (9) analyzed several drugs from plasma, with direct injection of the plasma onto their multidimensional system. It is worth noting that they employed a 'consumable' inexpensive pre-column, which they simply replaced when fouled, typically after several hundred injections. In Figure 15.3, an analysis of tricyclic antidepressants from blood plasma is shown. Filtration was the only sample treatment prior to injection and no carry-over between injections was seen. It is seen that excellent efficiency and separation of the drug from the matrix can be achieved in a relatively short (less than 20 min) analysis time. These authors also demonstrated the direct injection and identification of drugs in plasma by using multidimensional high performance liquid chromatography (HPLC) with photodiode array detection (10).

More recently, Carda-Broch et al. described the analysis of cardiovascular drugs from urine by using micellar HPLC (11). These authors present extensive method development information, including resolution maps showing micelle and mobile phase composition. In effect, this may be viewed as a multidimensional technique within a single column. Typical separations of beta-blockers, using three different mobile phase compositions, are shown in Figure 15.4. Although analysis times are somewhat long, excellent resolution can be obtained, and tremendous effects on the selectivity are seen.

Stopher and Gage used size-exclusion chromatography (SEL) coupled to reversed phase HPLC for the direct injection of plasma in the analysis of an antifungal agent, voriconazole (12). Their system consisted of three columns, i.e. first a size-exclusion

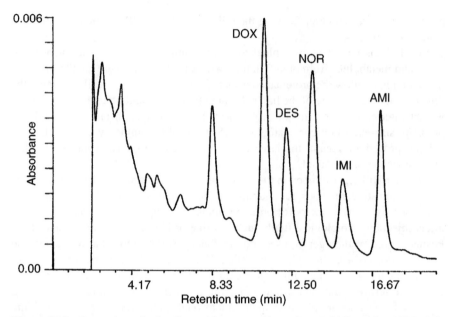

Figure 15.3 Separation of tricyclic antidepressants by using multidimensional LC–LC. Peak identification is as follows: DOX, doxepin; DES, desipramine; NOR, nortryptylene; IMI, imipramine; AMI, amitryptyline. Adapted from *Journal of Chromatography*, **507**, J. V. Posluszny *et al.*, 'Optimization of multidimensional high-performance liquid chromatography for the determination of drugs in plasma by direct injection, micellar cleanup and photodiode array detection', pp. 267–276, copyright 1990, with permission from Elsevier Science.

column, primarily used for de-salting followed by a pre-column, followed by a C_{18} analytical column. The pre-column was used for a final clean-up, plus re-focusing of the analytes prior to separation on the analytical column. This system allowed the direct injection of approximately 0.5 mL of plasma. The limit of quantitation of the method was found to be 5 μg/mL, with quantitation by internal standard. Figure 15.5 shows an analysis of voriconazole with an internal standard. Resolution between these very similar compounds is excellent, with a relatively short analysis time and little matrix interference.

Another system based on an SEC pre-column was employed by Earley and Tini for the analysis of a nonpeptidic inhibitor of human leukocyte elastase as part of a drug discovery project (13). The analytical column used here was octadecylsilane, although the authors commented that, depending on the particular application, any stationary phase could be used as the analytical column. In their application, a fraction of the effluent from the SEC column was transferred to the analytical column by the use of a C_{18} collection column that served to pre-concentrate and focus the analytes. They noted that frequent solvent flushing was required to avoid build-up of impurities on the collection column. A chromatogram obtained from a 50 μL plasma standard injection is shown in Figure 15.6. This separation provides an excellent

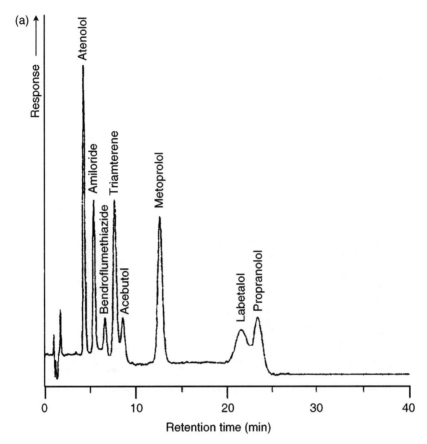

Figure 15.4   Separation of mixtures of beta-blockers by using micellar HPLC, employing the following mobile phases: (a) 0.12M SDS, 5% propanol, 0.5% triethylamine; (b) 0.06 M SDS, 15% propanol; (c) 0.11 M SDS, 8% propanol. Adapted   from *Journal of Chromatographic Science*, **37**, S. Carda-Broch *et al.*, 'Analysis of urine samples containing cardiovascular drugs by micellor liquid chromatography with fluorimetric detection', pp. 93–102, 1999, with permission from Preston Publications, a division of Preston Industries, Inc.

(*continued to p. 414*)

example of the ability of LC–LC to provide on-line sample clean-up along with high resolution of closely related structures.

Multidimensional LC has also been used to determine ursodeoxycholic acid and its conjugates in serum (14). These compounds are used in the treatment of cholesterol gallstones, hepatitis and bilary cirrhosis. These authors employed a traditional (10 × 4 mm) pre-column and a micro-bore (35 × 2 mm) analytical column that were interfaced by using a six-port switching valve.

Drug discovery applications also provide insights for the eventual determination of the compounds in a forensic or toxicological setting; this is another area in which multidimensional chromatography allows the direct injection of bodily fluids. For

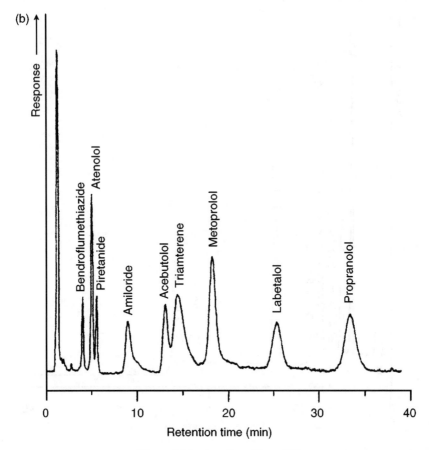

Figure 15.4   (*continued to p. 415*)

example, Szuna and Blain (15) described the determination of a new antibacterial agent by multidimensional HPLC using a guard column and a reversed phase analytical column. Mulligan *et al.* (16) combined liquid–liquid extraction with multidimensional HPLC for the analysis of a platelet aggregating factor receptor antagonist, again, by direct injection of bodily fluids. These two applications from drug discovery also demonstrate the applicability of multidimensional LC to a variety of toxicologically related problems.

15.4 GAS CHROMATOGRAPHY–GAS CHROMATOGRAPHY

GC–GC has typically been employed for complex samples or those requiring additional chemistry, such as chiral recognition, to be employed along with classical GC separation. Typical GC–GC systems employ multiple capillary columns connected

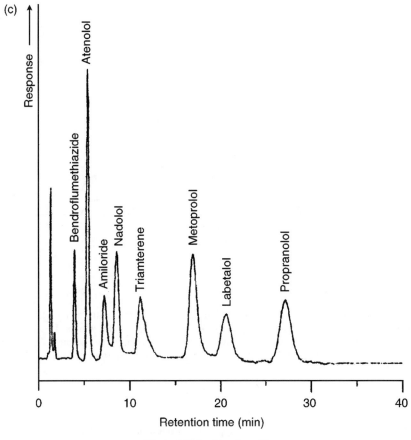

Figure 15.4 (*continued*)

with a switching valve or a cryogenically cooled concentrator. The multidimensional separation may also be combined with on-column or large volume injection. Applications in forensic and toxicological analysis are presented here, while instrumental and theoretical details can be found in Chapters 3 and 4 of this volume.

Recently, multidimensional GC has been employed in enantioselective analysis by placing a chiral stationary phase such as a cyclodextrin in the second column. Typically, switching valves are used to heart-cut the appropriate portion of the separation from a non-chiral column into a chiral column. Heil *et al.* used a dual column system consisting of a non-chiral pre-column (30 m × 0.25 mm × 0.38 μm, PS-268) and a chiral (30 m × 0.32 mm × 0.64 μm, heptakis(2,3-di-*O*-methyl-6-*O*-*tert*-butyldimethylsilyl)-β-cyclodextrin) (TBDM-CD) analytical column to separate derivatized urinary organic acids that are indicative of metabolic diseases such as short bowel syndrome, phenylketonuria, tyrosinaemia, and others. They used a FID following the pre-column and an ion trap mass-selective detector following the

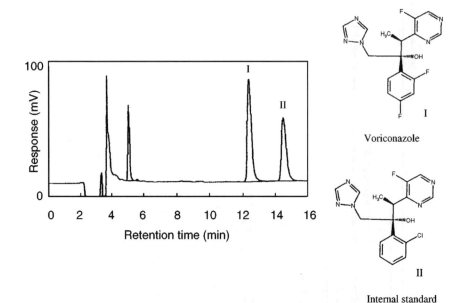

Figure 15.5 Separation of Voriconazole and an internal standard by using SEC–HPLC. Adapted from *Journal of Chromatography, B* **691**, D.A. Stopher and R. Gage, 'Determination of a new antifungal agent, voriconazole, by multidimensional high-performance liquid chromatography with direct plasma injection onto a size exclusion column', pp. 441–448, copyright 1997, with permission from Elsevier Science.

chiral column (17). Using the same approach, Podebrad *et al.* (18) determined several chiral amino acids that are indicative of maple syrup urine disease, which is a disorder involving several amino acids that leads to mental retardation and neurological damage. In addition, Kaunznger *et al.* used this approach in the analysis of 2-1 and 3-hydroxy aekanoic acids from the biomembranes of bacteria (19).

Figure 15.7 shows the chiral separation of several methyl esters of hydroxy alkanoic acids using multidimensional GC. The high resolving power of the achiral column is seen in chromatogram (a), which presents the separation of single enantiomers of three acids. The chiral selectivity of the second column is illustrated in chromatogram (b), which shows the separation of three enantiomeric pairs. The resolving power of capillary GC is also illustrated, as the selectivity of the chiral column is quite low, and yet the enantiomers are still resolved. The analysis time is relatively long, and could perhaps be lessened by using a shorter achiral column, as there is excess resolution of the three acids.

Figure 15.8 shows the multidimensional GC analysis of urinary acids, following lyophilization and derivatization by methyl chloroformate. In this figure, chromatogram (a) shows the complexity of the urine matrix and the need for a second separation dimension. A heart-cut is taken over a small range at about 45 min. The

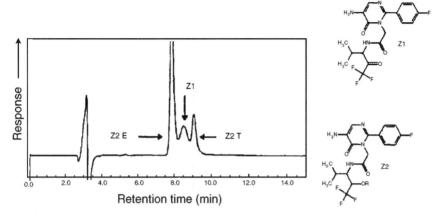

Figure 15.6 Chromatogram of a plasma standard of human leukocyte elastase inhibitors obtained by using LC–LC. Adapted from *Journal of Liquid Chromatography and Related Technologies*, **19**, R. A. Earley and L. P. Tini, 'Versatile multidimensional chromatographic system for drug discovery as exemplified by the analysis of a non-peptidic inhibitor of human leukocyte elastase', pp. 2527–2540, 1996, by courtesy of Marcel Dekker Inc.

second dimension separation (chromatogram (b) shows the GC/MS analysis (mass 84) for pyroglutamic acid enantiomers, separated on a chiral column. This allows the study of both enantiomers, which may be toxicologically significant, with a minimum of matrix interference. Again, these analysis times are relatively long, although high resolution in the first dimension is needed due to the complexity of the sample matrix.

By using a similar approach, bornane congeners, which indicate the presence of toxaphene, were analyzed by de Geus *et al.* (20). This presented a special analytical problem, as there are hundreds of congeners, many of which exist as racemates, with each of them having different toxicological properties. They also may be present in numerous biological and non-biological matrices. Samples of hake liver and dolphin blubber underwent Soxhlet extraction (pentane–dichloromethane (1:1)) and column chromatography to isolate the toxaphene compounds. Multidimensional GC was performed by using an Ultra-2 (24 m × 0.2 mm × 0.33 μm) pre-column, followed by an OV-1701 column (25 m × 0.25 mm × 0.15 μm), spiked with 10 wt% by TBDM-CD. Again, a heart-cutting technique was employed.

The enantioselective determination of 2,2′,3,3′,4,6′ -hexachlorobiphenyl in milk was performed by Glausch *et al.* (21). These authors used an achiral column for an initial separation, followed by separation of the eluent fraction on a chiral column. Fat was separated from the milk by centrifugation, mixed with sodium sulfate, washed with petroleum ether and filtered. The solvent was evaporated and the sample was purified by gel permeation chromatography (GPC) and silica gel adsorption chromatography. Achiral GC was performed on DB-5 and OV-1701 columns, while the chiral GC was performed on immobilized Chirasil-Dex.

(a)

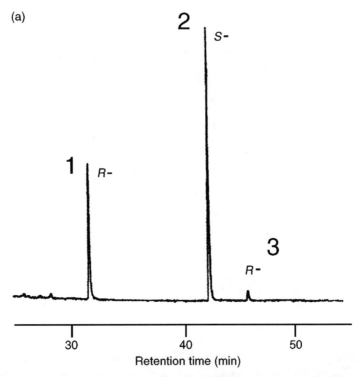

Figure 15.7 Chromatographic separation of chiral hydroxy acids from *Pseudomonas aeruginosa* without (a) and with (b) co-injection of racemic standards. Peak identification is as follows: 1, 3-hydroxy decanoic acid, methyl ester; 2, 3-hydroxy dodecanoic acid, methyl ester; 3, 2-hydroxy dodecanoic acid, methyl ester. Adapted from *Journal of High Resolution Chromatography*, **18**, A. Kaunzinger *et al.*, 'Stereo differentiation and simultaneous analysis of 2- and 3-hydroxyalkanoic acids from biomembranes by multidimensional gas chromatography', pp. 191–193, 1995, with permission from Wiley-VCH. (*continued p. 419*)

Jayatilaka and Poole analyzed fire debris by using headspace extraction followed by multidimensional gas chromatography (22). Since petroleum products are commonly used to start 'suspicious' fires, the samples are very complex, with misidentification of the propellant being common when analyzed by classical methods. In addition because of the complexity of the samples, pattern recognition by humans or by computers must be used to identify the samples, thus causing errors. Multidimensional gas chromatography can be used to simplify the final pattern recognition. For example, a second-dimension polar stationary phase can be used to separate aliphatic and aromatic portions of a heart-cut from a complex chromatogram on a non-polar column.

The use of heart-cuts to analyze a sample of 90% evaporated gasoline is illustrated in Figure 15.9. Each of the lettered chromatograms illustrates the further separation of mixture components which is not possible on a single column. The initial

(b)

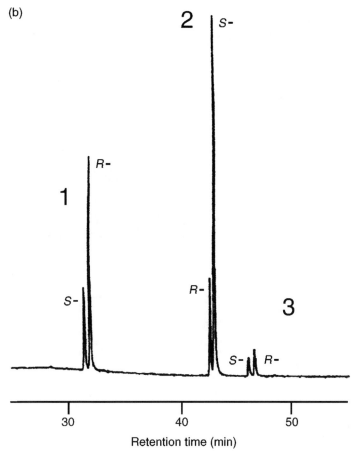

Figure 15.7   (*continued*)

separation is performed on a non-polar DB-5 (5% phenyl polydimethylsiloxane) column, while and the second separation is camed out on a more polar DB-225 (50% cyanopropylphenyl methyl polysiloxane) column, which allows highly selective separation of the aromatic components. This presents an excellent example of target compound analysis from a very complex matrix. Note that the individual targets which are not resolved in the first dimension are mostly well resolved in the second. This is also an excellent example of the use of multiple heart-cuts within a single analytical run, opening up the possibility of performing numerous analyses with a single injection.

Multiple cryogenic traps between the pre-column and the analytical column were developed in 1993 by Ragunathan *et al.* (23). These authors found that a multiple trap configuration provided the necessary resolution and automated sampling for simultaneous GC/IR and GC/MS determination of very complex mixtures. A

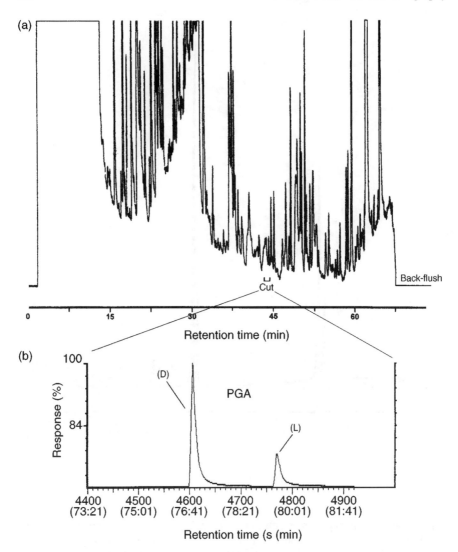

Figure 15.8 Multidimensional GC–MS separation of urinary acids after derivatization with methyl chloroformate: (a) pre-column chromatogram after splitless injection; (b) Main-column selected ion monitoring chromatogram (mass 84) of pyroglutamic acid methyl ester. Adapted from *Journal of Chromatography*, *B* **714**, M. Heil *et al.*, 'Enantioselective multidimensional gas chromatography–mass spectrometry in the analysis of urinary organic acids', pp. 119–126, copyright 1998, with permission from Elsevier Science.

separation of unleaded gasoline from their system is shown in Figure 15.10. The chromatographic system consisted of a DB-1701 pre-column, followed by the parallel traps, and then a DB-5 analytical column, mounted inside an HP 5890 gas chromatograph. Customized valves were used to reduce extra-column band broadening that might be caused by the traps and tubing. This technique could be easily applied

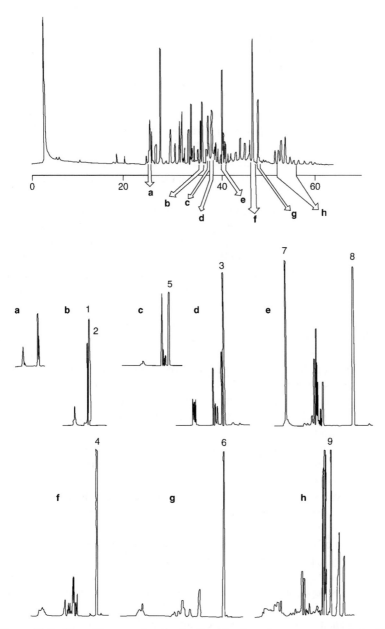

Figure 15.9 Use of heart-cutting for the identification of target compounds in 90% evaporated gasoline. Peak identification is as follows: 1, 1,2,4,5-tetramethylbenzene; 2, 1,2,3,5-tetramethylbenzene; 3, 4-methylindane; 4, 2-methylnaphthalene; 5, 5-methylindane; 6, 1-methylnaphthalene; 7, dodecane; 8, naphthalene; 9, 1,3-dimethylnaphthalene. Adapted from *Chromatography*, **39**, A. Jayatilaka and C.F. Poole, 'Identification of petroleum distillates from fire debris using multidimensional gas chromatography', pp. 200–209, 1994, with permission from Vieweg Publishing.

to the fingerprinting of gasoline and other flammable materials. The first chromatogram (Figure 15.10(a)) shows the GC-infrared chromatogram from the first dimension. There are approximately 110 peaks, which are significantly overlapped. Four heart-cuts ($A-C$) were taken. On the second chromatogram (Figure 15.10(b)), the GC/MS analysis of heart-cut C is shown. Twelve overlapping peaks that were initially trapped were resolved into over 30 separated peaks. The combination of non-destructive IR detection following the first stage and MS detection following the second stage presents especially powerful third and fourth dimensions, thus allowing spectral confirmation of peak identities and identification of overlaps.

The fragrance compounds commonly used in cosmetics and household products are some of the most common causes of contact dermatitis. These products often contain complex matrix interferences such as emulsifiers, thickeners, stabilizers, pigments, antioxidants and others, thus making the analytes of interest difficult to analyze. Tomlinson and Wilkins have applied multidimensional GC, coupled to infrared and mass spectrometry, to the analysis of common irritants from these complex mixtures (24). These authors modified a commercially available GC–IR–MS system to accommodate the second analytical column. By using an intermediate polarity (Rtx-1701) initial column and a variety of second columns, they obtained separations of a wide variety of fragrances. In order to evaluate the re-injection efficiency of their system, they used a Grob test mixture (25). This was injected on to the intermediate polarity column, and then the entire chromatogram was re-focused into a trap and re-injected onto an Rtx-5 non-polar column. They estimated a re-injection efficiency of about 85% between the two columns. The Grob test mixture chromatograms are shown in Figure 15.11. The first chromatogram (Figure 15.11(a)) shows a Grob test mix separated on the first-dimension column. The entire chromatogram is cryogenically trapped and then re-injected onto the second column, to give the chromatogram shown in Figure 15.11(b). This provides an excellent means for assessing the efficiency of the interface between the two columns. Thorough reviews of multispectral detection methods, such as GC–IR–MS, have been provided by Ragunathan *et al.* (26) and Krock *et al.* (27).

Authenticity evaluation has recently received increased attention in a number of industries. The complex mixtures involved often require very high resolution analyses and, in the case of determining the authenticity of 'natural' products, very accurate determination of enantiomeric purity. Juchelka *et al.* have described a method for the authenticity determination of natural products which uses a combination of enantioselective multidimensional gas chromatography with isotope ratio mass spectrometry (28). In isotope ratio mass spectrometry, combustion analysis is combined with mass spectrometry, and the $^{13}C/^{12}C$ ratio of the analyte is measured versus a CO_2 reference standard. A special interface, employing the necessary oxidation and reduction reaction chambers and a water separator, was used employed. For standards of 5-nonanone, menthol and (R)-γ-decalactone, they were able to determine the correct $^{12}C/^{13}C$ ratios, with relatively little sample preparation. The technical details of multidimensional GC–isotope ratio MS have been described fully by Nitz *et al.* (29). A MDGC–IRMS separation of a natural *cis*-3-hexen-1-ol fraction is

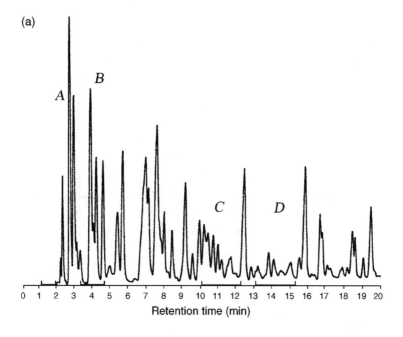

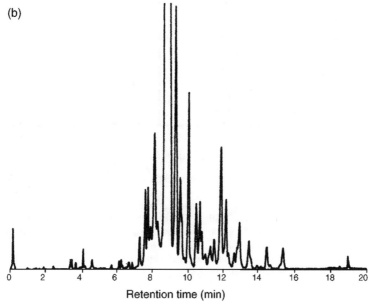

Figure 15.10 Primary (a) and secondary (b) separation of unleaded gasoline, where (a) shows the IRD chromatogram, and (b) shows the MSD total ion chromatogram of heart cut *c*. Adapted from *Analytical Chemistry*, **65**, N. Ragunathan *et al.*, 'Multidimensional gas chromatography with parallel cryogenic traps', pp. 1012–1016, copyright 1993, with permission from the American Chemical Society.

(a)

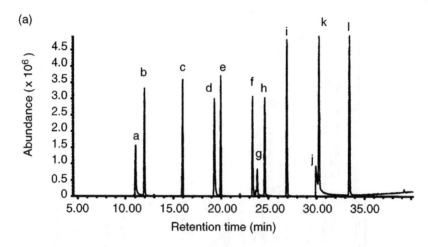

(b)

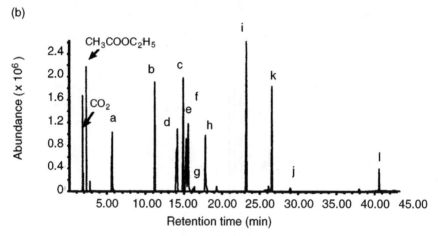

Figure 15.11 (a) Total ion chromatogram of a Grob test mixture obtained on an Rtx-1701 column, and (b) re-injection of the entire chromatogram on to an Rtx-5 column. Peak identification is as follows: a, 2,3-butanediol; b, decane; c, undecane; d, 1-octanol; e, nonanal; f, 2,6-dimethylphenol; g, 2-ethylhexanoic acid; h, 2,6-dimethylaniline; i, decanoic acid; methyl ester; j, dicyclohexylamine; k, undecanoic acid, methyl ester; l, dodecanoic acid, methyl ester. Adapted from *Journal of High Resolution Chromatography*, **21**, M. J. Tomlinson and C. L. Wilkins, 'Evaluation of a semi-automated multidimensional gas chromatography–infrared–mass spectrometry system for irritant analysis', pp. 347–354, 1998, with permission from Wiley-VCH.

shown in Figure 15.12. In the heart-cut(shown as an inset), taken from 25–30 min, several of the components are further resolved. In earlier studies, also related to authenticity determination, Mosandl *et al.* have described methods for the determination of several chiral natural product (30).

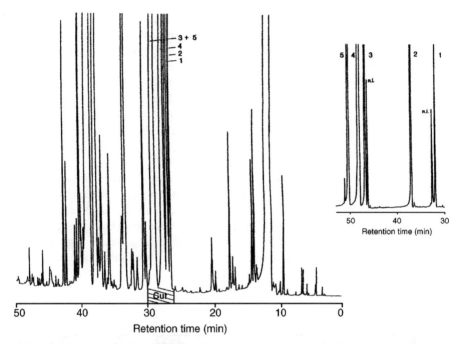

Figure 15.12 GC–GC chromatogram of a natural *cis*-3-hexen-1-ol fraction. Peak identification is as follows: 1, ethyl-2-methylbutyrate; 2, *trans*-2-hexenal; 3, 1-hexanol; 4, *cis*-3-hexen-1-ol; 5, *trans*-2-hexen-1-ol. Adapted from *Journal of High Resolution Chromatography*, **15**, S. Nitz *et al.*, Multidimensional gas chromatography–isotope ratio mass spectrometry, (MDGC–IRMS). Part A: system description and technical requirements', pp. 387–391, 1992, with permission from Wiley-VCH.

Comprehensive two-dimensional gas chromatography, originally proposed by Schomburg (31) and developed by Phillips and co-workers (32–35), in which the effluent from a traditional analytical column is sampled into a short, narrow-bore, thin-film second column, also shows promise in the analysis of the complex mixtures commonly found in forensic analysis. This technique offers a very rapidly obtained second dimension and high peak capacity which is necessary for complex mixtures, with thousands of peaks being possible in a single chromatogram. The method has been applied mostly in the petroleum and the environmental industries. Several authors have recently reported the use of comprehensive two dimensional gas chromatography on petroleum related samples (36–38).

Although comprehensive two-dimensional gas chromatography has not been applied to any great extent in forensic analysis, the technique shows great promise when samples or sample matrices are complex. For example, when oil is spilled into waterways, assigning responsibility for the economic and environmental damage is often difficult. Gaines *et al.* employed comprehensive two-dimensional GC in the forensic analysis of samples collected at oil-spill sites and were able to obtain results which were comparable to those obtained by classical methods (39). This article also provides an excellent description of the principles of comprehensive two-dimensional GC as applied to a forensic problem. The instrument consists of a typical gas chromatograph, modified by the addition of a thermal modulator as a switching valve after the first column. The effluent is switched into the second column at regular intervals. A flame-ionization detector (FID) was used for detection and the entire system required microprocessor control. The second dimension allowed these workers to separate compounds by both volatility and polarity in a single run, thus enabling over 500 separate peaks to be resolved, and in this way they were able to make a probable match as between various oil spill samples and a marine diesel fuel standard.

Comprehensive two-dimensional GC has also been employed for the analysis of pesticides from serum, which, although not strictly a forensic analytical 'problem', provides an example of the promise of this technique to forensic applications, such as the analysis of drugs of abuse (40). Two-dimensional gas chromatograms of a 17-pesticide standard and an extract from human serum are shown in Figure 15.13. The total analysis time of about 5 min, high peak capacity and the separation of all

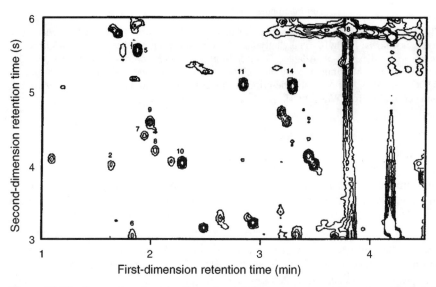

Figure 15.13 Comprehensive two-dimensional GC chromatogram of a supercritical fluid extract of spiked human serum. Peak identification is as follows: 1, dicamba; 2, trifluralin; 3, dichloran; 4, phorate; 5, pentachlorophenol; 6, atrazine; 7, fonofos; 8, diazinon; 9, chlorothalonil; 10, terbufos; 11, alachlor; 12, matalaxyl; 13, malathion; 14, metalochlor; 15, DCPA; 16, captan; 17, folpet; 18, heptadecanoic acid. Adapted from *Analytical Chemistry*, **66**, Z. Liu *et al.*, 'Comprehensive two-dimensional gas chromatography for the fast separation and determination of pesticides extracted from human serum', pp. 3086–3092, copyright 1994, with permission from the American Chemical Society.

17 of the components are the main features here. However, it was noted that one of the internal standards (heptadecanoic acid) co-eluted with major interferences, so even with two-dimensional separation, complete separation of the analytes of interest from a complex matrix is still difficult. The sample preparation used in this case was typical solid phase extraction. These authors also showed reproducibility and linear ranges that were easily competitive with those from traditional GC methods.

Solid phase micro-extraction (SPME) (41, 42) has also been employed by Gaines *et al.* (43), along with comprehensive two-dimensional GC in the analysis of trace components from aqueous samples. This combination fills the need for a rapid, high sensitivity and high resolution analysis of complex mixtures. These authors examined the analysis of oxygenated and aromatic compounds from water. While these are not strictly forensic analytes, they do provide effective models for other applications. As described above, Gaines and co-workers employed a two-column scheme, with the first analytical column being non-polar and essentially separating compounds by volatility, while the second (fast) column separated analytes of similar volatility by their polarity. In this way, they were able to demonstrate the low ppb analysis of various gasoline components spiked into water. The adaptation of this method to the analysis of volatile and semi-volatile components from water, fire debris, biological material and other forensic matrices would seem to be reasonably straightforward.

15.5   ON-LINE SAMPLE PREPARATION

Although on-line sample preparation cannot be regarded as being traditional multidimensional chromatography, the principles of the latter have been employed in the development of many on-line sample preparation techniques, including supercritical fluid extraction (SFE)–GC, SPME, thermal desorption and other on-line extraction methods. As with multidimensional chromatography, the principle is to obtain a portion of the required selectivity by using an additional separation device prior to the main analytical column.

The coupling of supercritical fluid extraction (SFE) with gas chromatography (SFE–GC) provides an excellent example of the application of multidimensional chromatography principles to a sample preparation method. In SFE, the analytical matrix is packed into an extraction vessel and a supercritical fluid, usually carbon dioxide, is passed through it. The analyte matrix may be viewed as the stationary phase, while the supercritical fluid can be viewed as the mobile phase. In order to obtain an effective extraction, the solubility of the analyte in the supercritical fluid mobile phase must be considered, along with its affinity to the matrix stationary phase. The effluent from the extraction is then collected and transferred to a gas chromatograph. In his comprehensive text, Taylor provides an excellent description of the principles and applications of SFE (44), while Pawliszyn presents a description of the supercritical fluid as the mobile phase in his development of a kinetic model for the extraction process (45).

Slack *et al.* have provided an example one of of the many types of forensic and related analyses that can be performed by using SFE–GC (46). These authors analyzed several explosives from water using on-line SFE–GC and describe several of the method development considerations. The SFE stage is essentially a non-selective elution of the components of interest from the matrix as a group. This is followed by selective elution using capillary gas chromatography. Figure 15.14 shows one of the typical method development problems, i.e. the temperature of the trap between the first (SFE) and second (GC) dimensions. Analyses of several explosives, using different trapping temperatures, are shown. It is seen that as the temperature of the trap is decreased, the response for the earlier eluting compounds is increased, while little

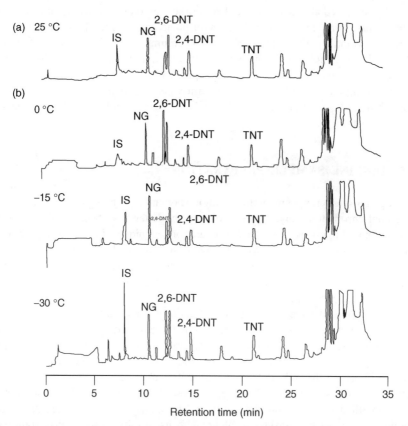

Figure 15.14 Separation of explosives extracted from water by using SPE–SFE–GC at several SFE trapping temperatures. peak identification is as follows: NG, nitroglycerin; 2,6-DNT, 2,6-dinitrotoluene; 2,4-DNT, 2,4-dinitrotoluene; TNT, trinitrotoluene; IS, 1,3-trichlorobenzene. Adapted *Journal of High Resolution Chromatography*, **16**, G. C. Slack *et al.*, 'Coupled solid phase extraction supercritical fluid extraction–on-line gas chromatography of explosives from water', pp. 473-478, 1993, with permission from Wiley-VCH.

change is seen for the later-eluting components. As automated systems become more common, on-line sample preparation systems of this type will see tremendous growth as a forensic and toxicological analysis technique in the near future.

15.6 CONCLUSIONS

The applications of multidimensional chromatography presented here show that such techniques provide great promise in the solution of the complex problems involved in forensic and toxicological analysis. These include complex matrices such as urine, blood and natural products, and difficult analytes such as drugs, aromas and natural products. When compared to environmental analysis, forensic and toxicological analysis using multidimensional chromatography has received relatively little attention, although the possibilities are many and the potential is bright. The environmental and pharmaceutical methods described in this present chapter could be readily adapted to forensic problems. In particular, multidimensional chromatography offers the forensic scientist high resolution, high sensitivity and short analysis times.

ACKNOWLEDGEMENT

The author gratefully acknowledges the influential work of Professor John Phillips, which has touched nearly all areas of multidimensional gas chromatography.

REFERENCES

1. T. A. Brettell, K. Inman, N. Rudin, and R. Saferstein, 'Forensic science', *Anal. Chem.* **71**: 235R–255R (1999).
2. R. H. Liu and D. E. Gadzala, *Handbook of Drug Analysis: Applications in Forensic and Clinical Laboratories,* Oxford University Press, New York (1997).
3. S. B. Karch (Ed.), *Drug Abuse Handbook,* CRC Press, Boca Raton, FL, USA (1998).
4. G. R. van der Hoff, R. A. Baumann, P. van Zoonen and U. A. Th Brinkman, 'Determination of organochlorine compounds in fatty matrices: application of normal phase LC clean-up coupled on-line to GC/ECD', *J. High Resolut. Chromatogr.* **20**: 222–226 (1997).
5. E. A. Hogendoorn, G. R. van der Hoff and P. van Zoonen, 'Automated sample clean-up and fractionation of organochlorine pesticides and polychlorinated biphenyls in human milk using NP-HPLC with column-switching', *J. High Resolut. Chromatogr.* **12**: 784–789 (1989).
6. G. R. van der Hoff, A. C. van Beuzekom, U. A. Th Brinkman, R. A. Baumann and P. van Zoonen, 'Determination of organochlorine compounds in fatty matrices. Application of rapid off-line normal-phase liquid chromatographic clean-up', *J. Chromatogr.* **754**: 487–496 (1996).
7. R. Barcarolo, 'Coupled LC–GC: a new method for the on-line analysis of organchlorine pesticide residues in fat', *J. High Resolut. Chromatogr.* **13**: 465–469 (1990).

8. C. G. Chappell, C. S. Creaser and M. J. Shepherd, 'On-line high performance liquid chromatography–multidimensional gas chromatography and its application to the determination of stilbene hormones in corned beef', *J. High Resolut. Chromatogr.* **16**: 479–482 (1993).

9. J. V. Posluszny and R. Weinberger, 'Determination of drug substances in biological fluids by direct injection multidimensional liquid chromatography with a micellar cleanup and reversed-phase chromatography', *Anal. Chem.* **60**: 1953–1958 (1988).

10. J. V. Posluszny, R. Weinberger and E. Woolf, 'Optimization of multidimensional high-performance liquid chromatography for the determination of drugs in plasma by direct injection, micellar cleanup and photodiode array detection', *J. Chromatogr.* **507**: 267–276 (1990).

11. S. Carda-Broch, I. Rapado-Martinez, J. Esteve-Romero and M. C. Garcia-Alvarez-Coque, 'Analysis of urine samples containing cardiovascular drugs by micellar liquid chromatography with fluorimetric detection', *J. Chromatogr. Sci.* **37**: 93–102 (1999).

12. D. A. Stopher and R. Gage, 'Determination of a new antifungal agent, voriconazole, by multidimensional high-performance liquid chromatography with direct plasma injection onto a size exclusion column', *J. Chromatogr.* **691**: 441–448 (1997).

13. R. A. Earley and L. P. Tini, 'Versatile multidimensional chromatographic system for drug discovery as exemplified by the analysis of a non-peptidic inhibitor of human leukocyte elastase', *J. Liq. Chromatogr.* **19**: 2527–2540 (1996).

14. S. J. Choi, C. K. Jeong, H. M. Lee, K. Kim, K. S. Do and H. S. Lee, 'Simultaneous determination of ursodeoxycholic acid and its glycine-conjugate in serum as phenacyl esters using multidimensional liquid chromatography', *Chromatographia* **50**: 96–100 (1999).

15. A. J. Szuna and R. W. Blain, 'Determination of a new antibacterial agent (Ro 23-9424) by multidimensional high-performance liquid chromatography with ultraviolet detection and direct plasma injection', *J. Chromatogr.* **620**: 211–216 (1993).

16. T. E. Mulligan, R. W. Blain, N. F. Oldfield and B. A. Mico, 'The analysis of Ro 24-4736 in human plasma by multidimensional reversed phase microbore HPLC/UV', *J. Liq. Chromatogr.* **17**: 133–150 (1994).

17. M. Heil, F. Podebrad, T. Beck, A. Mosandl, A. C. Sewell and H. Böhles, 'Enantioselective multidimensional gas chromatography-mass spectrometry in the analysis of urinary organic acids', *J. Chromatogr.* **714**: 119–126 (1998).

18. F. Podebrad, M. Heil, S. Leib, B. Geier, T. Beck, A. Mosandl, A. C. Sewell and H. Böhles, 'Analytical approach in diagnosis of inherited metabolic diseases: maple syrup urine disease (MSUD)–simultaneous analysis of metabolites in urine by enantioselective multidimensional capillary gas chromatography–mass spectrometry (enantio-MDGC–MS)', *J. High Resolut. Chromatogr.* **20**: 355–362(1997).

19. A. Kaunzinger, M. Thomsen, A. Dietrich and A. Mosandl, 'Stereodifferentiation and simultaneous analysis of 2- and 3-hydroxyalkanoic acids from biomembranes by multidimensional gas chromatography', *J. High Resolut. Chromatogr.* **18**: 191–193 (1995).

20. H.-J. De Geus, R. Baycan-Keller, M. Oehme, J. de Boer and U. A. Th Brinkman, 'Determination of enantiomer ratios of bornane congeners in biological samples using heart-cut multidimensional gas chromatography', *J. High Resolut. Chromatogr.* **21**: 39–46 (1998).

21. A. Glausch, J. Hahn and V. Schurig, 'Enantioselective determination of chiral 2,2′,3,3′,4,6-hexachlorobiphenyl (PCB 132), 'in human milk samples by multidimensional gas chromatography/electron capture detection and by mass spectrometry', *Chemosphere* **30**: 2079–2085 (1995).

22. A. Jayatilaka and C. F. Poole, 'Identification of petroleum distillates from fire debris using multidimensional gas chromatography', *Chromatographia* **39**: 200–209 (1994).

23. N. Ragunathan, K. A. Krock and C. L. Wilkins, 'Multidimensional gas chromatography with parallel cryogenic traps', *Anal. Chem.* **65**: 1012–1016 (1993).

24. M. J. Tomlinson and C. L. Wilkins, 'Evaluation of a semi-automated multidimensional gas chromatography–infared–mass spectrometry system for irritant analysis', *J. High Resolut. Chromatogr.* **21**: 347–354 (1998).

25. K. Grob Jr., G. Grob and K. Grob, 'Comprehensive quality test for glass capillary columns', *J. Chromatogr.* **156**: 1–20 (1978).

26. N. Rugunathan, K. A. Krock, C. Klawun, T. A. Sasaki and C. L. Wilkins, 'Multispectral detection for gas chromatography', *J. Chromatogr.* **703**: 335–382 (1995).

27. K. A. Krock, N. Ragunathan, C. Klawun, T. Sasaki and C. L. Wilkins, 'Multidimensional gas chromatography–infrared spectrometry–mass spectrometry', *Analyst* **119**: 483–489 (1994).

28. D. Juchelka, T. Beck, U. Hener, F. Dettmar and A. Mosandl, 'Multidimensional gas chromatography coupled on-line with isotope ratio mass spectrometry (MDGC–IRMS): progress in the analytical authentication of genuine flavor components', *J. High Resolut. Chromatogr.* **21**: 145–151 (1998).

29. S. Nitz, B. Weinreich and F. Drawert, 'Multidimensional gas chromatography–isotope ratio mass spectrometry (MDGC-IRMS). Part A: system description and technical requirements', *J. High Resolut. Chromatogr.* **15**: 387–391 (1992).

30. A. Mosandl, U. Hener, U. Hagenauer-Hener and A. Kustermann, 'Direct enantiomer separation of chiral γ-lactones from food and beverages by multidimensional gas chromatography', *J. High Resolut. Chromatogr.* **12**: 532–536 (1989).

31. G. Schomburg, 'Practical limitations of capillary gas chromatography', *J. High Resolut. Chromatogr. Chromatogr. Commun.* **2**: 461–474 (1979).

32. Z. Liu and J. B. Phillips, 'Sample introduction into a 5 μm i.d. capillary gas chromatography column using an on-column thermal desorption modulator', *J. Microcolumn Sep.* **1**: 159–162 (1989).

33. Z. Liu and J. B. Phillips, 'High-speed gas chromatography using an on-column thermal desorption modulator', *J. Microcolumn Sep.* **1**: 249–256 (1989).

34. Z. Liu and J. B. Phillips, 'Large-volume sample introduction into narrow-bore gas chromatography columns using thermal desorption modulation and signal averaging', *J. Microcolumn Sep.* **2**: 33–40 (1990).

35. Z. Liu and J. B. Phillips, 'Comprehensive two-dimensional gas chromatography using an on-column thermal modulator interface', *J. Chromatogr. Sci.* **29**: 227–231 (1991).

36. G. S. Frysinger and R. B. Gaines, 'Comprehensive two-dimensional gas chromatography with mass spectrometric detection, (GC × GC/MS) applied to the analysis of petroleum', *J. High Resolut. Chromatogr.* **22**: 251–255 (1999).

37. J. Blomberg, P. J. Schoenmakers, J. Beens and R. Tijssen, 'Comprehensive two-dimensional gas chromatography, (GC × GC) and its applicability to the characterization of complex, (petrochemical), "mixtures"', *J. High Resolut. Chromatogr.* **20**: 539–544 (1997).

38. G. S. Frysinger, R. B. Gaines and E. B. Ledford-Jr, 'Quantitative determination of BTEX and total aromatic compounds in gasoline by comprehensive two-dimensional gas chromatography (GC × GC)', *J. High Resolut. Chromatogr.* **22**: 195–200 (1999).

39. R. B. Gaines, G. S. Frysinger, M. S. Hendrick-Smith and J. D. Stuart, 'Oil spill source identification by comprehensive two-dimensional gas chromatography', *Env. Sci. Technol.* **33**: 2106–2112 (1999).

40. Z. Liu, S. R. Sirimanne, D. G. Patterson-Jr, L. L. Needham and J. B. Phillips, 'Comprehensive two-dimensional gas chromatography for the fast separation and

      determination of pesticides extracted from human serum', *Anal. Chem.* **66**: 3086–3092 (1994).

41.   J. Pawliszyn, *Applications of Solid Phase Microextraction,* Royal Society of Chemistry, Cambridge, UK (1999).

42.   J. Pawliszyn, *Solid Phase Microextraction: Theory and Practice,* Wiley-VCH, New York (1997).

43.   R. B. Gaines, E. B. Ledford-Jr and J. D. Stuart, 'Analysis of water samples for trace levels of oxygenated and aromatic compounds using headspace solid-phase microextraction and comprehensive two-dimensional gas chromatography', *J. Microcolumn Sep.* **10**: 597–604 (1998).

44.   L. T. Taylor, *Supercritical Fluid Extraction,* John Wiley & Sons, New York (1996).

45.   J. Pawliszyn, 'Kinetic model for supercritical fluid extraction', *J. Chromatogr. Sci.* **31**: 31–37 (1992).

46.   G. C. Slack, H. M. McNair, S. B. Hawthorne and D. J. Miller, 'Coupled solid phase extraction–supercritical fluid extraction–on-line gas chromatography of explosives from water', *J. High Resolut. Chromatogr.* **16**: 473–478 (1993).

Index